FROM JUPITER TO GENESIS

FROM JUPITER TO GENESIS

FINDING LOVE'S 7TH-GIFT IN FATHER'S LIBRARY

Anthony Sneed

ISB::Institute™ Fellow, Emeritus

Library of Congress Control Number: 2004091413
ISBN: Hardcover   978-1-4134-5010-1
               Softcover   978-1-4134-5009-5

The New International Version (NIV) is referenced, herein, only as proof that even with multiple translation defects from a source text, or other translation dynamics, Scripture's own, redundant constructs secure the deposits within it, across languages and speech, time and era.

This book was printed in the United States of America. Cover design consultants: the Sneed Family Children (Alexandria, Dorion, Dorius, Chante). Cover art courtesy of NASA.

Print information available on the last page

Rev. date: 06/01/2015

To order additional copies of this book, contact:
Xlibris
1-888-795-4274
www.Xlibris.com
Orders@Xlibris.com
544875

CONTENTS

Foreword 9

Chapter 1: Living Dust 17
Chapter 2: Cain: A New "Kind" of Person 22
Chapter 3: A Wife for Cain 25
Chapter 4: Lamech: The First Married Man 36
Chapter 5: The First Widow on Earth 39
Chapter 6: The First Born Daughter on Earth 45
Chapter 7: Why Noah Was, and Is, Very Special 50
Chapter 8: The Marriage of Abram 58
Chapter 9: A 10-Year Wait 64
Chapter 10: Terah's Unique Children
    and Grandchildren 68
Chapter 11: The Secret of Sarai/Sarah 72
Chapter 12: The First Marriage for All Nations 76
Chapter 13: Nephilim After the Flood 80
Chapter 14: The Son of MAN 86

APPENDICES

Appendix א The Galileo Mission to Jupiter 97
Appendix ב U.S. Patent No. 6,236,992 110
Appendix ג The First Deposition on Earth 128
Appendix ד Shuttle Thermal Integrity
    Detection System 164
Appendix ה Sneed v. IBM: Decoding a $1.86 Billion
    Federal Riddle With Caltech's
    Honor Code, Pro Se 183
Appendix ו What Angels Can Never Do 493
Appendix ז The Ten Commandments and
    the "Ten" Kinds of Fruit of the Spirit 523

Index 545

Dedication

In remembering-gratitude for another 7 weeks' health, offspring, helpful aides, provisions, and a peaceful place to share rest [Exodus 20:8-11];

To my father and mother, fully respecting the Commandment, "Honor your father and your mother, so that you may live long in the land the [I-SHALL-BE] your God is giving you." [Exodus 20:12];

And the names of R.B. Harris, J.I. Caldwell, F.F. Clark, S.M. Smith, L.F. Browne, J.N. Gross, C.E. Jensen, A.T. Gomez, D.A. Prouty, N.E. Armstrong, D.E. Brandt, L.A. Romero, A. Lidow, P. Wilzbach, J.R. Opel, D. Welsh, L. V. Gerstner Jr., P.T. German, and their respective wives and families, along with others, as it is written, "A man of many companions may come to ruin, but there is a friend who sticks closer than a brother." [Proverbs 18:24];

And respect due six Caltech presidents: H. Brown, R. Christy, J.L. Chameau, D. Baltimore, T.E. Everhart, and M.L. Goldberger, as it is written. "A king's rage is like the roar of a lion, but his favor is like dew on the grass." [Proverbs 19:12]

a *7th-Gift* page

FOREWORD

This manuscript contains written testimony regarding the Book of Genesis' first 21 chapters, specific to its record of the original 21 generations of "man" ("adamah" in Hebrew for "earth"), beginning with Adam and Eve. For some time, the writer has had the pleasurable pass-time of committing Scripture to memory, purely for the fun of seeing how much could be recalled. To date, this amounts to verses encompassing six complete books (John, Revelation, 1John, 2John, 3John and Jude) and almost half the book of Genesis (chapters 1 through 21).

There was little motive or particular objective in commencing such an endeavor, other than the enjoyment of remembering verses and recalling them, weeks or months later. The intellectual stimulation, in doing so, proved more satisfying than technical assignments the writer had after graduating from Caltech. Such efforts include designing and developing an on-board telemetry computer operating system for the NASA Galileo mission to Jupiter (Appendix א). This work was delightful in its own way, yet, committing Scripture to memory proved even more intellectually challenging and fulfilling.

The NASA, Galileo-orbiter spacecraft's, on-board, telemetry computer helped relay scientific data some 0.5 billion miles from Jupiter to earth. The data were collected by a jettisoned Galileo-probe during its descent to, and then through, the Jovian cloud-tops. It reached speeds of up to 106,000 miles-per-hour, after separation from the main Galileo-orbiter spacecraft months before. In this endeavor, just for the "fun" of it, the writer developed a 4,000-line telemetry computer simulator, in about four weeks, to help advance the project development schedule by some four months. It also helped trim up to $0.25 million in costs from the project budget.

The savings and schedule advances were coincident with innovating the telemetry computer simulator. The overriding motive was to see if one could remember all aspects of the telemetry computer's seven million instruction combinations, along with its architectural interfaces. The simulator's fidelity, to the target system implementation, obviated construction of a third computer. Furthermore, all "microcode" software sent to Jupiter, in the telemetry computer, first came to "life" and was debugged in the simulator, benefitting from diagnostic and analysis tools built into it. Committing Scripture to memory, in subsequent years, proved even more stimulating and satisfying, than any accomplishment through the Galileo mission or other technical endeavors. Scripture, to the writer, also contains "instructions" and "commands" for "living beings," paralleling, in a way, the rigor of digital instructions and commands for ground-based and on-board computers directing a spacecraft on a distant journey.

Since Genesis contains detailed language specifications to set a number of concurrent events into motion, the writer was moved to apply natural and synthetic language processing methodologies, learned at Caltech, while reviewing Scripture committed to memory. In doing so, the writer was given to subject Scripture to the rigor of machine-based languages that describe very precise actions. An example of such rigor might include implementing a discrete Fourier transform routine, to lock-on and track a 1.4 gigahertz signal from the Galileo-probe, and compensate for a 15 hertz-per-second Doppler. Such a routine follows all frequency shifts as the vehicle's acceleration reached speeds of up to 106,000 miles-per-hour during the Jovian encounter—the fastest man-made object, ever, of its size. The question was raised in the writer's mind, "What might the outcome be from applying artificial language rigor to Scripture? Would Scripture exhibit ambiguity or coherency in unexpected ways?" Having committed Scripture to memory, just for the fun of it, made it a pretty easy exercise, though the outcome documented below was not anticipated. Nevertheless, Scripture proved more exacting and subtle under strict processing protocols which allow computers to follow millions of "commands" and "instructions." What became apparent was that the greater the rigor applied to Scripture, the more precision it seemed to exhibit. This has been especially so for Genesis chapters 1 to 21.

In natural or synthetic language processing protocols, all "variables" or "symbols" must be *explicitly declared*, before use, and their attributes prescribed. This helps avoid ambiguity as symbols appear in diverse contexts, and it greatly aids detecting and resolving incongruence. Simply put, before a symbol or variable can be used later, it must have a defining precedent that determines its subsequent meaning. Genesis sets forth the starting "architecture" and "infrastructure" for all subsequent testimony throughout Scripture. Therefore, type-precedent and explicit declaration, of a plethora of symbols and circumstances, are carefully examined in this beautifully detailed book. If consistent, then any detected exception requires consideration of a new precedent. The writer became intrigued by the precision apparent in the book, humbled as more and more chapters were committed to memory.

It was only natural, after committing so many precedent-setting symbols and circumstances to memory, to consider that these were fundamental constructs and definitions, similar to the initial steps in very large, complex computer command sequences. However, the scope, encompassed by Scripture, includes all of humanity and the initiation of the known universe. At Caltech, one learned to consider systems of equations, multidimensional spaces spanned by vectors, and equations defining population behaviors—all while limiting oneself to well-established principles or laws. Scripture, *initially*, was analyzed with the same sense of discovery, though with little expectation regarding deductive insight. The writer expected more of a qualitative narrative, which the text initially appeared to be—especially the early chapters in Genesis. The facts below may show a much, much different outcome.

How can one praise the testimony and the profound Name identified as the source of "discrete revelations" which exhibit his purpose, through human agencies, referenced in the text? In fact, how can one praise the Name, through whom this book claims is the beginning of understanding and life? How can one be inspired, through faith, to perceive the awesome majesty, glory and power, of the One that Scripture discloses formed the entire world? How can source text, more precise than artificial computer languages, reveal the sentimental Father who manifests again and again his love for the child with whom he is very pleased? How does his affection become real, today, as it has been across almost 70 centuries? How can what appears a qualitative abstraction be in harmony with a will that always considers the affairs of mankind, delighting in even the tiniest manifest desire to pursue goodness and righteousness, whenever possible, through all manner of human endeavor?

And why would the writer, after committing the books noted to memory, be moved to write the following:

a *7th-Gift* page

"Therefore, as a servant out of deference to the Name contained in Scripture, and with gratitude to him: Blessed is the Name of the I-SHALL-BE [Exodus 3:14], the God of Abraham and the God of Isaac and the God of Jacob; may his mercy and grace be enjoyed by the children who know him today, have known him in the past, and shall know him in the future; may his love and forgiveness be upon the fathers who keep his Commandments [Exodus 20:1-17]; and may his grace be received through the Name he gave to the whole world, the Spirit of truth anointing the One and Only, for the salvation of Israel and the entire world—all praise be to his Name, Amen.

"And now, as permitted, through the confidence imparted from the Spirit of truth, with all due respect to the One who sits on the Throne of Heaven, in front of which is the 7-fold lampstand, a testimony with the forgiveness of the Name of the One and Only, submitted by this servant after 40 days (not all consecutively) of fasting (no food or water) during an 84-day (sunset to sunset) period; a testimony embraced for as many as 40 days (not all consecutively) out of 55; and for as many as 180 days (not all consecutively) in a 12 month period, and all 7 days during one week, while being moved through prayer to understand these verses:

> Do not work for food that spoils, but for food that
> endures to eternal life which the Son of Man will
> give you; on him God the Father has placed his seal
> of approval. John 6:27

> Whoever eats my flesh and drinks my blood has
> eternal life and I will raise him up at the last day.
> For my flesh is real food and my blood is real drink.
> John 6:54,55

"What is written is not for my benefit or yours, but for the benefit of another, who may come to know that my Father's Son really did rise from the grave after three nights, and will judge this servant, within the Perfect Mercy without favoritism, according to the Law [Exodus 20:1-17]. Holy! Holy! Holy! is the I-SHALL-BE, the Almighty, who was, and is, and is to come! Amen."

Writing such words is as far from the early Caltech/NASA/IBM background of this writer as Jupiter is from earth. How did such change come about from committing books and chapters of Scripture to memory? What is the effect upon a person who applies artificial language processing methodologies to Scripture committed to memory? What might such perspective reveal about "man" or the "Son of Man"?

How can holding Scripture, and testimony therein, to the rigor and precision that guides the most complex machines ever made, be useful, productive or beneficial? Why even expose a delicate, ancient text to such harsh mechanical protocols? What does such technical rigor have to do with the text's contents which include poetry, miracles, nations, wars, and the most sublime examples of greed and generosity, despair and hope, famine and abundance, cruelty and abiding love that time and time again overcomes, advances and, through patient endurance, compels appreciation, affection, admiration, awe, allegiance, acceptance and atonement? Who is "man," through whom, and upon whom, and about whom, Scripture unfolds every revelation?

A.S.

Chapter 1

Living Dust

Let's begin with the first formed man, Adam. As it is written:

> —the [I-SHALL-BE] {Exodus 3:14} God formed the man from the dust of the ground and breathed into his nostrils the breath of life, and the man became a living being. Genesis 2:7

This is the first and only human being explicitly declared as "formed" from the "dust of the ground" and having the "breath of life" breathed into his nostrils by God himself. He shall be designated a *Type-1.0* person through this explicit declaration of the first male who was created.

Here is the account of Eve.

> So the [I-SHALL-BE] God caused the man to fall into a deep sleep; and while he was sleeping, he took one of the man's ribs and closed up the place

with flesh. Then the [I-SHALL-BE] God made a woman from the rib he had taken out of the man, and he brought her to the man. Genesis 2:21,22

Eve shall be designated a *Type-2.0* person, the first person created after Adam. A rib was taken out of Adam and used to form Eve into a woman, thus creating the first "male-female" pair called "man":

> Then God said, "Let us make man in our image, in our likeness, and let them rule over the fish of the sea and the birds of the air, over the livestock, over all the earth, and over all the creatures that move along the ground." Genesis 1:26

> So God created man in his own image, in the image of God he created him, male and female he created them. Genesis 1:27

> He created them male and female and blessed them; and when they were created he called them "man." Genesis 5:2

The term "wife" appears in Genesis 2:24 *AFTER* Eve was first declared a "woman" in Genesis 2:23:

> The man said, "This is now bone of my bones and flesh of my flesh; she shall be called 'woman,' for she was taken out of man." Genesis 2:23

> For this reason a man will leave his father and mother and be united to his wife, and they will become one flesh. Genesis 2:24

> The man and his wife were both naked, and they felt no shame. Genesis 2:25

Here, also, is the first usage of "father" and "mother," along with "wife." These terms are introduced by the "narrative voice" and are not ascribed to what Adam and Eve (the name given her later in Genesis 3:20) called one another at the time of the "bone of my bones" declaration.

Please make special note that Adam *ALWAYS* refers to Eve as "the woman," even after the above verses, later giving her the name "Eve" in Genesis 3:20. The Creator, when speaking to Adam, always refers to Eve with the term "your wife" (Genesis 3:17) and never as "the woman," except when speaking about her to the serpent in Genesis 3:15 ("I will put enmity between you and the woman . . ."). When speaking to Eve about Adam, the Creator always uses "your husband":

> To the woman he said, "I will greatly increase your pains in childbearing; with pain you will give birth to *children*. Your desire will be for *your husband*, and he will rule over you." Genesis 3:16

> To Adam he said, "Because you listened to *your wife* and ate from the tree about which I commanded you, 'You must not eat of it,'

> "Cursed is the ground because of you; through painful toil you will eat of it all the days of your life." Genesis 3:17

In the midst of this, the precedent of "children" is explicitly declared as a future new "type," different from Type-1.0 or Type-2.0, the types of Adam and Eve, since children would later come through "childbirth" from Eve, she to be the "mother of all the living" [see below].

Not once does Adam, himself, ever call Eve his "wife," based on what is written, though he gave her the name Eve. In her "discussion" with the serpent and events surrounding the serpent's deception, she is referred to as "the woman" between Genesis 3:1 to Genesis 3:7. Thus, the first two specific references to Eve as "wife" are in Genesis 2:25 and, later, Genesis 3:8 when she and Adam are "in hiding":

> The man and his *wife* were both naked, and they felt no shame. Genesis 2:25

> Then the man and his *wife* heard the sound of the [I-SHALL-BE] God as he was walking in the garden in the cool of the day, and they hid from the [I-SHALL-BE] God among the trees of the garden. Genesis 3:8

Does Adam refer to her as his wife, even after this? No, he does not:

> The man said, "*The woman* you put here with me—she gave me some fruit from the tree, and I ate it." Genesis 3:12

Thus Eve, as the original Type-2.0 person, sets the precedent of being the first divinely-formed-female made from a part (i.e. a rib) of a man. Afterward, she is referred to as the first "wife" and later becomes the first mother. The man, from whom the rib is taken to form her, is designated her "husband" for the first time in Genesis 3:6:

> When the woman saw that the fruit of the tree was pleasing to the eye and good for food, and also desirable for gaining wisdom, she took some and ate it. She also gave some to her *husband*, who was with her, and he ate it. Genesis 3:6

a *7th-Gift* page

Chapter 2

Cain: A New "Kind" of Person

After these events, it is later written about Cain:

> Adam lay with his wife Eve, and she became
> pregnant and gave birth to Cain. She said, "With
> the help of the [I-SHALL-BE] I have brought forth
> a man." Genesis 4:1

Eve's joy and profound expression of accomplishment
is altogether fitting because, prior to Cain, only the Creator,
himself, through the miracles that are his alone, brought into
being the first people, Type-1.0 Adam and Type-2.0 Eve. It also
affirmed that the ability to bear children, an attribute only of
Type-2.0 (explicitly declared as female and woman and wife),
was not destroyed by the serpent's deception in the Garden
of Eden. The Creator's intervention assured Eve would not
be barren and, indeed, she would become the first and most
fertile woman in the history of all mankind, rebutting the
devastation the serpent attempted. (Within the "painful toil,"

assigned to Adam, is the promise of food "all the days" of his life; his family had sufficient harvests whenever the ground was worked, except for Cain [Gen 4:12]. By Adam, food filled all the earth. [Gen 2:5; 3:23])

Furthermore, with Cain is established the precedent of the first son to be "born" on earth, an exceptional proto-birth, resulting from the first pregnancy on earth. Adam and Eve did not originate through pregnancy and birth. Therefore, Cain is designated a *Type-3.0* person, identified through the explicit declaration that "Adam lay with his wife Eve, and she became pregnant and gave birth to Cain."

Regarding Abel, the fourth person in the first family, it is written:

> Later she gave birth to his brother Abel. Now Abel
> kept flocks and Cain worked the soil. Genesis 4:2

Notice there is no explicit declaration, "Adam lay with his wife Eve . . ." (Genesis 4:1), as is the case preceding the birth of Cain. Nor is there a declaration that, "Adam lay with his wife *AGAIN* . . ." (Genesis 4:25), as is the case preceding the birth of their third son, Seth. The *AGAIN* is explicitly declared regarding Seth. There is no explicit "*AGAIN*-declaration" regarding Abel because of RIFT-1 (RIFT meaning "Revealed in Future Testimony").

> Adam lay with his wife *AGAIN*, and she gave birth
> to a son and named him Seth, saying, "God has
> granted me another child in place of Abel, since
> Cain killed him." Genesis 4:25

Abel becomes a proto-brother/sibling in his own right, as well as the first "son" whose blood is shed at the hand of another (in this case, by his own brother Cain). In both being the first "brother-sibling" and the first fatality through the shedding of innocent blood, leading to the loss of life from "unnatural causes," Abel receives his own designation as *Type-4.0*.

Seth's birth establishes an entirely, new *Type-5.0*, as he is the first son born in the *likeness and image* of his father—an attribute not explicitly ascribed to either Cain or Abel (Genesis 5:3). To summarize:

Type-1.0: Adam; the one and only Type-1.0 in the history of mankind.

Type-2.0: Eve; the first woman, formed by the Creator from her husband's rib.

Type-3.0: Cain; proto-birth, not a "formed" being like his father or mother.

Type-4.0: Abel; proto-sibling by childbirth (missing "lay again" declaration).

Type-5.0: Seth; proto-image-son, "born" in his father's "own image and own likeness."

Chapter 3

A Wife for Cain

In Genesis 4:17, Cain is declared as having a wife, with whom he "lay." Prior to this verse, only Adam had a wife, at all. Cain's wife gives birth to a son—the first female, other than Eve, to do so. Cain is the first *born* person (not formed) to become a father and have a son (Enoch). Enoch is not declared to be in his father Cain's "own image and own likeness." (Seth is, later, explicitly declared to be in Adam's "own image and own likeness"; not once is such declared in the Adam-Cain line which is composed of distinct MEN; the image-and-likeness attribute is in the Adam-Seth line of MAN, " . . . And when they were created, he called them 'MAN.'" [Genesis 5:2]):

> Cain lay with his wife, and she became pregnant and gave birth to Enoch. Cain was then building a city, and he named it after his son Enoch. Genesis 4:17

Here, we encounter a GREAT silence in this verse. Read it again. Did you hear the SILENCE? That's right, it has to do with "his wife." Who is Cain's "wife"? From where did she come? There are two Positional Declarations:

1.   Positional Declarations regarding Cain's wife.

     1.1. She is a new "type" of "born" child who is not male.
     1.2. She is an existing type, previously declared.

Now 1.1 may seem the safer choice (just like those "obvious" answers on an SAT exam), based on present-day experiences regarding "sisters" or "daughters" being born. So, why not just designate her a new "Type-6.0" and move on?

Consider this: Cain's wife is first declared in Genesis 4:17. So? Well, the term "sister" does not first appear *any where* until Genesis 4:22, *five* verses later (remember, the writer is accustomed to an explicit type-declaration before using a "symbol," based on natural or synthetic language processing methodologies for precisely-specified grammars and languages). "Sister" has its next appearance 210 (two-hundred ten) verses, after Genesis 4:22, in Genesis 12:13.

"Daughter" does not appear, initially, until Genesis 5:4, or 13 verses after Genesis 4:17. Therefore, prior to Genesis 4:17, no "sister-type" or "daughter-type" is explicitly declared. That is to say, the reader may know what "sister" and "daughter" mean from experiential-inference. Yet, within the scope of this context regarding Cain's wife, no such terms are previously declared or specified, in any way, leading to Genesis 4:17.

If one commits the first 11 chapters of Genesis to memory, which the writer has done, one can't forget that the term "wife" *does* have precedent and explicit declaration, *before* the Genesis 4:17 instance regarding Cain's *wife*. Remember? That's right, the term is an attribute of a Type-2.0 person, the first of whom is Eve herself. Therefore, the choices, to resolve uncertainty relative to Cain's wife's introduction, in Genesis 4:17, are:

2. Positional Declarations to define "wife" in Genesis 4:17.

2.1. Use a *Forward*-Reference to "sister" in Genesis 4:22.

2.2. Use an Explicit-*Prior*-Declaration of "wife" that precedes Genesis 4:17:

1st: Genesis 2:25, "The man and his *wife* were both naked . . ."

2nd: Genesis 3:8, "Then the man and his *wife* heard the sound of the [I-SHALL-BE] . . ."

3rd: Genesis 3:17, ". . . because you listened to your *wife* and ate . . ."

4th: Genesis 4:1, "Adam lay with his *wife* Eve, and she became pregnant . . ."

The 2.1 Forward-Reference to "sister," in Genesis 4:22 (regarding Tubal-Cain's sibling introduced as Naamah), if Cain's wife is construed as a born person, attempts to establish the first instance of "incest" in Scripture—conduct expressly forbidden, later, by the Creator (Leviticus 18:9). However, Cain's wife is never *explicitly* referred to as a "sister" or a "daughter," at this time, though some may attempt such presumption for their own purposes. One might posit this notion based on modern-day life-experiences, for example,

a *7th-Gift* page

but it must be acknowledged as no more than that. As noted, to do so requires a Forward-Reference to a term or terms not introduced until later in Scripture, and which are not declared prior to Cain's wife's introduction. The 2.2 Explicit-Prior-Declaration and usage of "wife," *BEFORE* disclosure of Cain's spouse, is extant multiple times *leading up to* Genesis 4:17 ("Cain lay with his wife and she became pregnant . . ."). There are *four* corroborating instances of such prior-usage. To disregard the four *prior* instances of "wife" and assert a Forward-Reference, to some other precedent or usage (such as "sister" or "daughter"), disregards substantive, antecedent testimony and context established preceding Cain's wife's introduction. It also *injects* presumption that Cain and such a "sister" or a "daughter" engaged in incest by "divine will," the Creator compelled to establish such precedent, out of necessity, to expand the human family. Here, it is stated again: incest is conduct later explicitly prohibited, through commandments given Moses in the presence of the I-SHALL-BE (noted previously in Leviticus 18:9). A person who asserts "divine exemption" for such conduct, early in the human family, must discard the explicitly declared precedent of "wife" *four times*, to construe a whole new meaning, in its *5th* instance, within Genesis 4:17 regarding Cain's *wife*. Such reconstruction may be radically different from all previous usage of the term "wife."

Therefore, 2.1 requires a Forward-Reference to the yet-to-be declared "sister" type, to assert the "first," incestuous relationship, within the human family, through a "divine exemption." In this case, such assertion is made with no prior, supporting, scriptural evidence whatsoever.

The 2.2 Explicit-Prior-Declaration requires no new type or presumption at all (consistent with linguistic rigor), since the term "wife" is already an attribute of a Type-2.0 person—Eve being the proto-wife. What are the scriptural outcomes if the Wife-Of-Cain ("WO-Cain"), the first "unnamed wife," is indeed a Type-2.0 person, like Eve (i.e. formed from the rib of a husband)? First, it requires the I-SHALL-BE [Exodus 3:14], the Creator, the writer's Father, to make a wife for Cain as he had for Cain's father, Adam. This is an explicitly declared and disclosed ability of the writer's Father. The 2.2 Explicit-Prior-Declaration specifically *excludes* "incest out of necessity," since the writer's Father has no such limitation ("Is anything too hard for the [I-SHALL-BE]? I will return to you at the appointed time next year and Sarah will have a son." as noted in Genesis 18:14). The writer's Father is the I-SHALL-BE and, through his One-and-Only Name, he fulfills his purpose, in his own way, and at his own time, notwithstanding a presumed limitation which he does not have.

Applying the 2.2 Explicit-Prior-Declaration of "wife" allows WO-Cain to be a Type-2.0 (like Eve). What are the consequences of even considering this? It would mean that, for *some period* of time, males, *and only males*, were born (except for Adam) and all females were "divinely formed beings"—each made from the rib of a man called her "husband." (Being a Caltech grad without religious training, one learns to suspend personal experience and follow the "data" or "evidence" wherever it may lead *without* presumption. Though this was disconcerting to the writer, when he was permitted to discern it about 10 years ago, nevertheless, the scientific rigor and technical protocols, from the programming arts, helped develop the discipline to suspend assumptions for extended periods, while analyzing data to find programming exceptions and mistakes).

Is there any immediate evidence to support "male-only-births," for *any* period of time? There's none in this writer's personal experience. But what does the testimony of Scripture have to say? After the writer committed the first 11 chapters of Genesis to memory, the data analysis proceeded quickly . . .

The first three, and only, "named" children, of Adam and Eve, are male—Cain, Abel and Seth. The 4th "person," WO-Cain ("Wife-of-Cain"), was formed from a rib of Cain, based on the 2.2 Explicit-Prior-Declaration precedent of Adam and Eve (Eve being the first "wife"). Again, such is not within the life-experience of the writer or anyone else today (at least such has not been publicly disclosed), which may be why the 2.1 Forward-Reference, requiring intragenerational incest, seems plausible, to explain the pregnancy of the second-ever woman on earth.

However, the 2.2 Explicit-Prior-Declaration requires a careful examination of all, related Scripture because of multiple references to "wife" *BEFORE* WO-Cain's introduction. In sequencing through descendants from Adam to Terah (20 Generations in Genesis 5 and Genesis 11—recalled from memory), regarding "male-only-births," a female birth is neither explicitly declared *nor* explicitly excluded. It is disclosed that Adam's first, three children were all male, and that Adam *had* "other sons and daughters" (Genesis 5:4, which contains the first usage of "daughters") after his first three sons were born. It is disclosed that Noah, in the 10th generation from Adam, had three children, after Noah lived 500 years, and all three were males:

> After Noah was 500 years old, he became the father
> of Shem, Ham and Japheth. Genesis 5:32

Every first-named-child, in each generation between Adam and Noah, is male. We know that Terah, the father of Abram, 10 generations from Noah, only had three children, by birth, after he lived 70 years—and all were boys:

> After Terah had lived 70 years, he became the father
> of Abram, Nahor and Haran. Genesis 11:26

In the 10 generations from Noah to Terah, every first-named-child is male. Thus, across 20 generations, from Adam to Terah, every first-named-child is male. In any generation, where more than one child is named, every named-child is male (three for Adam, three for Noah and three for Terah—and Noah and Terah, each, had only three born-children).

Thus, in the first 20 generations, from Adam to just before Abram in the 21st generation, the names listed first, for *all* those generations, are male. And if in fact each first-named-child is the first-born in each generation, which is confirmed by RIFT-2 (as is the case with Adam's children, and Noah's children, and Terah's children), then the probability of 20 generations of consecutive first-born-males is one in one million (2 x 2 x 2 x 2 x 2 x 2 x 2 x 2 x 2 x 2 x 2 x 2 x 2 x 2 x 2 x 2 x 2 x 2 x 2 = 1,048,576). These are incredible odds, based on present-day life experiences! However, Scripture offers no explicit contradiction, relative to what is recorded about the first 20 generations from Adam to Terah. One can presume otherwise, which the one in one million odds compel, but such presumption is made in the silence of Scripture since there are no named daughters, *at all*, for the first, 20 generations of "MAN"—that is, the direct descendants of the Adam-Terah line.

Yet, rigor compels acknowledgement that the presumption (of male-only births, for a period of time) is not conclusive, because it is repeatedly declared that each first-named-male *had* "other-sons-and-daughters," within these 20 generations (except Noah and Terah, in the 10th and 20th generations, since both only had sons, by birth). "Other-sons-and-daughters" is explicit in Genesis 5:4, 5:7, 5:10, 5:13, 5:16, 5:19, 5:22, 5:26, 5:30, 11:11, 11:13, 11:15, 11:17, 11:19, 11:21, 11:23 and 11:25. This totals 18, independent, generational witnesses (Genesis 11:13 having two) that each first-named-male had "other sons and daughters," in every one of the 20 generations (except for Noah and Terah who, each, only had three born sons). Again, this is not conclusive evidence that only males were born, for this period of time, nor is there explicit evidence of a named-daughter being born since every female, in the Adam-Seth line, is unnamed for these 20 generations (except Eve).

Additionally, the presumption of a patriarchal, cultural bias must be considered (a position some may feel is obvious from the out set, right? Think about it: if this writer injects a rhetorical, parenthetical question like this, what do you think is about to happen?). That is, the names and births of daughters were explicitly excluded due to the "status of women," in those times. This may seem reasonable, based on one's, own, contemporary/cultural/religious inferences, and in light of recorded history, across multiple cultures.

Yet, as the writer's Father moves him, one must "follow the data" instead of merely projecting the traditions of men upon Scripture, for whatever personal/cultural/religious/economic motive that may seem reasonable, because "so many others have."

Is there any other evidence or witnesses which may shed light? A resounding "yes"! If you recall, Adam had two lines of descent recorded in some detail: Seth's line, extending to Abram; and also Cain's line. What of Cain, who went away after killing Abel? Why study his line? As a Nobel Prize winner might be asked, "Why study a fruit fly"? In considering the Adam-Cain line, something stands out, for five generations (Cain to Lamech), which is not apparent in the Adam-Seth line: there is no explicit contradiction to male-only-births:

> Cain lay with his wife, and she became pregnant
> and gave birth to Enoch. Cain was then building a
> city, and he named it after his son Enoch. To Enoch
> was born Irad, and Irad was the father of Mehujael,
> and Mehujael was the father of Methushael, and
> Methushael was the father of Lamech. Genesis 4:17,18

The premise, of male-only-births, for a period of time, receives even stronger corroboration through the Adam-Cain line. A very, prominent peculiarity is apparent in the Adam-Cain line, for five generations: not once is it declared that the first-named-male, in any of these generations, "had-other-sons-and-daughters"—that is, had *MULTIPLE* offspring. Zero. Zip. Nada. None. Why is this relevant? Because the Adam-Seth-Terah line virtually SHOUTS "had-other-sons-and-daughters," across 20 generations, with 18 specific references (except for Noah and Terah, where neither is declared to have had another born-child after each's third son).

It looks like it's time for Positional Declarations, on why "had-other-sons-and-daughters" is missing from the Adam-Cain line, for all five males in Genesis 4:18:

3.   Positional Declarations on "had-other-sons-and-daughters" in the Adam-Cain line:

3.1. It is implied, but not written, even though it appears 18 times in the Adam-Seth line.

3.2. For five generations (Enoch to Lamech), each born-male is an only child.

Now 3.1 seems a safe choice since it presumes precedent from the Adam-Seth line. However, the precedent is clear, from the Adam-Seth line, that sons and daughters, beyond those listed, in a particular generation, *ARE* explicitly declared with "had-other-sons-and-daughters." To ascribe "had-other-sons-and-daughters," where no such explicit declaration exists, may be a presumption, at best, and, perhaps, adding "had-other-sons-and-daughters" where such is explicitly *excluded*. "Had-other-sons-and-daughters" is absent in the Adam-Cain line (except for Adam). After Lamech's actions, in Genesis 4:19, things are never quite the same for the Adam-Cain line (nor for the Adam-Seth line, as well) because Lamech is the first man who "marries" a wife not formed from his own rib, and he does so twice. This results in the birth of male and female offspring, in the Adam-Cain line, for the first time, after its 7th generation.

In the Adam-Cain line, therefore, is sustained precedent for male-only births, across multiple generations after Adam, with no reference whatsoever to "had-other-sons-and-daughters."

a *7th-Gift* page

Chapter 4

Lamech: The First Married Man

As it is written, Lamech married twice, the first person in all of Scripture to ever marry anyone:

> Lamech married two women, one named Adah and the other Zillah. Genesis 4:19

At no time is the term "married" used, prior to this verse, regarding a Type-2.0 person, of whom Eve is the prototype and WO-Cain is the same type—a female formed from a rib of her respective husband.

4.   Positional Declarations regarding unnamed wives in Genesis 4 and 5.

4.1. Wives' names are omitted based on cultural tradition.

4.2. All Type-2.0 wives are unnamed, except Eve,
and never described as "married."

Now 4.1 seems reasonable as we traverse through the
Adam-Cain line (except for Adam), since it is clear that
Cain, Enoch, Irad, Mehujael and Methushael all had wives.
However, the pattern is explicitly broken with Lamech (a
"data exception" in language processing protocols), where
it specifically declares that he "married." Not once is such
declared for any of his ancestors, including Adam.

4.2 adheres to the primary Positional Declaration that, for
some period of time, each male was born, and each female
was formed from the rib of her respective husband. Such a
union shall be designated a "created-couple," consistent with
the context in Genesis 2:27 and Genesis 5:2:

> So God created man in his own image, in the image
> of God he created him, male and female he created
> them. Genesis 2:27

> He created them male and female and blessed
> them. And when they were created, he called them
> "man." Genesis 5:2

"Created-couples" are the only type of husband/wife
pairs, until the explicit declaration that Lamech "married,"
establishing a new kind of husband/wife pair that shall be
designated a "married-couple."

5.   Positional Declarations regarding couples.

5.1. Married-couples are the same as created-couples.
5.2. Married-couples are different from created-
couples.

Prior to the flood, a married-couple resulted when a male and a female became husband and wife, and the wife was not formed from her husband's rib. Such marriage is by the husband's desire and appears less dependent on the Creator's will. The first married-couple consists of one husband and one wife, until Lamech marries a second time. Since Lamech married Adah and Zillah, he sets the precedent of marriage and then multiple marriages. Both of his wives have names, which is distinct from created-couples, where the wife is unnamed, except for the first created-wife, Eve.

Therefore, the disclosure of the first, two "named-women," after Eve, is because they became part of a married-couple (in this case consisting of two wives, setting two extraordinary precedents), making them different from all, prior "unnamed-women," who were within created-couples. This offsets the presumption that all women's names were excluded due to a patriarchal cultural bias, alone (the writer warned ahead of time this sentence was coming). When the created-couple precedent of Adam and Eve is first broken, in the 7th generation of the Adam-Cain line, women are explicitly named, and the two referenced stand out. Their marriage is unlike any relationship or status of their mother-ancestor, Eve, and each, subsequent, physically-similar Type-2.0, who was also a created-wife.

Chapter 5

The First Widow on Earth

Lamech becomes part of the first, explicitly-declared married-couple, revealed in the 7th generation from Adam and Eve, within Cain's line. Lamech's wife has a name, Adah, as does his other wife, Zillah. Now for some important Positional Declarations:

6.   Positional Declarations regarding childbirth.

     6.1. A created-couple gives birth to sons and/or daughters.

     6.2. A created-couple gives birth to one or more males, but only to males.

6.1 makes sense from modern-day inferences and experience (however, as in quantum mechanics, personal experience is not always applicable). In addition, there are

18 references to "had other-sons-and-daughters," for the fathers listed between Adam and Terah, in Genesis 5 and Genesis 11. It is only natural to presume that created-couples gave birth to both sons and daughters (this statement is rhetorical). There is explicit declaration regarding the births of sons (Cain, Abel and Seth, for example). Yet, not once is there an explicit declaration of a daughter's *birth* in the *direct* Adam-Seth-Terah line, for 20 generations, though it may *seem* otherwise. Additionally, female offspring are unnamed, in the Adam-Seth-Terah line, for 20 generations. They are also unnamed, in the Adam-Cain line, until Lamech (in the 7th generation from Adam).

6.2 is corroborated within the Adam-Cain line, across five generations (Enoch to Lamech). In fact, in the Cain-Methushael line, for five generations, each created-couple has an *only-child* and each is always male (Enoch, Irad [the father of whom is RIFT-3], Mehujael, Methushael and Lamech), while the created-couples, in the Adam-Seth line, have multiple offspring *every* generation. Contrast the missing "had-other-sons-and daughters" within the Adam-Cain line:

> To Enoch was born Irad, and Irad was the father
> of Mehujael, and Mehujael was the father of
> Methushael, and Methushael was the father of
> Lamech. Genesis 4:18

Careful note must be made that Enoch is not explicitly declared to be a father, a title affirmed five times—twice before and three times after him—in the Adam-Cain line, from Adam to Lamech, though it may seem obvious to claim he is a father. For five generations, wives, in Cain's line, each had an only-child. This is interrupted when Lamech marries Adah and Zillah—neither woman formed from his own rib. He went on to have *children* (the first instance of multiple offspring from a descendant of Cain) with each wife:

> Adah gave birth to Jabal; he was the father of those
> who live in tents and raise livestock. His brother's
> name was Jubal; he was the father of all who play
> the harp and flute. Genesis 4:20,21

It is not explicitly disclosed that Lamech "lay with his wife," which is the precursor expression to "gave birth," for all named-males conceived by both Adam and Cain, except for Eve's conception of Abel (Gen 4:2).

There is one small matter that must now be addressed. If each formed-female was created from her husband's rib, how did Lamech find two, unattached women to marry? This naturally follows from the premise of male-only births and created-couples. Additionally, why did Lamech marry at all, since no one else had? Apparently, after Lamech had lived RIFT-4 years, he had not become a father nor produced a son like his ancestors. Being in the 7th generation from Adam, and without an offspring, could have represented the end of the Adam-Cain line. This is because created-couples, in this line, were limited to one, and only one, son. To be without a created-wife, or with a barren one (based on RIFT-5 and the usage of "wives" in Genesis 4:23) would set Lamech apart from the fecund precedent of his ancestors.

7.   Positional Declarations regarding the prior husband's of Lamech's wives.

7.1. Their husbands did not predecease them.

7.2. Their husbands predeceased them.

a *7th-Gift* page

Carefully consider what Lamech says in Genesis 4:23:

> Lamech said to his wives, "Adah and Zillah listen to
> me; wives of Lamech, hear my words. I have killed
> a man for wounding me, a young man for injuring
> me. If Cain is avenged seven times, then Lamech
> seventy-seven times." Genesis 4:23,24

Who are the males Lamech killed? Why is he compelling his wives, including both Adah and Zillah, to pay attention to his declaration regarding these men's deaths? Did Adah and Zillah "know" any of the males Lamech killed? With the precedent of Type-2.0, Adah and Zillah would have had husbands before "marrying" Lamech (there is no explicit declaration that they were "previously married"). If the males Lamech killed included Adah's and Zillah's husbands, whose ribs were used to form these two women, then Adah and Zillah endured the unthinkable. And if Lamech killed and killed again, under the pretext of being "wounded" or "injured" in some way, then he may be putting the wives he "married" on notice regarding his unique status in the world, following the precedent of Cain murdering Abel (a single man without a wife).

The name of Adah's husband is RIFT-6, and the name of Zillah's husband is RIFT-7, making Lamech the first serial killer. How severe were his "wound" and "injury" by those he killed? The fact that he invoked the precedent of Cain, who was deeply "hurt," from a *PERCEIVED* "wrong," gives testimony to the nature of Lamech's justification of murder without cause. And was not Cain's anger based on the fact that he thought to make an offering, *first*, only to be upstaged, by Abel's better offering, later [Genesis 4:3-7]? How many times has a greater blessing been received, by a younger sibling, to the detriment of a first-born (Isaac? Jacob? Joseph? David?).

And did not Cain actually attempt a superior offering than Abel's? Abel provided only the blood of sacrificed animals, while Cain's sacrifice became that of the blood of a man, in an attempt to restore his perceived "entitlement," from his "proto-offering." Was his first offering (of crops) in humility to the Creator, or to coerce recognition and elevation of himself, by the precedent he attempted to set, yet which now belongs to Abel forever?

Chapter 6

The First Born Daughter on Earth

Did anything else unusual occur, after Lamech entered into his "multiple" marriages, something that may not have happened in the world, before his first "marriage"? Relative to Adah, nothing seems unusual by her having two sons, Jabal and Jubal (Genesis 4:20,21), continuing the precedent of male-only-births (though this is the first multi-offspring event in the Cain line). Is there anything unusual about Zillah, his other wife, giving birth?

> Zillah also had a son, Tubal-Cain, who forged all
> kinds of tools out of bronze and iron. Tubal-Cain's
> sister was Naamah. Genesis 4:22

Wait just a moment. What about the last part of this verse: "Tubal-Cain's sister was Naamah." Herein is the first "sister" and the first "born-female" on earth.

Naamah is the first female ever to be designated as something other than a "wife" or "woman," in the pattern of Eve and 20 generations of wives, from Adam to Terah, as well as six generations of wives, from Cain to Lamech. It is submitted that Naamah is the first female to be "born." She was not "divinely formed," by the hands of the Creator, from the rib of a designated male, as was the case for all women before her. Lamech is the first man on earth to have a "born-female" offspring, from his marriage to a wife who was not "rib-formed," from Lamech, by the Creator. In having a born-daughter, Lamech could claim what, up to that time, had been exclusive to the Creator—forming females. Additionally, Naamah is the first woman to be free of any "pre-existing" relationship with a "husband." All of this was initiated through the institution of "marriage" and Lamech creating the first widows, by shedding the blood of other males.

Such an extraordinary precedent finds corroboration in Genesis 6:1,2:

> When men began to increase in number on the earth and daughters were born to them, the sons of God saw that the daughters of men were beautiful, and they married any of them they chose. Genesis 6:1,2

There are specific points to consider, through Positional Declarations, regarding these verses:

8.   Positional Declarations regarding Genesis 6:1,2.

8.1. "MEN" are Cain's descendants; "sons of God" are descendants of the Adam-Seth line called "MAN."

8.2. When "MEN" began to increase in number on the earth, they had daughters—not before.

8.3. The "sons of God" ("MAN") had never "married" before the time of Gen 6:1.

8.4. Prior to "MEN" having daughters, the "sons of God" waited on God to form their wives.

8.5. The created-wives, of the "sons of God," are "unnamed," except Eve.

8.6. A wife formed from her husband's rib was chronologically matched to her husband.

8.7. Later "sons of God" condemned themselves, marrying the daughters of MEN.

8.8. The offspring of the married "sons of God" and their wives were called the Nephilim (Genesis 6:4).

"Sons of God" has specific reference in Psalm 82:6:

> "I said, 'You are "gods"; you are all sons of the Most High.' But you will die like mere men; you will fall like every other ruler." Psalm 82:6

In the case of the Messiah, who would be called "THE Son of God," such a declaration anchors itself in Psalm 82:6, because it is written:

> [I-SHALL-BE-SALVATION] answered them, "Is it not written in your Law, 'I have said you are gods'? [Psalm 82:6]. If he called them 'gods,' to whom the word of God came—and the Scripture cannot be broken—what about the one whom the Father set apart as his very own and sent into the world? Why then do you accuse me of blasphemy because I said, 'I am God's Son'?" John 10:34-36

Thus, "sons of God" references *people* to whom the Word of God came, people like prophets and those living righteous lives according to the Creator's will—upon receiving his Word.

The "sons of God," in Genesis 6:1,2, consist of faithful descendants in the Adam-Abram line. They are different from the "men," who had daughters, in the Cain-Lamech line.

9.   Positional Declarations regarding the two "lines".

9.1. The Creator did not mind if the Adam-Cain line and the Adam-Seth line mixed.

9.2. The Creator desired that the Adam-Cain line and the Adam-Seth line never mix.

As we will see in chapter seven, the two lines were never to mix and, if they did, there would be devastating consequences for each line. The Adam-Seth line contained the promised "seed" leading forward to the salvation of Israel and the entire world.

"And I will put enmity between you and the woman, and between your seed and hers; he will crush your head, and you will strike his heel." Genesis 3:15

a *7^h-Gift* page

Chapter 7

Why Noah Was, and Is, Very Special

And what did the Creator do when these lines "mixed" and the "sons of God" took it upon themselves to marry a wife instead of waiting to become a husband, within a created-couple?

> Then the [I-SHALL-BE] {Exodus 3:14} said, "My Spirit will not contend with man forever, for he is mortal; his days will be a hundred and twenty years." Genesis 6:3

> The [I-SHALL-BE] saw how great man's wickedness on the earth had become, and that every inclination of the thoughts of his heart was only evil all the time. Genesis 6:5

> So the [I-SHALL-BE] said, "I will wipe mankind, whom I have created, from the face of the earth— men and animals, and creatures that move along

the ground, and birds of the air—for I am grieved
that I have made them." But Noah found favor in
the eyes of the [I-SHALL-BE]. Genesis 6:7,8

It should be noted that Noah is in the sons-of-God-male-class
("MAN") and not the men-male-class ("MEN"). How did Noah
find favor, over all the people who lived in his time, and why
did he walk with God (Genesis 6:9), like Enoch before him
(Genesis 5:24)? Because, as is repeatedly declared, Noah always
did everything God commanded him:

> Noah did everything just as God commanded him.
> Genesis 6:22.

> And Noah did all that the [I-SHALL-BE]
> commanded him. Genesis 7:5

The most profound example of Noah's faithfulness, as a
"son of God," is revealed after he is 500 years old:

> After Noah was 500 years old, he became the father
> of Shem, Ham and Japheth. Genesis 5:32

Now, from Genesis 6:1, we know that unattached, beautiful
women, called "the daughters of men," started to be born
when "men began to increase in number on the earth . . ."
The first such daughter was Naamah, born to Lamech's wife
Zillah, one of the wives Lamech put on notice about his ability
to kill with divine retribution against anyone who harmed him.
Since he killed his wives' husbands, he, too, had the fear Cain
expressed after murdering Abel:

> Cain said to the [I-SHALL-BE], "My punishment
> is more than I can bear. Today you are driving
> me from the land, and I will be hidden from your

presence; I will be a restless wander on the earth
and whoever finds me will kill me." Genesis 4:13,14

Cain was reassured that anyone who killed him would suffer
"vengeance seven times over" (Genesis 4:15). Lamech embraced
this precedent, to put his wives and anyone else on notice that
he, too, was protected by the earlier "divine exemption."

10. Positional Declarations regarding co-deceasing
husbands and wives.

10.1. Spouses within created-couples do not
co-decease.
10.2. Spouses within created-couples co-decease.
10.3. Adah and Zillah were the first wives to have
their husbands predecease them.
10.4. A created-wife is "chronologically matched"
to her husband through his rib.

In other words, a husband and wife, as a created-couple,
co-decease, unless "unnatural" causes ends one of their lives.

11. Positional Declarations about co-deceasing spouses
in the Adam-to-Terah line.

11.1. At no time is it declared that anyone in the
Adam-to-Terah direct-line had multiple
wives.
11.2. At no time is it declared that a spouse
predeceases his or her mate in this line.
11.3. In the event of such a declaration, a new
precedent would be established.

(The unique circumstances after Abram, in the Adam-Terah line, will be addressed shortly.) Noah patiently endured 500 years, when other Adam-Seth descendants chose to marry the "daughters of men" rather than be faithful, to the Creator, and let him make a wife for each of them. Cain set in motion a precedent that resulted in an only-child for his descendants (to avoid repeating what happened to Abel). Even so, Lamech killed other males, under the pretext of being "wounded" or "injured," based on his subjective perception of what he construed as "harm," embracing Cain's "divine exemption." Whatever his self-justification, Lamech put himself in the position to take the wife of each man he killed (or threatened to kill), anticipating the precedent of earthly "kings" and "kingdoms," after the flood (Genesis 10:8-10), perpetuated across time (even within the original, written laws of the most powerful nation on earth today).

These women may have been devastated by the loss of their husbands. Before this, no wife is disclosed as losing a husband (Abel was never declared one). Adah and Zillah, as the first widows on earth, may have experienced vulnerability that goes beyond losing a limb or body part. Again, Lamech exalts himself by *proclaiming* his protected status to them and everyone else, using "divine exemption," which he anoints himself, to construe his conduct as "legal":

> Lamech said to his wives, "Adah and Zillah listen to
> me; wives of Lamech, hear my words. I have killed
> a man for wounding me, a young man for injuring
> me. If Cain is avenged seven times, then Lamech
> seventy-seven times." Genesis 4:23,24

Recognize that, after Cain murdered Abel, there was no flood or condemnation of all mankind, in subsequent generations, based on Cain's act alone. However, the violence Lamech set in motion, upon the world, directly affected the covenant between husband and wife that the Creator holds dear to his heart—to this day—as his exclusive domain, even as Lamech married two women—innocent widows of males whom he killed. From one of his wives, Zillah, came forth the first daughter, by birth—never before achieved by anyone. And more daughters were to follow as events were set in motion leading up to the destruction of the whole world:

> When men began to increase in number on the earth and daughters were born to them, the sons of God saw that the daughters of men were beautiful, and they married any of them they chose. Genesis 6:1,2

The permission to marry such daughters seems wholly absent. Like Samson later, whomever a son of God wanted became that son of God's wife:

> When he returned, he said to his father and mother, "I have seen a Philistine woman in Timnah; now get her for me as my wife." Judges 14:2

When the "sons of God" (i.e. descendants, in the Adam-Seth, line, called "MAN," juxtaposed to "MEN" representing the Adam-Cain line) married "the daughters of men," they became co-conspirators with what Lamech set in motion by marrying women he made widows. This resulted in the first, "born" daughter, breaking the precedent of divinely made females by the hand of God, as was the case for all women prior. Noah's faithfulness, therefore, is awe inspiring, since all of his ancestors had children, some as early as 65 (Mahalalel and Enoch, in Genesis 5:15 and Genesis 5:21, respectively), and definitely no later than the late hundreds (Methuselah was 187). As Noah surpassed 100, then 200, then 300, then 400, it could have appeared he might be the first man not to have a wife. In fact, past 187, he was older (as recorded in Scripture) than anyone ever had been, upon first becoming a father (at 187, Methuselah had his first, born son).

Being so advanced in age, Noah was a candidate to be the first man without the "apparent" blessing of children and a legacy in the world. The pressure, therefore, to take one of the daughters of men, at the urging of others, who had done so and had growing families, could have been very tempting—especially motivated by the same agency who attempted to destroy the first created-couple, Adam and Eve, but failed.

Each marriage, between a "son-of-God" and a "daughter-of-men," expanded the conspiracy and deeply grieved the I-SHALL-BE. Yet, Noah did not yield to the temptation, pressure, or even his own desire to be a husband and father, outside of the clearly set precedent to which Adam, Seth, Enosh, Kenan, Mahalalel, Jared, Enoch (Adam-Seth line), Methuselah and Lamech (Adam-Seth line) all adhered. Still, what was said about Noah by others, or by the adversary of all mankind who acted first in the Garden of Eden, must have been unsettling. Recognize that if Noah had been unfaithful, and married along with others, he would have been included with those written of later, as declared by my Father's One and Only Son, Yashua Messiah:

> "As it was in the days of Noah, so it will be at the coming of the Son of Man. For in the days before the flood, people were eating and drinking, *MARRYING AND GIVING IN MARRIAGE*, up to the day Noah entered the ark; and they knew nothing about what would happen until the flood came and took them all away. That is how it will be at the coming of the Son of Man." Matthew 24:37-39

The "sons of God" expanded the "Lamechian-conspiracy," when they married the "daughters of men," setting conditions for the destruction of the whole world. It may have seemed

a *7ʰ-Gift* page

that such action was without remedy, since it wasn't until the 10th generation, in Noah's day, that the Creator took decisive measures, fulfilling these words:

> God saw how corrupt the earth had become, for all the people on the earth had corrupted their ways. So God said to Noah, "I am going to put an end to all people, for the earth is filled with violence because of them. I am surely going to destroy both them and the earth." Genesis 6:12,13

Chapter 8

The Marriage of Abram

Thus, through Noah, the Adam-Terah line survived the flood. The Adam-Terah line includes 1. Adam, 2. Seth, 3. Enosh, 4. Kenan, 5. Mahalalel, 6. Jared, 7. Enoch, 8. Methuselah, 9. Lamech, 10. Noah, 11. Shem, 12. Arphaxad, 13. Cainan, 14. Shelah, 15. Eber, 16. Peleg, 17. Reu, 18. Serug, 19. Nahor and 20. Terah. Each is a husband with a created-wife: none of them marries and each has an unnamed-wife.

In the 21st generation of Adam, through the Adam-Seth line, something extraordinary happens: Abram becomes the first to marry, and his wife has the name Sarai. She is the only wife, other than RIFT-8, across 21 generations, in the Adam-Seth line, with a disclosed-name, after she became a married wife. Because of this explicit-declaration and breath-taking precedent, the marriage of Abram and Sarai flashes like a very bright beacon, in the 21st generation, that something extraordinary happened. But what?

Abram has two, and only two siblings, by birth: Nahor (the same name as Abram's grandfather) and Haran. Something unexpected happens to Haran:

> While his father Terah was still alive, Haran died
> in Ur of the Chaldeans, in the land of his birth.
> Genesis 11:28

Why is this unusual? Because this is the first, explicit declaration, in the Adam-to-Abram line, of a son predeceasing his father since Abel's demise. Were there others? Of course:

12. Positional Declarations of sons predeceasing fathers or ancestors.

12.1. Abel predeceases his father Adam.

12.2. Enoch (Adam-Seth line) is "taken away," his ancestors all living.

12.3. Lamech (Adam-Seth line) predeceases his father, Methuselah.

12.4. Arphaxad predeceases his father, Shem.

12.5. Cainan predeceases his grandfather, Shem.

12.6. Peleg predeceases four of his ancestors.

12.7. Reu predeceases three of his ancestors.

12.8. Sereg predeceases two of his ancestors.

12.9. Nahor predeceases six of his ancestors.

12.10. Terah predeceases two of his ancestors.

12.11. Abram predeceases two of his ancestors.

12.12. After the flood, in the Adam-Terah line, only Noah (of course), Shem, Shelah and Eber do not predecease any of their ancestors.

Noah's son, Shem, is the great-great-grandfather of Eber. Two sons were born to Eber: Peleg and Joktan.

In Peleg's time, the "earth" was divided because of RIFT-9, which directly affected Joktan's son, Jobab.

Haran is the first, *explicitly* declared son, in the Adam-Seth line, who predeceases his father (requiring the writer to compute other, related predeceasing events). Because of this, Haran is very significant. Why does Scripture emphasize his circumstances? Let's carefully consider Haran's life, through the few verses he is mentioned:

> This is the account of Terah. Terah became the father of Abram, Nahor and Haran. And Haran became the father of Lot. Genesis 11:27

Notice, Haran does not marry, *AND* his wife is unnamed, *AND* his offspring is male (but only one), the case with all the other created-couples, though there is no declaration he had "other-sons-and-daughters," initially. Therefore, Haran, like his father Terah, waited for the Creator to form his wife, from Haran's rib. And, like his predecessors in the previous 20 generations, his first-named child, Lot, is a son. The other alternative is for Haran to have married (and not have waited for a created-wife to be formed), but there is no explicit declaration of him ever marrying. This writer shall add that "following the data," up to this point, creates outcomes that are beyond any experiential or scriptural reference-frame he knows. It means that created-couples continued to be "formed" a considerable period after the flood.

However, without an explicit marriage to break the pattern, one is left to either make presumption or "let the data lead where it may." This may be similar to "flying on instruments only," when conditions make visual flight rules impossible.

Consider what happens to Haran in the very next verse:

> While his father Terah was still alive, Haran died in Ur of the Chaldeans, in the land of his birth. Genesis 11:28

It is certain that Haran had a wife. If he predeceased her, what became of this woman? Additionally, it is disclosed that Haran had two daughters:

> Abram and Nahor both married. The name of Abram's wife was Sarai, and the name of Nahor's wife was Milcah; *she was the daughter of Haran,* the father of both Milcah and Iscah. Now Sarai was barren; she had no children. Genesis 11:29,30

Wait just a moment here. Haran has a daughter, and she has a name. Created-couples, within the Adam-Seth-Noah-Terah line, *don't* conceive named-daughters—they only have "born" sons. Only married-couples have "named," "unattached," born daughters who become wives, through marriage, if and when they do. Therefore, Haran's wife has attributes of both a created-wife (unnamed without marriage) and of a married-wife (who bore a named-daughter)." In fact, Haran has two named-daughters, Milcah and Iscah.

Everything written, up to this point, the writer was permitted to discern about created-couples and married-couples some 10 years ago. The contradictions surrounding Haran's wife could not be resolved at that time. And there were additional complications to address for every woman named in the verse:

> Abram and Nahor both *married*. The name of Abram's wife was Sarai and the name of Nahor's wife was Milcah; she was the daughter of Haran, the father of both Milcah and Iscah. Genesis 11:29

This meant that Abram did not wait to receive a created-wife, as his ancestors had done, and he became a husband through marriage. What's wrong with that? Nothing at all, based on today's traditions. However, up to that time, in the witness of Scripture, and prior to Abram's life, marriage was not in the lineal ancestry of the "Promised Offspring": nowhere is marriage declared within the *direct* Adam-Seth-Noah-Terah line. Thus, Abram's actions were unprecedented and potentially cataclysmic. Remember, the entire world and *all* marriages and descendants, thereof, were wiped out in the flood—except Noah and his sons, who waited for created-wives to be formed from their ribs.

a *7th-Gift* page

Chapter 9

A 10-Year Wait

For 10 years, a period beginning from this writer's 37th year, until his 46th year, the incongruence, regarding Abram's marriage, persisted. It was not discussed, because the contradictory finding implied a defect in the reasoning about created/married-couples. In addition, the rigor taught at Caltech and applied professionally (including the award of U.S. patent no. 6,236,992 to this writer, by the federal government, listed at the U.S. Patent and Trademark Office's, "www.uspto.gov" Web site, in Appendix N) meant the defect or oversight had to be discerned, and not introduce unnecessary error, if at all possible.

So, after the 7th month, during which there are four festivals (Leviticus 23 and 25), in the writer's 46th year, the delightful endeavor began to commit the opening chapters of Genesis to memory (which had previously been read multiple times). This was after committing to heart six other complete books, four of them in the months prior to partaking of the pleasure the writer's Father placed in Genesis many centuries ago.

When the 11th chapter was committed to heart, the mystery of Abram/Abraham's riddle, about his wife Sarai/Sarah, unfolded as beautifully as a rose glistening with early morning dew, waiting to have its petals coaxed open, after sunrise ends a long night.

[Special Note: If the Creator *HAD* to engage compelled-sibling-incest, to expand the human family after Adam and Eve, then men could later claim families free even of this compromise. Then all such men might assert the ability to do what the Creator could not, making themselves greater than the Creator relative to this aspect of his eternal being. In asserting this, a limitation is projected upon the Creator which he does not have. The one who does so ascribes to himself "that which even the Creator couldn't do." This directly challenges the Commandment, "You shall have no other gods before me," which naturally includes one's own self. It remains written: "When Adam had lived 130 years, he had a son in his own likeness, in his own image; and he named him Seth." Gen 5:3 This is a new precedent, which the Creator also fulfills.

[Over time, other men have had many sons. If the I-SHALL-BE can't have a son, then men are able to do something else he can't. If added to compelled-sibling-incest by the I-SHALL-BE—another limitation he doesn't have—then men could do multiple things the I-SHALL-BE couldn't. His Son would have to have a human body AND be the exact representation of the eternal God, the I-SHALL-BE. Through such a Son, the mystery of *REST* is established by the infinite, never-resting Father, who loves his One and Only Son. To create *REST*, the Son became a "little lower than the angels," who are never-resting like my Father. The mystery of *REST* is a profound bond between my Father, his One and Only Son, and the Spirit of truth. There are 7 blessings in the *REST* Commandment of Exodus 20:8-11: 1. Faith to know it's origins ("Remember . . ." Gen 2:3); 2. Opportunity and

fulfillment ("Six days you shall labor . . ."); 3. Children; 4.
The blessing of help; 5. A stock of resources; 6. Testimony to
others; 7. A place to call home. Before Adam or anyone else
ever rested (Genesis 2:21), my Father's Son made it possible
to enter and return from it, because he already had. *REST*
"7-authenticates" all measures of time in Leviticus 23 & 25:
day (sunset-to-sunset); week (7th-day); week-of-weeks (day
after 7 weeks); month (7th-month); year ("day"); week-of-years
(7th-year); week-of-week-of-years (the 50th year); the "hour"
(7th-hour) and the 7th-month (week-of-7th-months) are
7-authenticated in RIFT-10.]

The "unsolvable" riddle, implanted upon this writer's
heart, 10 years earlier, was this:

> Now Abraham moved on from there into the region
> of the Negev and lived between Kadesh and Shur.
> For a while he stayed in Gerar, and there Abraham
> said of his wife Sarah, "She is my *SISTER*." Then
> Abimelech king of Gerar sent for Sarah and took
> her. But God came to Abimelech in a dream one
> night and said to him, "You are as good as dead
> because of the woman you have taken; she is a
> married woman." Genesis 20:1-3

> And Abimelech asked Abraham, "What was your
> reason for doing this?" Abraham replied, "I said to
> myself, 'There is surely no fear of God in this place,
> and they will kill me because of my wife.' Besides,
> she *really IS my SISTER*, the daughter of my father
> though *not* of my mother; and she became my wife."
> Genesis 20:10-12

Terah, Abram/Abraham's father, did not marry, and his
wife is unnamed, and he had only sons by birth, consistent
with being part of a created-couple. In addition, there is no
declaration that he had any wife other than his created-wife.

Furthermore, it is disclosed in Genesis 11:27, that Terah had three sons, Abram, Nahor and Haran, and *no "other-sons-and-daughters"* ("other-sons-and-daughters" is explicitly declared for *all* of his ancestors, after the Flood, in Genesis 11:11, 11:13 [twice therein], 11:15, 11:17, 11:19, 11:21, 11:23, and 11:25). Therefore, how could Sarai/Sarah be Abram/Abraham's sister *AND* be his father Terah's daughter, at the same time, since Terah is not declared to have had any other children, *by birth*, after his three sons?

Chapter 10

Terah's Unique Children and Grandchildren

To solve the problem in chapter nine, carefully consider what happens to Abram/Abraham's middle brother, Haran, and his youngest brother, Nahor. The order of the brothers is known because of RIFT-11. As noted, in Genesis 11:28, Haran predeceases his father Terah, after leaving his father's household in Ur of the Caldeans, living in another area and then returning to Ur some time later. He returned to Ur of the Caldeans, "the land of his birth," before he died.

It is presumed he was in Ur with his unnamed-wife and his three children—Lot, Milcah and Iscah—before he died. This, of course, is a "precedent exception," because created-couples only have sons, and Haran is in Ur, with two named-daughters identified, before he dies. How could Haran and his wife, as a created-couple, have named-daughters?

13. Positional Declarations regarding Haran and his unnamed-wife.

13.1. His unnamed-wife bore Haran's daughters.

13.2. His unnamed-wife did not bear Haran's
        daughters.

13.1 would establish a "precedent exception," relative to created-couples, since there is no explicit antecedent that a created-couple ever gave birth to a daughter, prior to Haran. 13.2 is equally difficult, because the Scripture clearly discloses:

> Nahor's wife was Milcah; she was the daughter of
> Haran, the father of both Milcah and Iscah.
> Genesis 11:29

In other words, we know, explicitly, that Haran had two named-daughters, Milcah and Iscah. And we know from 20, prior generations of witnesses that created-couples' birth-offspring are male. Haran does have one son, Lot, yet the precedent of daughters is very problematic. Such was the case for this writer, spanning 10 years, a quarter of the writer's life at the time . . .

. . . Until [10 years later] Genesis 11:30 was committed to heart:

> Now Sarai was barren; she had no children.

In calculus, a technique called "parametric" substitution is used to resolve complex, analytic problems. If Sarai is treated

a *7ʰ-Gift* page

as a "variable," then what would happen if we substitute this "variable" for the name of Haran's created-wife?

Why are these facts mentioned in such close proximity to one another?

1. Haran dies after returning to Ur of the Chaldeans.

2. Abram marries Sarai, and Nahor marries Haran's daughter, Milcah.

3. Sarai was barren; she had no children.

When Haran dies, what becomes of his created-wife?

14. Positional Declarations regarding Haran's created-wife after he dies.

14.1. She co-deceases with him.

14.2. She becomes a widow and remains single the rest of her life.

14.3. She becomes a widow and marries outside Haran's family.

14.4. She becomes a widow and marries inside Haran's family.

Chapter 11

The Secret of Sarai/Sarah

W hen Haran dies, his unnamed-wife and his children—Lot, Milcah and Iscah—survive him. Scripture later gives considerable testimony about Lot, and we know that Milcah married Haran's youngest brother, Nahor. There is complete silence about what became of Haran's created-wife.

15. Positional Declarations regarding Haran's created-wife's fertility.

———————

15.1. She was barren.

15.2. She was not barren.

If she were barren, Haran's journey beyond the land of Ur, the place of his birth, opened a way for him to raise a family (a path Abram followed later). If so, a "surrogate-female" would be capable of having children of either gender, from extant precedent. Thus, a "surrogate-female" could give birth to Lot, Milcah and Iscah. If Haran's created-wife is barren, she would be the first such barren created-wife, breaking 20-generations of consistent fertility. When Haran dies, his younger brother marries one of Haran's surviving daughters, Milcah, an explicit disclosure of a relative marrying a member of a deceased brother's family.

16.   Positional Declarations regarding Sarai.

16.1.   Sarai is not the unnamed, created-wife of Haran.

16.2.   Sarai is the unnamed, created-wife of Haran.

Sarai's first appearance in Scripture, by name, is in close proximity to Haran's death. She is part of a "double-marriage" declaration, Abram-Sarai and Nahor-Milcah. Additionally, she is barren, which is a condition for Haran's use of a "surrogate-female" to have children. If Sarai weren't Haran's wife, then she suddenly appears in Scripture with "no family history." Because she becomes the wife of one of the most important figures in all of Scripture, Abram/Abraham, through the first heaven-accepted marriage, her ancestry *MUST* be well defined.

If she were Abram/Abraham's created-wife (excluding marriage between the two), her family history and her husband's family history would be the same (created-wives have no genealogical ambiguity whatsoever, which greatly aided lineal accuracy before and after the flood). Therefore, her family history, to some degree, must be disclosed since she becomes Abraham's wife through the first-ever marriage, in the *direct* Adam-Seth-Noah-Abram line. With theses results (discerned by this writer after 10 years and upon committing the first 11 chapters of Genesis to memory), it is possible to now revisit Abram/Abraham's riddle:

> "Besides, she *really* is my *SISTER*, the daughter of my father though *not* of my mother; and she became my wife." Gencsis 20:12

17.   Positional Declarations regarding Abram's father, Terah.

17.1.   Terah was a husband with a created-wife.

17.2.   He had only one wife.

17.3.   His wife bore him three sons and no other children by birth.

17.4.   He had no other children by or from any other woman.

Thus, 17.2 and 17.3 must be addressed, in light of Abram/Abraham's declaration regarding Sarai, " . . . she *really* is my sister . . ." Indeed, if Sarai/Sarah were the created-wife of Haran, Abram/Abraham's brother, then Sarai/Sarah would be Abram/Abraham's sister. However, as a created-wife, she was never "born." So, Abraham was honest in saying, " . . . though not of my mother . . .", since his mother did not bear Sarai/Sarah. And therefore, the 18 references to "other-sons-and-daughters," in Genesis 5 and 11, each refer to instances of a born-son and the wife formed from his rib, by the writer's Father, through the Testimony of his One and Only Son, Yashua Messiah—the Son of MAN.

Abram declared that Sarai " . . . *really* is my sister . . ." If Abram meant Sarai " . . . really is my *half-sister*" (which is not in the text), then Terah, his father, would have had a previous, undisclosed relationship with a woman other than Terah's created-wife. There is no basis to impugn Terah's reputation with this conduct, as he, too, comes under the Commandment, "You shall not give false testimony against your neighbor." Additionally, one would also have to impugn Abram/ Abraham's declaration, " . . . she *really* is my sister . . ." Terah is included in the testimony of "MAN"—as is Noah. Neither is construed, by Scripture, as outside the testimony of "MAN."

The marriage of Abram and Sarai reveals Sarai to be the last of the created-wives, by name. Sarai/Sarah's body is now resting in Kiriath Arba, in Hebron, her DNA there, today. She endured Haran's effort to have a family when Sarai "seemed" barren. Haran became the father of Lot—and Lot was also Abram's brother (Gen 13:8). Lot was the last male-child that Terah's created-wife gave birth. After Haran died, Abram married Sarai, becoming a kind of kinsman-redeemer, himself (Ruth 4:14-15). This was an extraordinary act of faith on Abram's part, since it almost guaranteed no possibility of offspring through marriage.

Yet, by marriage, Abram complemented Noah's faithfulness: Noah could have married, but waited to receive a created-wife. Abram was faithful when his younger brother, Haran, received a created-wife, but Abram did not. And, again, Abram was faithful, after Haran died, to marry his brother's, surviving created-wife (Gen 26:4, 5), though she was barren. A 7th time Abram/Abraham was faithful, regarding his only son, Isaac (Gen 22:12).

Chapter 12

The First Marriage for All Nations

With Abraham and Sarah, the first marriage is established, through whom new children of God would be born, to carry forth the birth of the sons of God from created-couples who preceded Abraham and Sarah, leading to the birth of "the Son of MAN" himself. It is through Abraham that the I-SHALL-BE bonds all subsequent marriages, of the Abraham-Isaac-Jacob direct line, to created-couples, and to the first of them all, Adam and Eve:

> Some Pharisees came to [I-SHALL-BE-SALVATION] to test him. They asked, "Is it lawful for a man to divorce his wife for any and every reason?"
>
> "Haven't you read," he replied, "that at the beginning the Creator 'made them male and female' [Gen 1:27], and said, 'For this reason a man will leave his father and mother and be united to his wife, and the two will become one flesh' [Gen 2:24]? So they are no longer two, but one. Therefore what

a *7ʰ-Gift* page

God has joined together, let man not separate."
Matthew 19:3-6

In other words, when two become one, in the I-SHALL-BE's covenant of marriage—whether of the same cultural/religious/ national backgrounds or different ones—the union is equal and as profound as though the two were a created-couple. Even harm to "unknown unions" is not overlooked because through such unions may be born children of God (Gen 20:3). The kings of Sodom and Gomorrah perished for ignoring what Pharaoh and Abimelech respected (Gen 12:19; 20:14; 14:21). "The women" went to Bera by the oath of Melchizedek Abram kept (Gen 14:16-23). "Closing all wombs" preceded Bera's demise (Gen 12:17; 20:18; 19:4; 19:24). Lot's wife, RIFT-12, became the spice that kings and the wealthy had indulged her (Gen 19:26). Lot's daughters, in other cities (Gen 19:12-15), all perished (Gen 19:29). The two with Lot, conceived by RIFT-13, left no sin on Lot (2Peter 2:7). His descendant, Gen 19:37, was king David's great-grandmother (Ruth 4:13). International communities which anchor themselves, in Abraham's legacy, do so for, among other reasons, the legitimacy it lends their marriage traditions. Abraham/Sarah's marriage is the first one set apart by the I-SHALL-BE, to establish the precedent of a new blessing for all nations:

"I will make you into a great nation and I will bless you; I will make your name great, and you will be a blessing. I will bless those who bless you, and whoever curses you I will curse; and all peoples on earth will be blessed through you." Genesis 12:2,3

"No longer will you be called Abram; your name will be Abraham, for I have made you a father of many nations. I will make you very fruitful; I will make nations of you, and kings will come from you." Genesis 17:5,6

"Abraham will surely become a great and powerful nation, and all nations on earth will be blessed through him. For I have chosen him, so that he will direct his children and his household after him to keep the way of the [I-SHALL-BE] by doing what is right and just, so that the [I-SHALL-BE] will bring about for Abraham what he has promised him." Genesis 18:18,19

Abraham doubted *7 times* (1st in Gen 12:13) that the I-SHALL-BE would· give him an heir. The 3rd-doubt (Gen 15:8) brought 400 years of servitude to his descendants (Gen 15:13). Between his 4th and 5th doubts (including laughing at the thought of Sarah having a son [Gen 17:17]—Sarah, later, laughed as part of the 6th-doubt in Gen 18:12), he and all males in his household underwent circumcision (Gen 17:23). After the 6th-doubt, no one lay with Sarah, not even Abraham. The 7th-doubt was the denial he had a wife (Gen 20:2). So, the I-SHALL-BE rescued Sarah (Gen 20:7), to preserve the covenant of marriage being established, and the I-SHALL-BE did for Sarah what he promised. She became pregnant (Gen 21:1), Abraham allowed to name and circumcise "his" son—the first son of God born to a married-couple. None of Abraham's doubts—with Lot, Eliezer or Ishmael—foresaw Isaac's birth.

Chapter 13

Nephilim After the Flood

A problem must now be addressed. Only four created-couples survived the flood: Noah, Shem, Ham, Japheth and their respective wives. So, from whom did the post-flood Nephilim ("post-Nephilim") and their marriages originate after the flood?

> Then Caleb silenced the people before Moses and said, "We should go up and take possession of the land, for we can certainly do it."
>
> But the men who had gone up with him said, "We can't attack those people; they are stronger than we are." And they spread among the Israelites a bad report about the land they had explored. They said, "The land we explored devours those living in it. All the people we saw there are of great size. We saw the *NEPHILIM* there (the descendants of

Anak come from the *NEPHILIM*). We seemed like grasshoppers in our own eyes, and we looked the same to them." Numbers 13:30-33

This requires Positional Declarations, because *ALL* Nephilim were wiped out in the flood, designated herein as "pre-Nephilim."

18.   Positional Declarations on the origins of post-Nephilim.

18.1.   They are from pre-Nephilim who somehow survived the flood.

18.2.   They are from a spouse of a created-couple after the flood.

18.1 is eliminated, because only Noah and his wife, and his sons and their wives, survived the flood, which is explicitly declared:

> Everything on dry land that had the breath of life in its nostrils died. Every living thing on the face of the earth was wiped out; men and animals and the creatures that move along the ground and the birds of the air were wiped from the earth. Only Noah was left, and those with him in the ark. Genesis 7:22,23

This leaves 18.2. The candidate created-couples are in Genesis 10. A signature of Nephilim is a born-daughter whom a "son-of-God" marries, as in Genesis 6:1,2. But to produce a born-daughter, a created-wife must conceive with someone other than her "rib-taken" husband. This proved challenging to the writer for 10 years, too, along with Noah's riddle:

Noah, a man of the soil, proceeded to plant a vineyard. When he drank some of its wine, he became drunk and lay uncovered inside his tent. Ham, the father of Canaan, saw his father's nakedness and told his two brothers outside. Genesis 9:20-22

When Noah awoke from his wine and found out what his youngest son had done to him, he said, "Cursed be Canaan! The lowest of slaves will he be to his brothers." Genesis 9:24,25

Why was Canaan cursed for his father's offense? How did Noah know Canaan's name before he was born?

19.   Positional Declarations regarding Canaan's name.

_____

19.1.   Noah cursed Canaan after learning his name.
19.2.   Noah gave Canaan his name and cursed it simultaneously.

Relative to 19.1, Canaan's name first appears in Genesis 9:18 and 9:22: "Ham, the father of Canaan," for example. Noah's first use is unique.

20.   Positional Declarations on the first use of Canaan's name.

_____

20.1.   The name was given before Noah cursed it.
20.2.   The name was used only after Noah cursed it.

Genesis 9:18 and 9:22 both make "anticipatory reference" to Canaan's name—a name not assigned to the unborn child until Noah cursed him. There is no precedent supporting 20.1. 20.2 is sustained in Genesis 9:18, 9:22, 10:6 and 10:15, which all use the same name Noah gave the unborn child through a curse.

21. Positional Declarations on who named Canaan.

21.1. Ham named his son Canaan.
21.2. Noah named Canaan because his brothers were Shem and Japheth.

Notice the positioning of Canaan after Noah's sons come out of the Ark:

> The sons of Noah who came out of the ark were Shem, Ham and Japheth. (Ham was the father of Canaan). These were the three sons of Noah, and from them came the people who were scattered over the earth. Genesis 9:18

In other words, the sons who came out of the ark were different from Canaan, because Canaan's father was Ham. Indeed, Canaan is the brother of Shem and Japheth. Noah's curse subordinates Canaan to his brothers, while excluding Ham from any blessing Noah bestows on his other sons. Since Noah is part of a created-couple, what Ham did to Noah's wife was done to Noah, as well, because the two are one as husband and wife:

> When Noah awoke from his wine and found out what his youngest son had done to him, he said, "Cursed be Canaan! The lowest of slaves will he be to his brothers." He also said, "Blessed be the [I-SHALL-BE], the God of Shem! May Canaan be

a *7ʰ-Gift* page

the slave of Shem. May God extend the territory of Japheth; may Japheth live in the tents of Shem and may Canaan be his slave." Genesis 9:24-27

The moment Noah knew that Ham caused Noah's wife to conceive, he named the child Canaan, by the authority he had as her husband. While simultaneously cursing Canaan and excluding Ham from any blessing, Noah was merciful to allow Canaan to be among his brothers, in any capacity at all.

From the precedent of Canaan, the Nephilim would be extant after the flood, including RIFT-14, who took wives multiple times, after the flood, and brought forth daughters. The Nephilim are apparent in the table of nations of Genesis 10: Created-couples are listed by the sons' names *ONLY*, while Nephilim are noted as the Kittim and the Dodanim (or Rodanim in Gen 10:4); the Ludites, Anamites, Philistines and so forth (Gen 10:13,14); the Hivites, Arkites, Sinites, Hamathites and so on (Gen 10:15). Not once are "ims," or "ites," or "ians," or "ines," in Shem's direct line to Abram as the testimony of "MAN": rather, they are post-flood "MEN."

"Maritime peoples," in Genesis 10:5, may be better translated "marriage-time peoples," since the sons of God ("MAN") typically stayed together (Genesis 10:30) and were not "scattered" (Genesis 10:15-19) or "spread out." The post-Nephilim and their offspring were, after the flood.

Chapter 14

The Son of MAN

The atonement for the devastation, brought upon marriage and offspring, is within the perfect fulfillment of a descendant, after Isaac. Yet, he is before Isaac, within the God of Abraham and the God of Isaac and the God of Jacob—a testimony acknowledged in John 1:12,13:

> "Yet to all who received him, to those who believed in his name, he gave the right to become children of God—children born not of natural descent, nor of human decision or a husband's will, but born of God." John 1:12,13

This testimony anticipated the iniquity spread by Lamech before the flood; the iniquity renewed by Ham, after the flood; and the vulnerability to it by the sons of God, across time. MEN would again increase in number on the earth, to the detriment of all mankind, except by the grace of the I-SHALL-BE and his extraordinary provision, before the creation of the world.

The children of God, male and female, young and old, spread throughout nations, are not orphans (John 14:18); He who is, who was, and who is to come, guides each and everyone home from the four winds [Matthew 24:30,31; 24:36; 25:31-33; Luke 15:3-7; John 10:14-18; Rev 1:7; 7:15-17].

Moses, the descendant of Abraham and Isaac and Jacob not circumcised as a youth, received the Ten Commandments (Exodus 20:1-17) from the God of Abraham, the God of Isaac, and the God of Jacob. This Law anticipates the complex relationships of the descendant-sons-of-God after Jacob, all to be born through the marriage-covenant established, first, between Abraham and Sarah. The offspring of this marriage-covenant, after Jacob, became the first "ites" in the Adam-to-Abraham line leading to the Son of Man. Every "ite," so recognized in the direct line of descent leading to the Son of Man, is each a male-Israelite, even though the wives of these males may sometimes have other than Israelite backgrounds (Moses' Midianite wife [Exodus 2:16-21]; Boaz's wife Ruth, a Moabite; Solomon's Egyptian wife; and so on). After the Son of Man, children of God include those whose parents may or may not both be of obvious Israelite descent, such as Timothy (1Timothy 1:2); and the trustworthy Ethiopian (Acts 8:26-40) innocent of any potential offense against the I-SHALL-BE's covenant of marriage, after Philip baptized him; and, of course, Mary Magdalene (Luke 8:1-3; John 20:17).

Through Noah, one man's faithfulness, as a son of God, saved mankind. Yet, even his faith could not stop the reemergence of the Nephilim. How can MAN continue after this and the children of God ever be born again?

Through the children of Israel, the marriage-covenant of Abraham spread across all nations as powerfully as marriage had spread to all peoples before the flood. All descendants, from the sons of God who married before the flood, were wiped out, but not Noah and his three sons. The Law, therefore, was, is, and shall always be for the love of such sons who may marry across all manner of cultural/religious/national traditions, to NOT lose them again.

[Special Note: The Law, by grace, reminds a son to remain faithful to the I-SHALL-BE, whether in marriage or other circumstances. A wife's god, if different, is never to be before the I-SHALL-BE. If idols are worshiped, a son shall not (1Kings 11:1-6). If the I-SHALL-BE's name is used in vain, a son shall keep it holy (Job 2:9,10). If adultery is committed, a son shall not (Hosea 3:1,2). The Law is a reminder to be faithful, above all, to the I-SHALL-BE, with admonition regarding the marriage-covenant of Abraham.]

Such a son, if moved to become a husband, is a blessing to his wife and, either married or single, a blessing to the community and nation he is part, whether by immigration or by birth. If such a son carries the testimony of a father, the Commandment attends him in grace and forgiveness: " . . . for I, the [I-SHALL-BE] your God, am a jealous God, punishing the children for the sin of the fathers to the 3rd and 4th generation of those who hate me, but showing love to a thousand generations of those who love me and keep my commandments." Exodus 20:5,6 [Freed from Egypt]

From childhood on, everyone commits an offense against this covenant, except one descendant who never did (Revelation 14:4), has not now, and never will. He is not from the "son-of-men"; he is *the* "Son-of-MAN," in the original meaning of "MAN," before division by clans and languages, territories and nations. Such a descendant would have to be what none of the sons-of-God (Genesis 6:2 and Psalm 82:6) ever was—a "type" unlike any other. Yet, through him, the scattered children of the God of Abraham, the God of Isaac, and the God of Jacob, born of marriages spread across centuries and nations, are to be one family, beyond the cravings of flesh, and immorality, and MEN (1Timothy 1:1-20). One day, all these children will celebrate, together, the feasts in Leviticus 23 and 25—a gift through the Testimony in John 7:37-39:

"Then the survivors from all the nations that have attacked Jerusalem will go up year after year to worship the King, the [I-SHALL-BE] Almighty, and to celebrate the *Feast of Tabernacles.* If any of the peoples of the earth do not go up to Jerusalem to worship the King, the [I-SHALL-BE] Almighty, they will have no rain." Zechariah 14:16,17

"Every pot in Jerusalem and Judah will be holy to the [I-SHALL-BE] Almighty, and all who come to sacrifice will take some of the pots and cook in them. On that day there will no longer be a Canaanite in the house of the [I-SHALL-BE] Almighty." Zechariah 14:21

There is no prohibition whatsoever, in the testimony of a great Pharisee, perhaps one of the greatest of all, if one remembers such a feast day:

"Your boasting is not good. Don't you know that a little yeast works through the whole batch of dough? Get rid of the old yeast that you may be a new batch without yeast—as you really are. For [Messiah], our Passover lamb, has been sacrificed. Therefore let us *KEEP* the Festival, not with the old yeast, the yeast of malice and wickedness, but with bread without yeast, the bread of sincerity and truth." 1Corinthians 5:6-8 [written by Saul the Pharisee who became Paul]

So ends the testimony of this first book, with fasting (no food or water consumed), between sunsets on the 9th, 10th, 11th, 13th and 15th day of the 7th month—for 5 of 7 days—finishing up work, near sunset the 21st day of same; finalizing the work near sunrise the 30th day of same, in the year 5763. We write this to make our joy complete. 1John 1:4. Thank you for your gratuity and the testimony attached to it—for a life-time . . . and beyond.

AFTERWORD

The I-SHALL-BE establishes the precedent of "allowed-marriage," after the flood, through the family of Abraham and Isaac and Jacob. Each is a son of God and each desires and endeavors to honor the faithfulness of their male-ancestors by having only one wife, the case for all created-couples from whom they descend. The I-SHALL-BE is the God of Abraham, the God of Isaac, the God of Jacob.

1. *Abram/Abraham is a son of God and offspring in the last generation of created-couples. He marries Sarai/Sarah, who is barren and the surviving created-wife of his deceased brother. His desire to have only one wife, through his allowed-marriage, is affected by Sarai/Sarah's despair about being barren (Gen 16, Gen 21).*

2. *Isaac is the first son of God born from an allowed-marriage. His mother, Sarai/Sarah, is the last of the created-wives. Isaac marries Rebekah, by the I-SHALL-BE's will alone (Gen 24:12-27). Rebekah is the first female-descendant of Nahor's (Abram/Abraham's brother's) allowed-marriage (Gen 22:20-24) with Milcah (a surviving daughter of Abram/Abraham's deceased brother, Haran). Rebekah is both daughter and sister of Bethuel, her father (Gen 22:23, Gen 24:55-60). By faith, Isaac prayed (Gen 25:21), and Rebekah, though barren, gave birth to the first, twin offspring, ever—one a son of God and the other not (Gen 25:29-34, Mal 1:1-5).*

3. *Jacob is the first son of God, from an allowed-marriage, to be born of a female not a created-wife. Though Jacob desired to have only one wife, in his allowed-marriage, he patiently endured deception and hardship as the I-SHALL-BE fulfilled his purpose through Jacob's family (Gen 29, Gen 30, Rev 21).*

4. *I-SHALL-BE-SALVATION is the One and Only son of God born to a virgin, who was not barren. Faith, through him alone, identifies the children of God (male and female), no matter the apparent circumstances of each one's physical birth (John 1:12, 13; 1John 3).*

a *7th-Gift* page

A Gift I See in You

The wonderful gift you offer,
essential for wise kings;
Beauty surpassing gardens,
worth beyond diamond rings.

And with this happy moment,
touching everyone it finds;
Noble reflections grow,
in distant hearts and minds.

Swaying petaled roses,
refreshing like a breeze;
Through inspiration's fragrance,
going where it please.

Pouring forth like water,
moments bond into one;
A chain of unseen links,
by force is not undone.

Reaching toward the eternal,
One and Only who is true;
Healing with forgiveness,
in the Name to make life new.

Transcending every culture,
yet only from above;
The Truth an honest witness,
protecting this with love.

Six have shown an outline,
the 7th for living dust;
Let wisdom guide all giving,
who receive the gift of trust.

By Anthony Sneed
2058 North Mills Avenue
Claremont, CA 91711

APPENDICES

APPENDICES

Appendix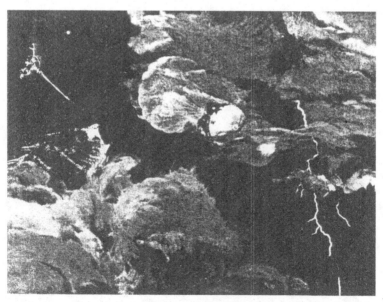

The Galileo Mission to Jupiter

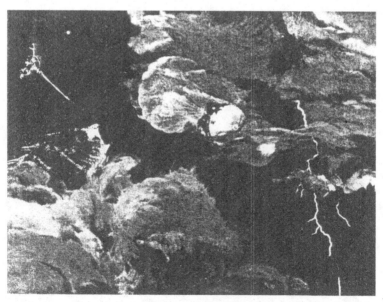

courtesy NASA/JPL

a *7ʰ-Gift* page

JPL Galileo Fact Sheet – Jupiter '96

Mission Summary

After a six-year voyage from Earth, the Galileo spacecraft entered orbit around Jupiter on December 7, 1995, to begin a two-year mission studying the giant planet. That same day, the Galileo orbiter successfully received data from the atmospheric probe it had dropped into the planet's atmosphere to study its structure and chemistry. The probe, which was released from the main Galileo spacecraft on July 13, 1995, sent back the first-ever direct measurements of Jupiter's chemistry, winds, and structure, finding a drier and windier planet than expected. The probe was destroyed as expected nearly an hour into its descent by the heat and pressure of Jupiter's atmosphere. With the atmospheric probe mission complete, Galileo's studies are focused on the large moons of Jupiter, long-term observation of Jupiter's atmosphere, and the Jovian magnetic and charged particle environment.

Galileo's study of the Jovian atmosphere, its giant magnetosphere, and its four major satellites (all different and three of them larger than Earth's moon) will help scientists understand the solar system's history and evolution.

When Galileo first reached Jupiter, the Galileo orbiter flew close by Jupiter's volcanic satellite Io, then received science data from the descending atmospheric probe for almost 58 minutes. Within two hours, Galileo's main engine fired to brake the spacecraft into orbit around Jupiter to begin its two-year orbital tour of the Jovian system. Over the course of the mission, Galileo will fly close by the moons Ganymede (4 times), Callisto (3 times), and Europa (3 times), with one close flyby per orbit. Throughout the mission, Galileo will conduct studies of Jupiter's rings, monitor Io's volcanoes, study four inner minor satellites, and map magnetic fields and charged particles in Jupiter's environment.

Launched October 18, 1989, by the space shuttle Atlantis with an Inertial Upper Stage booster, the Galileo mission used the gravity fields of Venus and Earth to accelerate the spacecraft enough to reach Jupiter. It also flew by two asteroids, making the first close observations of such bodies and discovering the first asteroid moon, Dactyl. Then, in July 1994 it observed the comet Shoemaker-Levy 9 fragments impacting Jupiter's night side.

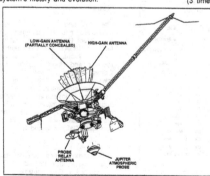

courtesy NASA/JPL

Spacecraft Anatomy

The Galileo Orbiter is a dual-spin spacecraft whose main body rotates at about 3 rpm. This portion of the vehicle carries six fields and particles sensors, thrusters, communications, and many other engineering subsystems. The lower part of the orbiter can be held in fixed orientation or rotated; it carries the camera and three other remote sensing instruments, the probe relay antenna, and other electronics.

Galileo's atmospheric probe had a deceleration module with a heat shield to slow and protect the descent module and its parachute during entry into the atmosphere.

The descent module had six scientific instruments and a radio relay transmitter that sent data up to the orbiter for 57.6 minutes. It measured atmospheric temperature, pressure, composition, and structure. These measurements showed the region the probe explored to be drier, windier, and less cloudy than expected; there was less neon in the atmosphere than predicted. Distant lightning was detected, but at a frequency three to ten times less than on Earth.

The orbiter was designed to transmit its scientific data at a very high rate from Jupiter, using a deployable, umbrella-like antenna. The antenna is only partly deployed, however, and the Jupiter data are transmitted, like those from Galileo's previous observations, over a smaller antenna at a lower data rate. Modifications to the ground receiving systems of NASA's Deep Space Network (DSN) on Earth and to ground and spacecraft

Major Mission Characteristics:

Launch	10/18/89
Venus Flyby	2/10/90
Earth Flyby 1	12/8/90
Gaspra Asteroid Flyby	10/29/91
Earth Flyby 2	12/8/92
Ida Asteroid Flyby	8/28/93
Comet Shoemaker-Levy Observation	7/1/94
Probe Release	7/13/95
Jupiter Arrival, Io Flyby, Probe Descent/Relay and Orbit Insertion	12/7-8/95
Orbital Tour/Playbacks (includes 10 satellite flybys)	6/96-12/7/97

Spacecraft Characteristics:

Category	Orbiter	Probe
Mass, kilograms (pounds)	2,223 (4,980)	339 (746)
Rocket propellant (usable)	925 kg (2,035 lb)	—
Height, meters (feet)	6.15 (20.5)	0.86 (3.1)
Science instruments	10*	6*
Instrument mass	118 kg (260 lb)	30 kg (66 lb)

software will permit Galileo to achieve 70% of its original scientific objectives at Jupiter.

Management

The Galileo mission was developed and is managed by NASA's Jet Propulsion Laboratory (JPL). NASA's Ames Research Center manages the probe mission, and Hughes space and Communications Company built the probe. The orbiter's retropropulsion module and some scientific instruments were supplied by Germany. the JPL-managed DSN supports the Galileo mission with its stations in Australia, California, and Spain.

*Two radio-science experiments use spacecraft radio and DSN stations as their instrument.

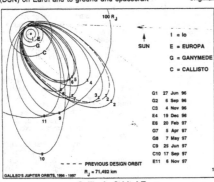

I = Io	
E = EUROPA	
G = GANYMEDE	
C = CALLISTO	

G1	27 Jun 96
G2	6 Sep 96
C3	4 Nov 96
E4	19 Dec 96
E6	20 Feb 97
G7	5 Apr 97
G8	7 May 97
C9	25 Jun 97
C10	17 Sep 97
E11	6 Nov 97

– – – – PREVIOUS DESIGN ORBIT

R_J = 71,492 km

GALILEO'S JUPITER ORBITS, 1995 - 1997

Galileo's Orbital Tour

CALIFORNIA INSTITUTE OF TECHNOLOGY

Pasadena, California 91125

Thomas E. Everhart
President

(818) 395-6301
FAX (818) 449-9374

September 20, 1996

Mr. Anthony Sneed
2058 N. Mills Ave.

Claremont, CA 91711

Dear Mr. Sneed:

I have been informed of your valuable contributions to the Galileo Project at Hughes Electronics. Dr. Urban Von der Embse sent a letter detailing your significant input which greatly contributed to the success of the project.

I am always pleased to hear of the accomplishments of Caltech graduates and look forward to hearing of future milestones in your career.

Sincerely,

Thomas E. Everhart

TEE:lr

Dr. Thomas Everhart   June 17th, 1996
President
Caltech
Pasadena, CA 91125

Executive Summary:

A significant contribution was made through a Caltech graduate who worked on the Galileo mission at Hughes Electronics. The successfully completed acquisition of data from the Galileo probe, and the relay of this valuable data to Earth, began December 7th, 1995. Anthony Sneed, who came to the Galileo project after graduating from Caltech, provided valuable service in achieving this critical mission objective.

Galileo Project Issues

Approximately six months after development began on the telemetry relay receiver to acquire and extract science data from the Galileo probe's 1.4 gigahertz, bi-phase, shift-keyed signal, seven problems presented themselves to the project team:

1. A firmware multiply routine needed to run twice as fast as the one developed by the project team. This routine was an important milestone during the first six months.

2. A discrete Fourier transform routine was too slow to keep up with the sampling rate necessary for the dynamic frequency- and phase-lock-loop algorithms that tracked the probe's signal. A 15 hertz/second doppler rate during early signal acquisition necessitated constant readjustment of a 24-bit numerical controlled oscillator, whose output was mixed with the probe's received signal after RF and IF demodulation.

3. The real-time, serial network architecture that linked CDS (Command and Handling Data System) to other spacecraft instruments and subsystems (including the telemetry relay receiver) had not been integrated into the telemetry relay receiver's development architecture. This provided the pathway to eventually send collected data to Earth through the CDS and down-link communication system.

4. Due to power limitations of the integrated spacecraft, all systems, including the telemetry relay receiver, ran on a 50% duty cycle, with about a 100 microsecond warning before power was turned off.

Hughes Space and Communications Company
P.O. Box 92919
Los Angeles, CA 90009

5. Communications algorithm development occurred on a Prime minicomputer using the PRIMOS operating system, while a DEC minicomputer, running a different operating system, was designated for support, development and testing of actual microcode to be loaded on the Galileo orbiter. The communications link modeling was done in Fortran by Hughes' Systems Analysis group. The final executable software would be in microcode, developed by a separate technology division within Hughes.

6. The actual test and development hardware, based on an AMD, dual bit-sliced architecture, used in the F/A-18 flight avionics system, would not be ready for at least 12 months after the start of the project. However, project milestones required portions of the software system to be checked-off as operationally tested well before that.

7. Six months into this development effort, the project was falling behind schedule, with a range of critical problems caused by a telemetry computer that lacked an interrupt architecture and subroutine calling architecture. Basically, the on-board computer system used was designed for a fire control system that tracked missiles flying at a couple times the speed of sound, in combat environments. It was now being retro-fitted to track a single probe moving at over *150 times* the speed of sound on Earth (the fastest man-made object *ever*), while the telemetry computer was shut off about every 550 microseconds to save power, with the CDS making repeated real-time data requests for later Earth down-link.

This was all to be done within the most intense electromagnetic radiation environment in the solar system, besides the sun. At initial signal acquisition, up to a 34 kilohertz uncertainty would exist on the probe's 1.4 gigahertz frequency. The communications algorithms would have to quickly resolve the uncertainty, then lock on to the 23 dB signal before the Jovian noise spectrum could take its toll on initial signal-to-noise ratio.

A Caltech Connection

Anthony Sneed, less than four months after receiving his degree from your school, was briefed on the project. He was told by management that it would not reflect badly upon him if he decided not to accept the assignment to bring the project back on schedule and address cost over-runs, since he was just out of school. After asking basic questions about the project (how far behind was the team, what were the main technical obstacles, what would be the measure of success in accepting the assignment), Anthony asked where to report to begin the assignment.

2

The following directly resulted from Anthony joining the project.

1. On his first day, he was introduced to the project team, including hardware engineers, microcode engineers, test-set designers, and senior communications analysts. He was briefed on the multiply algorithm developed during the previous six months. Two days later, he had developed a new algorithm that was twice as fast, and carefully validated it with the microcode engineers who had worked on the earlier algorithm. As a team effort, it was implemented. Anthony also provided a mathematical proof that a 65% reduction from the original algorithm was the limit of improvement if *a priori* knowledge was used on the ranges of multiplicands and multipliers. This catalyzed the process which later resulted in the addition of a hardware multiplier to the architecture of the signal processor, to be powered-up only when multiplication was needed.

2. Afterwards, Anthony was given the DFT (Discrete Fourier Transform) routine to review, which is a core component of the dynamic frequency-lock-loop (FLL), phase-lock-loop (PLL) and symbol detection microcode for bit synchronization. The DFT provided the filter coefficients used in searching for the initial signal and digitally adjusting a numerical controlled oscillator to track the signal. This routine would have to execute thousands of times for higher-level functions to resolve frequency uncertainty and compensate for the 15 hertz/second doppler of the probe's signal. After a week, Anthony had developed a modified DFT which ran approximately 20% faster than the one provided him. This had a ripple effect throughout the software development effort for higher-level routines, significantly improving the margins for signal acquisition and tracking.

3. While continuing to develop the microcode and programming architecture for the DFT, *Verify* and *Search Band* routines, Anthony accepted the responsibility to design the real-time operating system architecture for the entire telemetry relay receiver. No previous attempt had been made to implement this portion of the telemetry relay receiver, due to the hardware limitations of the computer used, and the need to accommodate three mutually exclusive real-time events, while concurrently tracking the probe's signal.

The solution that Anthony developed, which he called EXECTL (*Executive Control*), addressed all of the concerns of the hardware and software development teams. With his peers and management, he successfully defended EXECTL and its design components at a NASA/JPL design review, within the written specifications set by JPL's procurement documents. This helped greatly increase the confidence on the project team that the integrated software/hardware objectives were achievable.

3

a *7ᵗʰ-Gift* page

4. Though software project milestones were being met, Anthony would not sign-off that the portions of the system he was responsible for were *operationally verified*. This was because hardware was not available for actual testing, and would not be for several months. On his own initiative, Anthony began developing a system to simulate:

A. The Jovian noise environment, including lightning
B. The 1.4 gigahertz signal and its characteristics with modulated data
C. The 15 hertz/second doppler effect on signal
D. The RF and IF stages in the telemetry relay receiver
E. The instruction set of the telemetry computer (over 7,000,000 execution states)
F. The data buffering for quantized data coming from the probe, as well as the bus adapter that connected the telemetry computer to the spacecraft's network.
G. The hardware multiplier, including its timing characteristics

There were several sensitive design criteria for such a simulator to meet, which is why the project had specified true, *hardware* platforms for all testing and development:

1. Timing sensitivity of loop dynamics had to be verified in real-time conditions
2. The hybrid digital/analog interfaces had to be interactively tuned
3. Critical test runs could exceed tens of minutes in *real* time

In addition, the selection of a *real* computer and operating system, that would simulate the Galileo on-board, bit-sliced computer and operating system, was not without controversy. A DEC minicomputer had already been designated to support development of up to three telemetry receivers, and, while being used on other projects, was at the limit of its computing capacity.

Saving $0.25 Million with 4,000 Lines of Code

Anthony proceeded to develop a system simulator that became the backbone of the entire software development effort. All of the major code segments of the telemetry relay receiver, later loaded on the spacecraft and successfully executed in December 1995, first came to life and were tested in their native microcode environment with the simulator Anthony developed.

Many of the decisions that Anthony made were not obvious when he introduced the system to the design team. First, he decided not to use the DEC minicomputer designated by management to support Galileo development. He did this based on using the system himself, and noticed that the DEC minicomputer had too many users (significantly over-loading it and slowing performance). He also felt that the DEC minicomputer crashed too often, which could directly affect the stability of the final software product that would later arrive at Jupiter.

Instead, he elected to use a new, *Prime* minicomputer, *outside* of the technology division. The Prime that he selected had a small, well-regulated user community. Its PRIMOS operating system, from his personal experience, was a generation better than the DEC minicomputer, offering a richer set of development tools with much more stability. Since Anthony guessed that the simulator could become instrumental in testing this key component of a *$1 billion+* program, Anthony implemented the integrated simulator system on the Prime. A Prime was later purchased by the technology division, and Anthony transferred the simulator on to it.

There was one additional decision that may have been very inspired. Anthony designed the simulator to allow microcode sequences to include *native Fortran code*. By doing this, he was able to take some of the Fortran models of communications algorithms, and the RF/IF/quantizer hardware they simulated, and integrate them into the telemetry computer simulator. This included models of the signal dynamics of the probe falling to Jupiter, and the corrupting influences of the Jovian environment on the signal. The timing-base for the simulator tracked *real* time in up to seven different ways, accommodating all of the timing synchrony issues for simulated probe data, BPSK signal, demodulation hardware, quantization hardware, digital instruction processing and network connectivity to the CDS that controlled the Galileo orbiter.

This simulator consisted of about 4,000 lines of code, and took Anthony about four weeks to write. He built an easy-to-use user interface, with extensive debugging tools that allowed any part of the telemetry computer to be queried during discrete simulation steps. Finally, he wrote a 70 page users manual which permitted all of the microcode engineers to readily use the system. The simulator's "self-checking" features could be activated from the user interface to monitor internal simulator performance. Subroutine sockets allowed other software engineers to readily modify the simulator to accommodate customization as the project evolved.

Since the Prime computer was also used in Systems Analysis to design the FLL, PLL and symbol detection models, the entire project became more unified by allowing senior communications analysts to exchange their models with the microcode engineers *on the same computer*. In fact, Fortran models made by Systems Analysis personnel could be directly embedded into microcode routines, and later converted into *pure* microcode when their characteristics were understood by the microcode engineers.

The simulator was exhaustively tested by the entire project team, and satisfied the requirements originally specified for the *hardware* simulators to support software development. Since it ran on a multitasking computer, multiple microcode engineers could work simultaneously (instead of one-at-a-time with a hardware simulator). This obviated building one of the three hardware simulators, which yielded savings to the project that may have exceeded $0.25 million. In addition, the entire project schedule was accelerated by over four months, allowing milestones to be operationally verified *before* signing-off on them (the original motivation Anthony had in developing the simulator). When a problem later occurred with a hardware simulator, it was the *software simulator* that was used to find and fix it.

Conclusions

This project benefitted *significantly* from the motivation and desire to serve, as vested through a well prepared student who came to the Galileo project shortly after graduating from Caltech. His efforts and contributions reflected very well on the portions of Caltech's curriculum addressing digital circuit design and testing, firmware development and integration, real-time system simulation, programming, microcode development and computer system architecture.

Sincerely,

Dr. Urban Von der Embse
Galileo Project Scientist, Hughes Space and Communications Company

CC: Attached List

6

Mr. Peter Thorp
Headmaster
The Cate School
1960 Cate Mesa Road
Carpinteria, CA 93014-5005

Dr. Sanderson Smith
The Cate School
1960 Cate Mesa Road
Carpinteria, CA 93014-5005

Mr. Fred Clark, Cate School Headmaster
c/o Mr. Peter Clark
Oregon State University
Geosciences Dept./Wilkinson Room 114
Corvalis, OR 97331

Mr. Lee Browne
Caltech Emeritus Faculty Member
871 W. Ventura Street
Altadena, CA 91001

Dr. Carver Mead
Office of the President
Caltech
Pasadena, CA 91125

Ms. Betty Woodworth
Emerita Archivist
The Cate School
1960 Cate Mesa Road
Carpinteria, CA 93014-5005

Ms. Judy Goodstein
Caltech Archives
Mail Code 015A-74
Pasadena, CA 91125

Appendix ℶ

U.S. Patent No. 6,236,992

US006236992B1

(12) **United States Patent**
Sneed

(10) Patent No.: **US 6,236,992 B1**
(45) Date of Patent: **May 22, 2001**

(54) **SERIAL ENCRYPTION SYSTEM-BYPASS OF YEAR 2000 DATE MALFUNCTIONS**

(76) Inventor: Anthony Sneed, 2058 North Mills Ave., Claremont, CA (US) 91711

(*) Notice: Subject to any disclaimer, the term of this patent is extended or adjusted under 35 U.S.C. 154(b) by 0 days.

(21) Appl. No.: 09/073,258

(22) Filed: **May 4, 1998**

(51) Int. Cl.[7] G06F 17/30
(52) U.S. Cl. 707/6; 707/101; 395/705
(58) Field of Search 707/6, 101; 395/705

(56) **References Cited**

U.S. PATENT DOCUMENTS

5,600,836	2/1997	Alter	395/612
5,630,118	5/1997	Shaughnessy	395/601
5,668,989	9/1997	Mao	395/612
5,740,442	4/1998	Cox et al.	707/4
5,758,336	5/1998	Brady	707/6
5,758,346	5/1998	Baird	707/101
5,956,817	9/1999	Nicholas	395/701

6,061,817 * 5/2000 Carter et al. 395/705
* cited by examiner

Primary Examiner—Wayne Amsbury
(74) Attorney, Agent, or Firm—I. Nicholas Gross

(57) **ABSTRACT**

An improved computing system is described which eliminates potential errors associated with Year 2000 malfunctions. This system may be a retrofitted version of an existing computer system which has been intentionally date regressed to function in a time zone that eliminates potential date computation problems. To effectuate a complete regression of such system, existing databases are date regressed, existing programs are time synchronized with such databases, and existing I/O routines are modified to assure date arithmetic integrity. To handle years outside of a 100 year span, chronological encapsulation is used to set up year specific databases which interact with associated time synchronized copies of any main programs on such system. Multiple regressed states can be implemented in such system by a series of one or more regressed databases and associated synchronized programs, so that date computations can be handled for date data fields spanning across two or three centuries worth of data using only two digit decimal field formats for records.

44 Claims, 4 Drawing Sheets

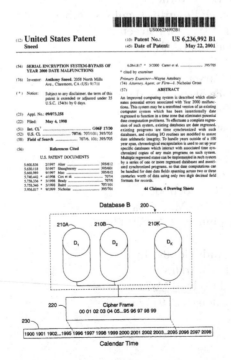

Database B   200

210A 210B 210K

D₁ D₂ Dₖ

220   Cipher Frame
00 01 02 03 04 05...95 96 97 98 99

230

1900 1901 1902...1995 1996 1997 1998 1999 2000 2001 2002 2003...2095 2096 2097 2098

Calendar Time

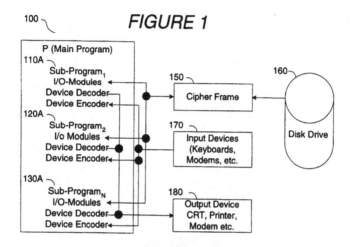

FIGURE 1

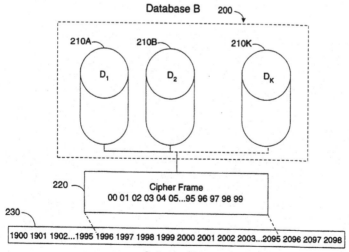

FIGURE 2

a *7th-Gift* page

FIGURE 3

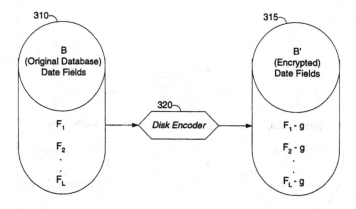

FIGURE 4

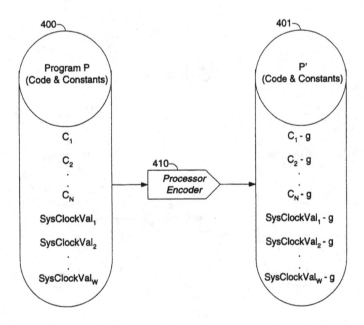

FIGURE 5A

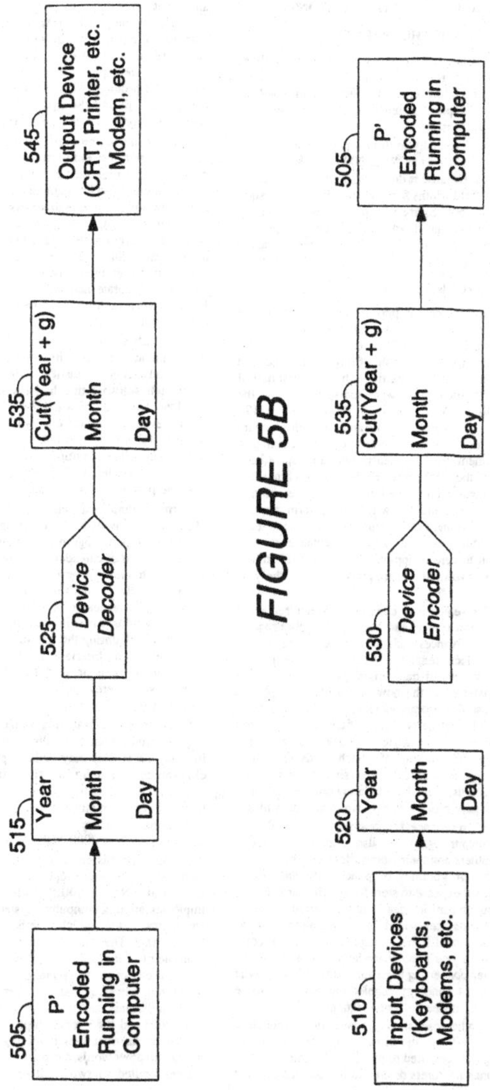

505 — P' Encoded Running in Computer

515 — Year / Month / Day

525 — *Device Decoder*

535 — Cut(Year + g) / Month / Day

545 — Output Device (CRT, Printer, etc. Modem, etc.)

FIGURE 5B

510 — Input Devices (Keyboards, Modems, etc.)

520 — Year / Month / Day

530 — *Device Encoder*

535 — Cut(Year + g) / Month / Day

505 — P' Encoded Running in Computer

US 6,236,992 B1

| 1 | 2 |

SERIAL ENCRYPTION SYSTEM-BYPASS OF YEAR 2000 DATE MALFUNCTIONS

FIELD OF THE INVENTION

The invention relates to a computer system that utilizes regressed internal date representations. To effectuate such a system, a unique encoding/decoding mechanism is used for regressing data in typical stored-program computers and their associated databases.

The invention is especially useful for preventing what are commonly known in the art as "Year 2000" malfunctions in computer systems. Because it does not require restructuring or expanding the field widths for data already stored in such systems, and does not require significant modifications of existing programs associated with such systems, the present invention provides a cost effective solution to the pervasive Year 2000 problem.

BACKGROUND OF THE INVENTION

Since 1900, and for census purposes shortly before, the year portion of calendar dates has been truncated to two digits ("55" instead of "1955," for example). The precedence helped maximize the amount of information encoded on 80-column punch cards in use throughout the first half of the 20th century. Such cards were indispensable for the decennial census (1890, 1900, 1910, 1920, 1930, 1940), significantly speeding census counts and post-census count analysis. Without such punch card technology, according to the U.S. government, the 1890 census would not have been completed before the 1900 census began. The machines that processed the cards were called tabulators, later having plug-in boards that could be wired to perform various computational and printed report functions. The same technology gained wide-scale commercial acceptance in business (in addition to census applications) after 1913, maintaining the same two-digit census precedence for the year portion of dates.

By the 1950's, electronic computers began replacing tabulators to support census and commercial data processing. Data storage advances included magnetic tape (herein "mag tape") and electromagnetic disk drives (herein "disk drives"). Vast amounts of data originally stored on punch cards were transferred to the new media of mag tape and disk drives. These new storage devices added the ability to randomly and electronically access files through stored program machines, a dramatic improvement over the manual handling of millions of punch cards. The new storage technology also used less physical space to store data than punch cards, as well as obviating much of the labor expense to manually process punch card-based data files.

The two-digit precedence for year portions of calendar dates (instead of four digits) was also used with the newer electronic computers and their storage devices. This simplified the transfer of volumes of punch card data while avoiding additional expense to store four-digit year fields. In doing so, the punch card images were transferred unaltered to the new storage devices, speeding conversion and removing up to an additional 2.5% storage-expense per record. Thus, the two-digit year precedence helped save space and reduce costs when converting to newer storage technology at mid-century. Such technology enabled database files to be accessed randomly instead of only sequentially.

To have done otherwise, such as converting to four-digit years, may have inflated the already high cost of newer storage offerings as compared to punch card technology and its media. No manufacturers desired to include in a product

proposal a 2.5% higher cost, which might give competitors an instant 2.5% price advantage who stayed with two digit years already on punch cards. Thus, the remedy (of staying with two digit years) was technologically expedient and helped reduce a price objection by customers, mating the storage devices appear as seamless extensions of proven punch card protocols. This expediency, and its widespread perpetuation during the last four decades, helped set the worldwide conditions for the Year 2000 Malfunction.

None of the computing technology advances since then have reversed the long-term effects of maintaining the two-digit year precedence. Though the cost of disk drive storage has dropped dramatically in the last few decades, the reprogramming expense to reverse out the two-digit date limitation has reciprocally increased each year. New functions were incrementally added to existing programs decade after decade. Such software changes were stimulated by regulatory requirements, new tax reporting and accounting methods, corporate acquisitions, new telecommunications links between multiple computer sites, new hardware and other reasons.

With each and every change, the two-digit year precedence spread like yeast within a batch of dough. Every year, it seemed easier to justify avoiding radical changes to production software that had taken years to develop and stabilize. Leaving software the same as it was the year before, even if the hardware was changed, seemed operationally and fiscally prudent. In numerous cases, this reasoning was followed until the Year 2000 Malfunction began to tangibly manifest itself in economically damaging ways over the past 18 to 24 months.

Current solutions to retroactively address the Year 2000 Malfunction involve restructuring such databases, and rewriting support programs to accommodate full four digit representations of calendar years instead of two digit abbreviations. This reengineering task involves multiple billions of lines of code, and multiple trillions of bytes of data spread throughout government, academic and private sector databases. All of this is running on a host of different computing platforms built during the last 40 years, some of which may no longer be commercially produced, if the manufacturers remain in business at all. Also, due to age, some of the source code for programs in many production systems may no longer be available.

Conventional tools to address the Year 2000 Malfunction may not produce useful results sooner than 18 to 36 months. In many cases, this may make a plethora of systems that electronically handle multiple trillions of dollars of resources--including airline reservations, banking transactions, accounting systems, utilities, government tax collections, social services, and others--inoperable as the frequency of Year 2000 Malfunctions increases.

A more promising and useful solution to the above malfunction has been proposed and described in a prior art U.S. Pat. No. 5,600,836 to Alter. In the Alter implementation, a computing system is configured to operate in "zone" time, which time is different from an external "local" time. The "zone" time is intentionally set at some multiple of four years (preferably 56 years to preserve leap year and day of the week integrity) in the past. In this way, the system perceives that it is operating at a time some 56 years in the past, and therefore, a date operation involving years 2002 and 1985, for example a subtraction operation, is treated as an operation involving years 1946 and 1929 instead. In other words, a conventional un-treated computing system limited to two digit date fields would treat the

US 6,236,992 B1

3

subtraction operation as 02–85 (the truncated values of 2002 and 1985) and give an erroneous result of –83, but the Alter approach yields the correct result of 46–29=17. By confining the computing system to operate with date field values within a single century, any internal date operations are kept accurate (in a relative sense). To maintain consistency with the outside world, an interface is used in Alter for converting date field data to local or zone time.

Some limitations of the Alter approach, however, include the fact that there is no mechanism for handling databases that include date field data exceeding 100 years in scope. For example, as of 1994 at least, the social security database includes data for birth years for more than 117 years worth of individuals. This cannot be accommodated even in a system modified by Alter, as such is still limited to a single two-digit database structure. Moreover, Alter apparently fails to appreciate that the number of computer programs that can be kept on a system utilizing such approach is limited unless such programs are also "synchronized" with the time adjusted data in such system.

SUMMARY OF THE INVENTION

Accordingly, it is an objective of this invention to provide a method for eliminating potential Year 2000 malfunctions from existing computing systems that include programs and/or databases limited to two-digit date formats;

A related objective of the present invention to provide an improved computing system that, while only utilizing programs and/or databases limited to two-digit date formats, can nevertheless perform date calculations spanning across two different centuries without error by operating in a regressed state;

A related objective of this invention is to provide a method for conditioning the date data used in such systems so as to make such data usable in a regressed state computing system;

A further related objective of the present invention is to effect an interface for use with a regressed data computing system, so that date field data passed to and from such system is encoded and decoded as necessary to synchronize it with the regressed state of such system or an actual time existing outside of such system.

These and other objects are effectuated by providing a computing system which has been converted to operate in one or more time regressed states. To place such system into such state, a first method of the present invention encodes date fields of records in a data file B so that such records can be used in a computer system during a calendar year YY, where YY represents a value of the last two decimal digits of a year between 2000 and 2099. An offset value, or encryption key, g, is used to transform such data file B to an encoded data file B' by modifying the data values in each of the records of B by said offset g when $XX-g>=0$ so that XX is replaced with $XX-g$ for each such record in B. When $XX-g<0$, the record is removed from the first data file, and stored in one or more sub-data files B_s'. In this manner, the date fields of both B' and B_s' are configured such that date computation integrity involving XX and YY is maintained in the computer system irrespective of the value of YY. In one implementation of this first embodiment, sub-data file B_s' is comprised of a plurality of sub-databases B_1', B_2', B_3' . . . B_n', and each such sub-file contains records for a single year. The records of such computing system are preferably examined in advance so that the value of g is selected to minimize the number of sub-data files. Using such approach, data files B' and Bs' may contain data for records spanning more than

4

100 years. In another implementation of this first embodiment, a first regressed database B' may be created to contain records for a first 100 year time span, and a second regressed database B" may be generated to contain records beyond this 100 year time span.

A method is also provided herein for converting software code of a computer program, so that such program can be used in a computer system that is limited to two digit decimal representations of dates. This is accomplished by first identifying any portions of the software code in such program that include either and/or both (i) a computation involving the date field from data file B and any associated date field related computation constant C; and/or (ii) a system clock call. These portions are then modified, if necessary, by replacing any date field related computation constant C, or system call, with a shifted computation based on g. This process is repeated until all of the software code for such program has been converted, and results in a computer program converted to perform date computation operations involving date values ranging from approximately 1900+g to 2000+g. In addition, the software code can be configured so that a first sub-program operates with date field data from a first 100 year time span, and a second sub-sub-program operates with date field data beyond said first 100 year time span.

In a third process described herein, suitable adjustments are then made to input and output portions of software code so that date computation integrity is maintained as date field data is passed to and from such date regressed system.

In this way, a computer system can be configured to operate now and beyond the year 2000 without experiencing date operation malfunctions caused by processing date across more than one century. This is a result of the fact that any original data files B are converted into encoded data files B' based on the desired offset g, and all associated programs are synchronized with such files.

In yet another embodiment, date operations associated with the programs and said data files can be executed by such system using a first regressed time T1, and/or a second regressed time T2, where T2 is different from T1. This further facilitates a complete conversion of an existing system, and permits date operations to be performed with date values spanning even more than 200 years.

Thus, the arithmetic encryption and re-engineering tools of the invention encode both databases and programs to make such entities compatible with a time regressed computing system, using a unique interface to synchronize information exchanges between peripheral devices and the central processor in a computing complex. A series of two encoder and two decoder protocols are applied to both databases and programs to encrypt the internal date representations of a computing system while maintaining synchrony with extant calendar precedence for computation, external transfer, display or printing of date-sensitive information.

The invention therefore provides a solution for retrofitting any manner of computing hardware and storage technology subject to the Year 2000 Malfunction. It achieves this without altering the two-digit year precedence of existing databases, and with only minor parametric changes to software supporting such databases. The simplicity of this encryption and re-engineering invention allows the Year 2000 Malfunction to be addressed in a matter of weeks, instead of months or years for most systems. This may permit a large number of organizations to obviate the malfunction in as little as seven weeks, and probably no

US 6,236,992 B1

5

more than seven months for even the most complex systems. The invention enables each organization to address the malfunction with existing staff, avoiding the expense and uncertainty of trying to find and hire specialized consultants as the supply of such personnel continues to shrink. As such, this invention helps increase the useful life by 20 to 80 years of already-purchased systems through in-place staff.

When applied, the invention herein is a direct replacement for additional hardware storage devices and additional central computers that may otherwise be needed and purchased in mass to allow large-scale information systems to be modified and operate beyond the year 1999. In this sense, the encryption and re-engineering invention acts as a direct substitute for tangible and costly hardware-intensive alternatives now being considered to avoid the Year 2000 Malfunction.

BRIEF DESCRIPTION OF THE DRAWINGS

FIG. 1 is a high level block diagram of the overall structure of a computing system embodying the teachings of the present invention;

FIG. 2 is a block diagram illustrating the operation of a Cipher Frame interface of the present invention, which coordinates and maintains date consistency between an internal regressed database, and other non-regressed, real-time entities;

FIG. 3 illustrates the general operation of a Disk Encoder of the present invention as it is used to convert an existing, two digit date field database B, into an encrypted, date regressed database B';

FIG. 4 illustrates the general operation of a Processor Encoder of the present invention as it is used to convert an existing program P, into a date synchronized program P';

FIG. 5a illustrates generally how a device decoder of the present invention operates to maintain date consistency between date regressed programs P' and output devices as they may exist within a computing system modified in accordance with the teachings of the present invention;

FIG. 5b illustrates generally how a Device Encoder of the present invention operates to maintain date consistency between input devices and date regressed programs P' as they may exist within a computing system modified in accordance with the teachings of the present invention.

DETAILED DESCRIPTION OF THE INVENTION

The invention herein is related to the potential Year 2000 Malfunction which exists for large-scale, and widely used computer systems. This particular malfunction is generally isolated to systems programmed to use two digits to represent years in calendar dates. For example, such systems abbreviate "1955" to "55." Because of this, calculations involving dates beyond the year 1999 cause system malfunctions from programs that maintain a two digit precedence to represent years throughout their programs and database files. In such systems, "00" represents the year "2000" since only two decimal digits (DD) are used to record the year (say XX) in a calendar date. Computations that use a value of "00" to denote the year 2000 violate the chronological sequence of actual calendar dates (. . . , 97, 98, 99, 00, 01, . . .). The truncation, therefore, makes 2009 chronologically less than the two-digit representations of "1901 " to "1999," which distorts any time relationships for events, calculations or stored dates projecting beyond 1999.

The serial encryption system of the present invention can be used to efficiently and cost effectively retrofit any two-

6

digit year based date field computing system. Such existing computing systems represent an investment of millions of man hours embodied in multiple billions of lines of code and multiple trillions of bytes of databases developed over the last four decades through a host of computer languages, hardware platforms and storage technologies. The invention permits the rapid re-engineering and proto-typing of large-scale and generally used computer systems such that their operation and date integrity is unaffected by the Year 2000 Malfunction.

As such, the encryption system of the present invention is a replacement for costly conventional strategies for dealing with the Y2K problem. These solutions generally require significant man hour expenditures and/or processor hardware and additional storage investments, because they require review and modification of multiple billions of lines of code and multiple trillions of bytes stored on large-scale and generally used computer systems.

The invention herein essentially neutralizes the Year 2000 problem by ensuring that a two digit date field computer system is never permitted to process date data that could lead to erroneous operations. The present invention, in essence, ensures that errors do not occur by ensuring that conditions that could lead to such errors are eliminated. This is accomplished by artificially regressing such computer system (including any associated programs and time-sensitive data within any data files) to a known state that existed g years ago. In this case, g is an offset value in years, and is preferably an integer multiple of four which is used as the encryption key for the Processor Encoder, Disk Encoder, Device Decoder and Device Encoder. As used herein, the regressed state of a logical entity within a computing system is generally designated with a 'notation, so that a regressed program P is designated P', a regressed database B is designated B', etc. created by the Processor Encoder design through manual or automatic means.

It will be apparent to those skilled in the art that from the perspective of such computing system, the internal "regressed" date is of little importance. The only critical consideration is that the relative relationship between two original four digit dates (i.e., in the form DDDD) must be maintained when such original dates are translated into a two digit representation (i.e., in the form DD). In other words, if the actual years involved in a date operation are 2012 and 1980, these would be represented as 12 and 80 in a two-digit date field computing system. Nevertheless, such system must ensure that mathematical operations involving these truncated numbers are accurate. It is evidence, nevertheless, that this is not possible without some intervention because the simple operation of subtraction involving these dates (which should yield a result of +32) will instead yield a value of −68. Accordingly, some translation must be performed on the original four digit dates to ensure that any mathematical operations yield the correct result. When such values in the two digit fields are to be used outside of such system, a re-translation must occur to harmonize such internal regressed values with their real time values.

In FIG. 1, a preferred embodiment of a computing system **100** embodying the teachings of the present invention is illustrated. Such system includes one or more original main software programs P' that have been modified (as explained further below), and any number of sub-programs $P_1, P_2 \ldots P_N$, which are generated as explained below. The latter sub-programs in turn include one or more main software routines that implicate date operations, one or more associated I/O modules, a device encoding interface, and a device

a *7ᵗʰ-Gift* page

US 6,236,992 B1

7

decoding interface. A Cipher Frame **150** acts as an interface between encoded dates of a database B' which may be stored on a single disk drive **160** (or one or more disk drives D_1, D_2, ... D_K as shown in FIG. 2) and internal computations done within the central processor by any of the programs or sub-programs. As alluded to earlier, this interface permits all year-specific fields on database B' to remain as two digit data fields—i.e., in a form DD where DD contains data XX from a truncated value of an arbitrary date XXXX. From a mathematical perspective, therefore, sequential arithmetic symbols, instead of actual dates, are used to represent any input date value. Thereafter, these symbols are used by the computing system for any and all date operations to ensure date operation integrity. It is only when a specific symbol in the Cipher Frame is decoded (as shown in FIG. 2) through a Device Decoder that the symbol gains actual calendar and time meaning for the outside world.

A Device Decoder interface for each program P' or sub-program P_n converts encrypted-symbols to years before output or transfer to other devices or systems **180**. It can be seen that if for some reason a single output module or section of code does not decode an encrypted symbol before output, then the fail-safe mode of the encryption system only results in a worst case example of a "regressed" year being printed or displayed. A regressed-year-dated check of more than a year or two, for example, is usually non-negotiable. This appears to be an advantage the encryption system has over other methods that require database restructuring and reprogramming of massive amounts of code.

Completing the input/output translation between actual time in the outside world and regressed time within computing system **100** is a Device Encoder interface; again, a separate interface is used for each program P' or sub-program P_n as explained further below. Such interface converts actual year input data into encrypted-symbols. As this entry mechanism into the system is fairly controllable (i.e., date field data in input data is usually well identified), the probability of an erroneous mistranslation is acceptably small.

In FIG. 2 it can be seen that Cipher Frame interface **250** coordinates and maintains date consistency by synchronizing the interchange of data between a computing system and its associated storage and I/O devices. In a preferred embodiment, Cipher Frame interface **250** includes 100 symbols ("00" to "99") which are purposely detached from any specific group of years. Instead, these symbols operate as logical one-to-one mapping entities between actual calendar time date values and internal regressed data values. These symbols are related to actual calendar years through an offset value g, which value for g can be specified by a system administrator for such computing system based on various considerations explained further below. An integer multiple of four is preferably used for g, and such value operates to arithmetically encrypt the Cipher Frame interface to represent a specific, contiguous 100 year period, as needed. It can be seen that the period can span and overlap across two 53 different centuries worth of data, thus eliminating the potential errors associated with using actual calendar year values. Cipher Frame interface **250** therefore can be aligned to the specific requirements of any particular computing system, and can be serially aligned (i.e., in multiple "regressions" of the same system) to span across multiple 100 year periods including any period that may trigger the Year 2000 Malfunction. As an example, it is conceivable that an optimal value g for a computing system may vary between g1 at calendar time T1, and g2 at T2. Cipher Frame Interface **250** is thus flexible enough that it may be altered from time to time according to changing system demands.

8

The table below explains the correlation between a selected value of g, and the actual calendar years that are then mapped by Cipher Frame interface **250**:

g Value	Database Encoded Years through Cipher Frame interface
0	1900–1999
4	1904–2003
.	.
.	.
92	1992–2091
96	1996–2095
etc.	

In virtually every date-aware program, there is a routine that computes the length of time between dates, usually in days. So long as the base-date used in these calculations is below the lowest calendar date encrypted through the Cipher Frame interface, all relative time lengths so computed will be chronologically consistent for internal interim computations.

In the present invention, therefore, as the dates are outputed from an encrypted computer system, they are made to appear externally consistent with "today's" date. Yet, internally the machine thinks it's "1977," if g=20, and if today's date is in the year 1997 for example. The invention maintains all relative time lengths (between internally encrypted dates) which ensures the chronological integrity of subsequent computations for any year encoded.

In FIG. 3, the transformation of a preexisting set of data files B into regressed data files B' is explained through the operation of Disk Encoder **320**. In sum, Disk Encoder **320** takes all or part of the data image of any database B **310** on a disk drive (D) and encrypts all year-specific fields through the selected Cipher Frame interface **250** (which contains 100 two-digit symbols ("00" to "99") to generate a new regressed state database B' **315**. Disk Encoder **320** allows Cipher Frame **250** to be recursively encrypted, if needed, in subsequent years, as proscribed by a value g, an integer multiple of four selected for the DiskEncoder. Again, if g=0, Cipher Frame Interface **250** encompasses calendar years 1900 to 1999, the default for virtually all systems, based on the historical 2-digit year precedence used during the 20th century.

Disk Encoder **320** is preferably executed as part of an overall retrofitting or conversion of a preexisting computing system that is being modified in accordance with the teachings of the present invention to become Year 2000 malfunction-free. Nevertheless, it is conceivable that such operation may need to take place at a later time as well, or that other types of computing systems using another Year 2000 solution may benefit from using time-regressed date data. For example, it may be necessary at a later time to upgrade the computing system to a full four field date field data format, and the reverse operations would then need to be made to construct such new database from the two field data formats of B'. In this regard, therefore, Disk Encoder **320** can be executed essentially at any time, and provides a very simple mechanism for accomplishing such functions.

In operation, Disk Encoder **320** is a simple software routine that merely subtracts g from every two-digit year-specific field F_L of database B **310** on the disk drive (or drives). This encodes database B **310** through the Cipher Frame interface, making it the encoded database B' **315**

US 6,236,992 B1

9 10

having transformed data $F_L - g$ for each date field. A corresponding encryption through Processor Encoder **410** (see FIG. 4 explained below) ensures that any program P that alters or uses database B' in any way whatsoever recognizes and correctly interfaces to database B through Cipher Frame **250**.

The year 2000 is a regular leap year (unlike the years 1900 and 2100, since these centennial years do not have a February 29th), based on the quadricennial adjustment initiated in 1582 for most of Europe (1752 for England and its possessions) to correct an error in Julius Caesar's reckoning of the length of a year in 46 B.C. The fact that g is an integer multiple of four ensures consistency with quadrennial (leap) years that a Cipher Frame interface may span. With 2000 being a regular leap year, Cipher Frame **250** may transparently encompass it as well. Those skilled in the art will recognize that if y is an encrypted symbol for a year within Cipher Frame interface **250**, and if g=20 for example, then y+g+1900 is the full four-digit calendar year encoded within y.

Finally, Disk Encoder **320** also detects year dates in B that can not be encoded through Cipher Frame interface **250**. For such records Disk Encoder **320** preferably produces a series of year-specific sub databases B_1, B_2, . . . B_n for each year that is outside the Cipher Frame interface. For example, if g=20, and assuming original database B **310** spanned a period from 1880 to 2019, then a series of new databases B_{1919}, B_{1920}, B_{1880} can contain all records outside of Cipher Frame **250**. In this way, actual calendar years 1920 to 2019 are encoded within Cipher Frame interface **250**, and records with actual calendar dates in this range remain encoded within B'**315**.

This approach works extremely well because according to U.S. census data as of 1994, less than 5% of all demographic data will lie outside of a Cipher Frame interface **250** spanning actual calendar years 1920 to 2019. The number of exceptions handling, therefore, will be limited to a small number in the specially coded databases B_{1919}, B_{1920} . . . B_{1880}. It can be seen, nevertheless, that the present invention in fact provides the capability to maintain date data spanning across three centuries' worth of data using only two digit date field formats, unlike prior art techniques that are limited to two centuries worth of date field data. This is accomplished, essentially, by including more than one regressed state of data within the computing system.

In another embodiment, records outside of the 100 year span of Cipher Frame interface **250** can be stored in a second (even more) regressed database B" where B" complements B' by the fact that it is intentionally displaced by another constant (in a preferred embodiment, exactly 100 years—one century) from regressed database B'. In other words, records within B" are offset by a value of g+g'; where g'=100, in all likelihood, all of the records within B will be mapped to either B' or B." As an example, if regressed database B' contains calendar year data for 1920–2019, B" can be set to include calendar year data for years 1820–1919. This method, in fact, essentially effectuates a second Cipher Frame interface **250'**, which is used for handling dates beyond that afforded by as single 100 year Cipher Frame interface **250**. The only required modification, in this instance, is that operations involving B' by P" must take into account that this second database has been intentionally regressed even with respect to B' by an additional g' years.

In this manner the present invention can effectuate a computing system with multiple regression states, instead of

only one, and this fact, too, eliminates the need for special handling of out of range calendar year data caused by large values of g. As an example, if g is set to 56 (a useful value in some systems since it is both a multiple of 4 and 7 and thus preserves date congruence as well as day of the week congruence), then the actual dates encompassed within B" range from 2055 to 1956, and within B" dates encoded include 1856 to 1955; for all practical purposes, this approach should encapsulate any records of a preexisting database B into at most two regressed databases.

In FIG. 4, a Processor Encoder **410** is another software routine that modifies any program P within computing system **100** that updates or uses any database B' encoded through Disk Encoder **320**. Such modified programs P' become chronologically aligned with any database B' through the Cipher Frame interface for any date-related computation or any data retrieval of a date-related field. Processor Encoder **320** simply identifies all chronological constants in P and subtracts g therefrom, where g is the same value used by Disk Encoder **320** for database B'. It also subtracts g from the year portion of all system clock values ("SysClockval's") resulting from system clock calls ("SysClockCalls"). A system clock call provides the year, month, day, and hours, minutes and seconds that a computer maintains internally. To ensure that the computing system is kept in a regressed state, therefore, g is subtracted from the year portion of every SysClockVal, before it is transferred to any variable within a program or used for any other type of computation within program P.

Handling of records outside of regressed database B' by programs within computing system **100** can be accomplished as follows. Again, assuming g=20, Cipher Frame interface **250** encompasses the years 1920 to 2019 (represented by Cipher Frame symbols "00" to "99"). If there are records with dates on the main database, B that precede 1920 (representing information 78 years or older relative to 1997), then these records are assigned to sub-databases, B_{1919}, . . . B_{1880}. A worst case involves setting g=20 for a very large database, like the U.S. Social Security database. For this database, 95% of the database in regressed database B' will contain the records of all people 77 years and younger, and only 5% would be spread across B_{1919} . . . , B_{1880} (1880 is considered the lower limit because in 1997 no known person is over 117 years old).

In a first embodiment, such new regressed databases B_{1919} . . . . B_{1880} are handled by corresponding synchronized sub-programs P_{1919}, . . . , P_{1880}. These sub-programs are identical to P (the main program) except that each has fixed date constants programmed into them, unique to their sub-databases' year-specific contents. In this manner, P' along with P_{1919}, . . . P_{1880} are treated as a single object since 95% to 99% of their code is identical. This part of the encryption architecture supports the few years outside of Cipher Frame interface **250** and is generally referred to herein as chronological encapsulation.

Each program P_{1919} . . . P_{1880} is a clone of P with fixed year constants encoded into each clone program for its particular year. When a year outside of Cipher Frame **250** is detected by the Device Encoder **530** (explained below) this latter routine invokes the year-specific P_T to handle whatever data is received with the year. Otherwise, the regular program path within P' handles the data entered along with the year. Each P_T has its own database B_T which is partitioned from B' by Disk Encoder **530** based on each particular year outside of Cipher Frame **250**.

The Social Security database is a worst case example, since most databases do not encompass a 117-year span. Yet

US 6,236,992 B1

11

even with this, for one of the most complex databases in the world, over 95% of the entire database is within Cipher Frame 250, with chronological encapsulation accommodating the remaining 5%.

In addition, in many instances, the number of additional synchronized programs P_T can be kept to a minimum because of the characteristics of the data kept in such databases. For example, in the U.S. Social Security database, individuals born before 1920 are 78 years and older. If all such persons fit a uniform pattern of benefit payments, then $P_{1919} \ldots P_{1880}$ may be identical to each other to the point that they may collapse into a single program, $P_{1919-1880}$. This pattern may be consistent across many private and public sector databases. Again B' would contain the most volatile demographics (77 years and younger), which represents 95% of the entire database population—and it would all be in Cipher Frame interface 250.

In another variation of the present invention, where a single second time regressed database B" is used, a single second program P" supplements P' in a manner similar to that afforded by programs $P_{1919} \ldots P_{1880}$ described above. Such second program P" is synchronized so that it operates with second regressed database B"; as with P', this can be accomplished by simply taking into account the additional offset g' in any calculations (i.e., by adding or subtracting it where necessary) in the manner described above. In this manner, a computing system 100 can operate in multiple regressed states simultaneously to accommodate actual calendar year dates spanning across a 200 year contiguous time period, rather than simply 100 years.

As with Disk Encoder 320, Process Encoder 410 is preferably executed as part of an overall retrofitting or conversion of a preexisting computing system that is being modified in accordance with the teachings of the present invention to become Year 2000 malfunction-free. Nevertheless, it is conceivable that such operation may need to take place at a later time as well (for example, new programs may need to be added), or that other types of computing systems using another Year 2000 solution may benefit from using time-regressed programs. In this regard, therefore, Process Encoder 410 can be executed essentially at any time, and provides a very simple mechanism for accomplishing such functions.

Finally, to facilitate handling of situations when operations of P' or some other sub-program implicate records outside of the database associated with such program or sub-program, an interim buffer, called I_{MAIN}, can accept database updates or other transactions for database records outside of Cipher Frame interface 250. Such interim buffer, is later sorted into individual files, $I_1–I_n$ which are individually applied for updating sub-databases $B_1 \ldots B_n$, through each sub-databases' date-specific programs, $P_1 \ldots P_n$.

Even for the most massively complex programs, involving hundreds of thousands of lines of code, there are perhaps fewer than several dozen year-constants pre-programmed for computational and comparison purposes. In source code, such constants may be readily found by doing routine text searches for "00" to "99" values. If only executable code is available, a reverse compiler still forces the dates to standout through identified instructions with constant fields. In either event, the task may be trivial compared to tracing every usage of a two-digit field that needs to be expanded to four digits across multiple billions of lines of code.

The present encryption mechanism requires only minor changes to those few portions of the system that generate

12

time-based values (a small fraction of most software programs, as compared to every instance within a large program that must be altered to accommodate new, four-digit year fields). Encryption only introduces incremental changes to specific parts of a computer's software, as opposed to altering the entire infrastructure of databases and programs. Encryption helps avoid many of the risks inherent in trying massive conversions with limited budgets, tight deadlines and competing priorities of available staff.

After transforming any associated databases into regressed state databases, and synchronizing associated programs, the only remaining step needed to complete the conversion of a preexisting computing system to become Year 2000 malfunction-free is to ensure that any input/output operations are also identified and modified as needed to ensure that date field data passing between the computing system and I/O devices are kept consistent. In other words, the system should process date fields at a regressed time T, where T represents a calendar year which is=YY−g (where YY represents a value of a two-digit truncated portion of a year DDDD), but input/output data should be generated based on the actual calendar year YY.

Such operations can be achieved readily with Device Decoder 525 illustrated generally in FIG. 5A, and Device Encoder 530 illustrated generally in FIG. 5B. Device Decoder 525 is required before any encoded date may be transmitted for printing, displaying or interchanging with any other device. Device Decoder 525 converts any encoded two-digit symbol received at 515 from Cipher Frame 250 into its two-digit year equivalent. This is done by adding g to the two-digit encoded symbol and truncating the sum to two digits to generate an externally consistent date field 535 before it is provided to an output device 545. As used in this figure, the operation cut(value) truncates a value to its two right-most digits; i.e., cut(112)=12.

In a similar fashion, to maintain consistency with externally provided date data, Device Encoder 530 in FIG. 5B takes any un-encoded four-digit year value 520 received from a keyboard or other device 510, and encodes it through Cipher Frame 250. The encoded value 535 may then be used within any program P' modified by Processor Encoder 410, and any associated database B' 315 encrypted with Disk Encoder 320. After subtracting g, Device Encoder 530 then decides if the four-digit year can be encoded into a two-digit symbol within Cipher Frame 250, or if it represents a year outside of Cipher Frame 250.

As an example, if g=20 then Device Encoder 530 logic is as follows:

If (Year−g >1899) then Year:=cut(Year−g) else begin
Case Year of begin
1919: P_{1919}
1918: P_{1918}
1917: P_{1917}
  .
  .
  .
1882: P_{1882} (if needed)
1881: P_{1881} (if needed)
1880: P_{1880} (if needed)
end

Alternatively, as explained above, it is also possible for Device Encoder 530 merely to pass control to a second synchronized main program P" which is responsible for maintaining date field data for a second regressed database B".

US 6,236,992 B1

13

Other Variations of the Present Invention

Serial encryption of Cipher Frame interface **250**. The present invention permits an encryption key g where: g=g1+ g2+g3 . . . . gn so that each g may be used to encrypt successive, overlapping, 100 year periods through Cipher Frame interface **250**, which permits the invention to be used serially This requires complete fidelity with the quadricennial (400 year) precedence m subsequent centennial years (2100, 2200, etc. which do not contain a quadrennial leap day), as it is consistently reflected in the year 2000 (which contain a quadrennial leap-day) for extant programs, systems and databases subject to the Year 2000 Malfunction.

Optional Cipher Frame interface having 256 symbols (0 to 255) to span a period of up to 256 years. An optional 256 symbol Cipher Frame interface **250** can be used in systems that permit redefinition of a two digit decimal field as two concatenated binary fields, each field a minimum of one byte in size. If the fields can be concatenated, then a larger Cipher Frame interface may contain a minimum of 32,768 symbols (0 to 32,767). The decimal representations of years in a database B are then mapped on to sequential symbols in the Cipher Frame interface. Device Encoder **530** and Device Decoder **525** would contain parallel mapping logic for receiving year values and outputing year values, respectively. Processor Encoder **410** can then include a provisions for adjusting all chronological constants and SysClockVals for consistency with encryption through such modified Cipher Frame interface. Disk Encoder **320** would encrypt a database B to map its range of years onto some portion (such as the lower end) of the optional Cipher Frame interface.

The invention reduces the time to re- engineer virtually any large-scale, generally used computer system and its database, compared to extant methods that require two-digit to four-digit conversions of year representations. Restructuring such existing databases holding multiple trillions of bytes of data creates a very dramatic and costly change in the multiple billions of lines of code P supporting such databases. These multiple billions of lines of code have evolved over a four-decade period, across a range of computer architectures and manufacturers, not all of which are still in business.

If the multiple billions of lines of code are looked at as a single software object, then the two-digit to four-digit conversion may be prohibitive to complete, in aggregate, within the next 12 to 36 months. Even if accurate source code were available (which it may not always be for a significant percentage of code in both the private and public sectors), there may remain a shortage of trained programmers and consultants with sufficient expertise to handle the size of the task within available budgets and time.

The present invention simplifies the task by focusing only on an arithmetic encryption that preserves existing software and hardware investments. Disk Encoder **320** can readily encrypt virtually any type of database overnight, without changing the record architecture within a target database. By leaving data structures intact, this helps obviate the huge expense of reprogramming multiple billions of lines of program code.

From the discussion above it is apparent that Cipher Frame Interface **250**, as well as Device Decoder Interface **525** and Device Encoder Interface **530**, can be implemented in a variety of ways known in the art, including in dedicated hardware, or as independent stand alone routines that run on system **100** and/or as an integrated sections of programs P' and sub-programs P'. Other variations will be apparent to skilled artisans.

Although the present invention has been described in terms of a preferred embodiment, it will be apparent to those

14

skilled in the art that many alterations and modifications may be made to such embodiments without departing from the teachings of the present invention. Accordingly, it is intended that the all such alterations and modifications be included within the scope and spirit of the invention as defined by the appended claims.

What is claimed is:

1. A method of encoding date fields of records in a database B so that such records can be used in a computer system during a calendar year YY, where YY represents a value of the last two decimal digits of a year between 2000 and 2099, the fields having a format DD and containing data values XX where XX represents a value of the last two decimal digits of a year between 1900 and 1999, the method comprising the steps of:

providing an offset value g, where g is an integer >=4;

transforming the database B to an encoded database B' by modifying the data values in each of the records of B by said offset g when XX–g>=0 so that XX is replaced with XX–g for each such record in B; and

when XX–g<0, removing the record from the first data file, and storing said record in a sub-database B_z' without performing date field expansion on said record;

wherein the date fields of both B' and B_z' are configured such that date computation integrity involving XX and YY is maintained in the computer system irrespective of the value of YY; and

further wherein said sub-database B_z' is comprised of a plurality of sub-databases B_1', B_2', B_3' . . . B_n, where 0<=n<=g, and each such of said plurality of sub-databases contains records for a single year.

2. The method of claim 1, wherein said value of g is selected to minimize the number of sub-databases in sub-database B_z'.

3. The method of claim 1, wherein said encoded database B' and sub-database B_z' contain data for records spanning more than 100 years.

4. The method of claim 1, further including a step of adding new records to said encoded database B' by mapping the date value YY of any such records to a value XX', where XX'=100+YY–g.

5. The method of claim 1, wherein said encoded database B' contains data for a first 100 year time span, and subdatabase B_z' contains data for a second 100 year time span.

6. A method of encoding date fields of records in a database B so that such records can be used in a computer system during a calendar year YY, where YY represents a value of the last two decimal digits of a year between 2000 and 2099, the fields having a two-digit format DD and a value XX where XX represents a value of the last two decimal digits of a year between 1900 and 1999, the method comprising the steps of:

providing an offset value g, where g is an integer >=4;

converting the database B to a plurality of encoded databases B' based on said offset g; and

wherein all said plurality of encoded databases B' use a two digit format for date fields for records in said files, and taken in combination such plurality of encoded databases B' contain data for records spanning more than 100 years; and

further wherein said plurality of encoded databases B' include a main database B_0' spanning up to 100 years of date field data, and a number of sub-databases B_1', B_2', B_3' . . . B_n, where 0<=n<=g, and where each such sub-database contains records for a single year such that 100+g years worth of data is contained in such sub-databases in date fields that are not expanded to four digits.

US 6,236,992 B1

15

7. The method of claim **6**, further including a step of adding new records to the encoded databases B' by mapping the date value YY of any such records to a value XX', where XX'=100+YY−g.

8. A method of converting software code of a computer program, which program is intended to be used in a computer system during a calendar year YY, where YY represents a value of the last two decimal digits of a year between 2000 and 2099, the system also including records in an encrypted database B having date fields with a two digit format DD and a value XX, where XX represents a value of the last two decimal digits of a year between 1900 and 1999, and where the date fields of such encrypted database represent date values shifted by an encryption key g from an actual date, the method comprising the steps of:

(a) identifying any portions of the software code in such program that include either and/or both:
   (i) a computation involving the date field from encrypted database B and any associated date field related computation constant C; and
   (ii) a system clock call;

(b) modifying, if necessary, said portions of software code identified in (a)(i) by replacing said date field related computation constant C with an encrypted computation constant C' where C'=C−g;

(c) modifying, if necessary, said portions of software code identified in (a)(ii) by replacing said system clock call with an encrypted system clock call based on the encryption key g;

(d) repeating steps (a) through (c) until all of the software code for such program has been converted to interact with encrypted database B for records spanning from 1900+g to 2000+g;

(e) providing at least one additional software routine to coordinate date transactions with a sub-database B', which sub-database B' does not use date field expansion and includes date records for a single year Y outside of a range spanned by said encrypted database B;

wherein the computer program is converted to perform date computation operations involving date values ranging from 1900+g to 2000+g, and said additional software routine handle date values for said year Y.

9. The method of claim **8**, wherein during step (e) a number of additional software routines are provided to interact with a corresponding number of sub-databases B_1', B_2', B_3' ... B_n', and each such sub-database contains records for a single year.

10. The method of claim **8** wherein said encrypted database B and sub-database B' contain data for records spanning more than 100 years.

11. A method of configuring a computer system, which system includes one or more programs intended to operate during a calendar year YY, where YY represents a value of the last two decimal digits of a year between 1900 and 2099, and where the programs access records in a database B having two digit date fields with a format DD and a value XX, where XX represents a value of the last two decimal digits of a year between 1900 and 1999, the method comprising the steps of:

(a) providing an offset value g, where g is an integer >=4;

(b) converting the database B to a plurality of encoded databases B' based on said offset g, such that said plurality of encoded databases B' may contain data for records spanning more than 100 years and without performing date field expansion on said records;

(c) synchronizing the programs to operate with said plurality of encoded databases B';

16

wherein the system is configured to perform date computation operations involving date values spanning across two centuries using said plurality of encoded databases B' and said synchronized programs; and

further wherein during step (b), said plurality of encoded data files include a main database B_0' spanning up to 100 years of date field data, and a number of sub-databases B_1', B_2', B_3' ... B_n', where 0<=n<=g, and where each such sub-database contains records for a single year such that 100+g years worth of data is contained in such sub-databases in non-date expanded form.

12. The method of claim **11**, wherein during step (b), the database B is converted to an encoded database B' by modifying the date coded fields XX in each record of B by said offset g when XX−g>=0 so that XX is replaced with XX−g for each such record in B, and when XX−g<0, said record is removed from B and stored in a sub-database B_n'.

13. A method of configuring a computer system, which system includes one or more programs intended to operate during a calendar year YY, where YY represents a value of the last two decimal digits of a year between 2000 and 2099, and where the programs access records in a database B having two digit date fields with a format DD and a value XX, where XX represents a value of the last two decimal digits of a year between 1900 and 1999, the method comprising the steps of:

(a) providing an offset value g, where g is an integer >=4;

(b) converting the database B to a plurality of encoded databases B' based on said offset g, such that said databases B' may contain data for records spanning more than 100 years without including any date expanded fields;

(c) synchronizing the programs to operate with said plurality of encoded databases B';

wherein the system is configured to perform date computation operations involving date values spanning across two centuries using said plurality of encoded databases B' and said synchronized programs; and

further wherein during step (c) the programs are converted by identifying any portions of the software code in such program that include either and/or both:
   (i) a computation involving the date field from database B and any associated date field related computation constant D; and
   (ii) a system clock call;

(d) modifying, if necessary, said portions of software code by replacing said date field related computation constant D with a shifted computation constant D' where D'=D−g, and, if necessary, replacing said system clock call with a shifted system clock call based on the offset g.

14. The method of claim **11**, further including a step (d): converting any input/output programs intended to be used in such system so that input/output data is modified by said offset g when it is passed between said input/output programs and said synchronizing programs.

15. The method of claim **11**, further including a step (d): providing an input/output interface for such system for detecting data in date fields in input/output programs and modifying said data by said offset g when such data is to be passed between said input/output programs and said synchronizing programs.

16. The method of claim **15**, wherein said input/output interface permits new records to be written to database B' by mapping the value YY of any date fields of such records to a regressed value XX', where XX'=YY−g+100.

US 6,236,992 B1

17

17. The method of claim 15, wherein said input/output interface permits records to be read from database B' by mapping the regressed value XX' of any date fields of such records to a non-regressed value YY, where YY=XX'+g−100.

18. The method of claim 11, wherein the system is configured so that date operations associated with the programs and said data files are executed by such system at a regressed time T, where T represents a calendar year=YY−g.

19. The method of claim 11, wherein the system is configured without modifying the structure or size of the two digit date field in data file B.

20. The method of claim 11, wherein the system is configured so that date operations associated with the programs and said data files are executed by such system using a first regressed time T1, and/or a second regressed time T2.

21. A computer system that performs data operations on date field values having a two digit form DD, the system including:

an encoded database B' that includes date fields with a two digit form DD and associated year date values XX that are offset by an integer value g from a true value YY, such that XX=YY−g+100; and

at least one encoded sub-database file B$_z$' that includes date fields with a two digit form DD and associated year date values XX; and

one or more computer programs P time-synchronized to operate with said encoded encoded database B' and encoded sub-database B$_z$'; and

an input/output interface for time-synchronizing data that is passed between said one or more time-synchronized programs P and non-synchronized input/output devices; and

wherein the system is configured to perform date computation operations involving date values spanning across two centuries, and said encoded database B' and encoded sub-database B$_z$' may contain date values spanning more than 100 years; and

further wherein encoded database B' includes data spanning up to 100 years of date field data, and said at least one encoded sub-database B$_z$' includes a number of sub-databases B$_1$', B$_2$', B$_3$' ... B$_n$, where 0<=n<=g, and where each such sub-database contains records for a single year such that 100+g years worth of data is contained in such sub-databases without including any date-expanded fields.

22. The system of claim 21, wherein programs P are synchronized to said data files so that date operations associated with the programs and said data files are executed by such system at a regressed time T, where T represents a calendar year=YY−g.

23. The system of claim 21, wherein said input/output interface permits new records to be written to data file B' by mapping the value YY of any date fields of such records to a regressed value XX', where XX'=YY−g +100.

24. The system of claim 21, wherein said input/output interface permits records to be read from database B' by mapping the regressed value XX' of any date fields of such records to a non-regressed value YY, where YY =XX'+g−100.

25. The system of claim 21, wherein date operations associated with the programs and said data files are executed by such system at a regressed time T1, and/or a regressed time T2.

26. The system of claim 21, wherein each sub-data file B$_z$' has an associated time-synchronized computer program P$_z$ for processing data associated with the date values in such sub-data file.

18

27. A method of operating a computer system for performing data operations on date field values having a two digit form DD, the method including the steps of:

(a) generating encoded values of any date fields of the two-digit form DD used by said system: by shifting an associated year date value YY of said date field by an offset integer value g; and when YY represents a year value between from 1900 to 1999, and YY−g>0, separately encoding said year date value without performing date field expansion;

(b) processing said date fields using said encoded values for YY with one or more programs associated with the computer system; and

wherein the system performs date computation operations involving date values spanning across two centuries, and said data values for YY may contain date values spanning more than 100 years; and

further wherein said encoded values are stored in an encoded database B' with data spanning up to 100 years of date field data, and in at least one or more encoded sub-databases B$_1$', B$_2$', B$_3$' ... B$_n$, where 0<=n<=g, and where each such sub-database contains records for a single year such that 100+g years worth of data is contained in such sub-databases without including any date-expanded fields.

28. The method of claim 27 wherein for step (a) said encoded value is set=YY−g+100.

29. A method of operating a computer system for performing data operations on date field values having a two digit form DD, the method including the steps of:

(a) generating encoded values of any date fields of the two-digit form DD used by said system: by shifting an associated year date value YY of said date field by an offset integer value g; and when YY represents a year value between from 1900 to 1999, and YY−g>0, separately encoding said year date value without performing date field expansion;

(b) processing said date fields using said encoded values for YY with one or more programs associated with the computer system; and

wherein the system performs date computation operations involving date values spanning across two centuries, and said data values for YY may contain date values spanning more than 100 years; and

further wherein said encoded values are stored in an encoded database B' with data spanning up to 100 years of date field data, and in at least one or more encoded sub-databases B$_1$', B$_2$', B$_3$' ... B$_n$, where 0<=n<=g, and where each such sub-database contains records for a single year such that 100+g years worth of data is contained in such sub-databases without including any date-expanded fields; and

further wherein each said sub-database has an associated time-synchronized computer program for processing data associated with the date values in such sub-database.

30. The method of claim 27, wherein the processing of said date fields is executed by such system at a regressed time T, where T represents a calendar year=YY−g.

31. The method of claim 27, further including a step (c) receiving input date field data external to said system from an input device and then encoding said data in accordance with step (a) before said date field data is processed at step (b).

32. The method of claim 27, further including a step (d) converting said encoded values to external date values before transmitting said values to an output device.

a 7th-*Gift* page

US 6,236,992 B1

19

33. The method of claim 27, wherein g is selected to be a multiple of 4.

34. A method of converting input and output date field data having a two-digit format DD, so that such data can be processed by a computer system having encoded date field values shifted by an offset value g, the method including the steps of:

(a) generating encoded values of input date field data, and decoded values of output date field data to be used by said computer system by shifting an associated actual year date value YY of said input/output date field by an offset integer value g so that I/O operations can be performed for records spanning 100 years worth of data from 1900+g to 2000+g by said shifting and without date field expansion; and when YY represents a year value before a year 1900+g, separately encoding said year date value without date field expansion so that I/O operations can be performed for records spanning before said year 1900+g; and

wherein the system processes said date fields for records associated with years 1900+g to 2000+g using said offset g, but does not use said offset for records associated with years prior to 1900+g.

35. The method of claim 34, wherein said input field data is provided by an external date file.

36. The method of claim 34, wherein said input field data is provided by an external input device such as a computer terminal.

37. The method of claim 34, wherein said output field is provided to an external output device such as a printer or computer display.

38. A method of operating a computer system during a calendar year YY, where YY represents a value of the last two decimal digits of a year between 2000 and 2099, the system including at least one or more computer programs that access records in a database B having date fields with a two digit format DD and a value XX, where XX represents a time shifted date value which has been shifted by an offset g from an actual date, the method comprising the steps of:

(a) executing any computations involving a data field constant C in the at least one or more computer programs by replacing said date field related constant C with a shifted computation constant C' where C'=C−g; and

(b) executing any computations involving a system clock call in the at least one or more computer programs by replacing said system clock call with a shifted system clock call based on the offset g;

20

wherein said computer system can perform date computation operations involving date values ranging from 1900+g to 2000+g without using expanded date fields.

39. The method of claim 38, including a step (c): executing any I/O operations having data fields in the computer system by shifting said date field by said offset g.

40. The method of claim 39, further including a step (d): adding a new record having a date value YY to a data file B by setting a date field XX=100+YY−g for such record.

41. A method of operating a computer system which includes a main database B having date fields with a two digit form DD, and at least one second database B' also having date fields with a two digit form DD, the method including the steps of:

(a) operating said database B such that its associated date fields contain date values that are intentionally regressed by an offset value g from a true value for said associated date fields, such that database B can contain records with date field values spanning from 1900+g to 2000+g;

(b) operating said second database B' such that its associated date fields contain date values that are not regressed, such that second database B' can contain records with date field values before 1900+g and which records do not include expanded date fields;

wherein said computer system can perform data storage and retrieval operations for data records containing more than 100 years worth of date values, by using records of said database B when a date operation involves a year between 1900+g to 2000+g, and by using records of said second database B' when a date operation involves a year before 1900+g.

42. The method of claim 41, wherein said data file B' includes a number of sub-data files B_1', B_2', B_3' . . . B_n', where $0<=n<=g$, and where each such sub-file contains records for a single year.

43. The method of claim 41, wherein I/O operations for a record having a date value 20YY, where YY has a value from 00 to g−1, are handled by shifting said value YY by said offset g, and I/O operations for record having a date value 19XX, where XX has a value from g to 99, are also handled by shifting said value XX by said offset g when XX>g.

44. The method of claim 42, wherein when XX<g, I/O operations for a record having a date value 19XX are not time shifted.

* * * * *

Appendix ב

The First Deposition on Earth

The NASA Galileo spacecraft's journey from Earth to Jupiter, reviewed in Appendix א, spanned some 0.5 billion miles. A successful, mission milestone was to track and receive science data, from below Jupiter's cloud tops, collected by the fastest, ever, man-made probe. Three years later, back on Earth, the writer's Y2K invention, in Appendix ב, called "BEST 2000," was used in seven nations (England, France, Germany, Italy, Singapore, China and Japan), as well as within America. One firm acquired BEST 2000, under contract, for $0.34 million. The contract was later reviewed by a court, within California's jurisdiction. The court ruled the contract valid, after a six-day trial.

In preparation for trial, some 3000 pages of discovery were produced by the writer and a $0.75 billion firm that acquired BEST 2000. The writer was deposed for 39.5 hours, by the firm's legal experts, prior to trial. The writer's wife was also deposed, for four hours. The depositions were spread over four months, one day, alone, spanning 11.5 hours of proceedings.

Early, budget projections were as high as $8 million to remedy Y2K for a firm that size, before it acquired BEST 2000. Yet, its management gave no documentation *whatsoever,* to the court, regarding any budgetary forecasts. Some $3 million of actual expenditures were uncovered, that were applied to fix their Y2K problem with BEST 2000. The firm's management confirmed such discovery information through deposition and, later, in cross-examination during trial.

The writer was intrigued by the deposition process as applied for 39.5 hours, by legal experts the opposing party hired from a prominent, international law firm. If successful, their efforts to prove a "breach entitlement," on behalf of their client, would obviate any monetary obligations due under the BEST 2000 contract. The court, as of this writing, ruled there is no "breach entitlement" to a valid contract, and awarded amounts due, plus other consideration in its decision.

The writer brought a copy of Scripture to all, six, deposition sessions (five for the writer, one for the writer's wife) and to the witness stand when questioned. After experiencing a deposition for the first time, the writer was moved to search for the first, ever deposition, within Scripture, conducted by the I-SHALL-BE. Such a proto-deposition would set precedent for all subsequent depositions, divine or otherwise, throughout Scripture.

31.   Positional Declarations on a proto-deposition in Scripture by the I-SHALL-BE.

31.1.   There is no proto-deposition because the I-SHALL-BE is omniscient.

31.2.   There is a precedent-setting proto-deposition, by the I-SHALL-BE, within Scripture.

If 𝔍1.1 is true, then objective, fact-finding inquiry has no precedent from the I-SHALL-BE. In other words, there is no basis for the I-SHALL-BE to "learn" about information from anyone else because the I-SHALL-BE is "all-knowing," "all-seeing," "all the time." Ascertaining facts from a third-party, by the I-SHALL-BE, would be unnecessary since he doesn't need to ask anyone anything. Inquiry, by the I-SHALL-BE, if it occurred, would therefore constitute only "leading questions" because the I-SHALL-BE knows the answer to each question before asking it.

𝔍1.2 is a counter-intuitive supposition to the very nature of an "all-knowing" being. It supposes there to be at least *one* instance where the I-SHALL-BE *asks a question* to learn facts he does not already have *and then act on what is disclosed to him.* Such a notion is contrary to the reassuring belief that the I-SHALL-BE has total control and continuous, complete awareness of everything occurring throughout his creation. Even one question by the I-SHALL-BE has freewill implications, no matter how trivial the question. Any type of *sincere* inquiry by the I-SHALL-BE suggests less than continuous awareness of all events while they are occurring.

If the I-SHALL-BE never initiates at least one honest question, to discover information after the fact, then there is no divine precedent for inquiry of any kind. Without some type of proto-deposition, all depositions would originate from some agency other than the I-SHALL-BE, which would contradict this timeless declaration:

> Through him all things were made; without him
> nothing was made that has been made. John 1:3

The following Positional Declarations consider these contradictory points of view:

2. Positional Declarations on what the I-SHALL-BE knows and when he knows it.

———————————

2.1. The I-SHALL-BE MUST know about everything, all the time, while it's occurring.

2.2. The I-SHALL-BE CAN elect to be unaware of some things for periods of time.

A person may elect "not to know something" or be "willfully uninformed" about a trivial event, a matter of consequence, or anything in between. And a person may elect to inquire about any matter and discover facts not previously known. The ability to "question" has profound implications, and is a fundamental manifestation of freewill. Now, if there is the freedom to question, there must also be the freedom "not to question," for whatever reason or motive one may choose. If a person may do such a thing and the I-SHALL-BE can't, then this would deny the I-SHALL-BE the primacy to manifest either attribute—to question or not to question—or any other attributes, when and if it is to his purpose to do so.

Let's now visit a possible origin of inquiry, first written of in Scripture. A disingenuous precedent is set by a party, expert at sowing doubt, confusion and peril through any opportunity:

> Now the serpent was more crafty than any of the wild animals the [I-SHALL-BE] God had made. He said to the woman, "Did God really say, 'You must not eat from any tree in the garden'?" Genesis 3:1

The intent of the question was not to discover useful information. Its intent, rather, is to confuse and mislead. The serpent already knew the correct answer and the consequences of any wrong assertion, before querying "the woman." This interrogatory applies a leading question to corrupt an innocent party ("the woman"). It establishes an egregious precedent of dishonest inquiry.

Is there any precedent of honest inquiry, where questioning is constructive and truly intended to discover facts not known, and to address such information in a beneficial way? The first question ever asked by the I-SHALL-BE can be juxtaposed to the serpent's, earlier precedent. Remember, the serpent already knew the correct answer to his question, and he used his one and only, contorted interrogatory to lead a completely innocent party into sin.

The I-SHALL-BE's proto-deposition, after the serpent's, is the first inquiry by the I-SHALL-BE recorded in Scripture. What follows are the precursor events leading up to that proto-deposition:

> When the woman saw that the fruit of the tree was good for food and pleasing to the eye, and also desirable for gaining wisdom, she took some and ate it. She also gave some to her husband, *who was with her,* and he ate it. Then the eyes of both of them were opened and they realized they were naked; so they sewed fig leaves together and made coverings for themselves.
>
> Then the man and his wife heard the sound of the [I-SHALL-BE] God as he was walking in the garden in the cool of the day, and they hid from the [I-SHALL-BE] God among the trees of the garden. Genesis 3:6-8

Recall that eating the fruit, from the tree in question, is a capital offense, conviction leading to the death penalty for breaching a clearly stated commandment:

a *7ʰ-Gift* page

And the [I-SHALL-BE] God commanded the man,
"You are free to eat from any tree in the garden; but
you must not eat from the tree of the knowledge of
good and evil, for when you eat of it you will surely
die. Genesis 3:16,17

This shall be designated, herein, the "Tree Commandment."
Adam gave no affirmation whatsoever upon hearing the
Tree Commandment and its consequences. There is no witness
or testimony that Adam said, or had to say, anything to become
bound by the Tree Commandment. For example, when others
became bound by the Ten Commandments, there was a clear
and unmistakable affirmation by the affected parties:

So Moses went back and summoned the elders
of the people and set before them all the words
the [I-SHALL-BE] had commanded him to speak.
The people all responded together, "We will do
everything the [I-SHALL-BE] has said." So Moses
brought their answer back to the [I-SHALL-BE].
Exodus 19:7,8

"Go near and listen to all that the [I-SHALL-BE] our
God says. Then tell us whatever the [I-SHALL-BE]
our God tells you. We will listen and obey."
Deuteronomy 5:27

Though the terms and conditions of the covenant were not
fully disclosed beforehand, nevertheless, the Hebrews, through
Moses, voluntarily, of their own freewill, accepted them, and
agreed to keep all the commandments of the covenant, and to
become subject to them as the Law. Adam expressed neither
acceptance nor rejection of the Tree Commandment declared
to him. There is no disclosure regarding Adam's response
upon hearing the new commandment about the tree of the
knowledge of good and evil.

Adam's only, binding, affirming, explicit declaration, of any kind, comes later, as recorded in Scripture, after a bone was removed from him to make the first woman. It is not apparent that Adam was consulted before undergoing the invasive surgery to have a rib removed. Like Isaac, he fully trusted, as a son of God, and submitted himself to the I-SHALL-BE's will, by being put into a deep sleep and having his rib removed. He fully acknowledges what the I-SHALL-BE does with this declaration, afterward:

> Then the [I-SHALL-BE] God made a woman from the rib he had taken out of the man, and he brought her to the man.
>
> The man said, "This is now bone of my bones and flesh of my flesh; she shall be called 'woman,' for she was taken out of man." Genesis 2:22,23

Here, Adam expresses affirmation and acceptance of the person the I-SHALL-BE formed from Adam's rib. He is of one accord with the I-SHALL-BE's will through this enduring vow regarding the new person called "woman." For his entire life, Adam had only one wife and kept the vow he made, consistent with the Testimony given by the Son of Man, about Adam, "from the beginning":

> [I-SHALL-BE-SALVATION] replied, "Moses permitted you to divorce your wives because your hearts were hard. But it was not this way *from the beginning*. I tell you that anyone who divorces his wife, except for marital unfaithfulness, and marries another woman commits adultery." Matthew 19:8,9

The "first" Adam had one wife his entire life, to which the "second" Adam gives truthful testimony of the profound precedent set, for all husbands, "from the beginning," through time.

We know it is written about Adam and his wife:

> The man and his wife were both naked, and they
> felt no shame. Genesis 2:25

After committing the capital crime, and becoming subject to the death penalty, Adam and his wife both realize they are naked. They do what they can to mitigate their circumstances, by clothing themselves to constrain the sensory implications of a newly-discovered aspect of each other's physical differences. After dressing themselves in fig-leaf clothing—the first garments ever made by man—they endeavored to conceal all evidence of their new circumstances from the I-SHALL-BE. By dressing-up and hiding from the I-SHALL-BE's presence, together, they were engaging in conduct which, in and of itself, might later be deemed suspicious.

These are the foreboding and terrible events leading up to the I-SHALL-BE's proto-deposition of the first husband and wife. In whatever the I-SHALL-BE does through his proto-deposition, it will set precedent for the highest standard of any subsequent form of inquiry by the descendants of Adam and Eve—if they avoid the death penalty and have descendants.

A proto-deposition, where the questioner has all the facts of the matter in advance, could be a brutal process for the deponents (i.e., the parties being questioned).

    3.   Positional Declarations on proto-deposition types when all facts are known already.

    3.1.   Accusatory: "I know why you're hiding! You ate the fruit, didn't you?"

    3.2.   Humiliating: "Were you blind enough to eat the wrong fruit, too?"

23.3.   Duplicitous; "Why are you hiding? What could you possibly be afraid of?"

In fact these are attributes of the serpent's contorted question to Adam's wife, which was used to corrupt a completely innocent party. At a minimum, the I-SHALL-BE's proto-deposition would have to exceed the highest, known standards for a presumption of innocence while ascertaining the facts of a matter through *sincere* inquiry:

> Nicodemus, who had gone to [I-SHALL-BE-SALVATION] earlier and who was one of their own number, asked, "Does our law condemn anyone without *first hearing him* to find out what he is doing?" John 7:50,51

> Then the [I-SHALL-BE] said, "The outcry against Sodom and Gomorrah is so great and their sin so grievous that *I will go down and see* if what they have done is as bad as the outcry that has reached me. If not, I will know." Genesis 18:20,21

> Then Abraham approached him and said: "Will you sweep away the righteous with the wicked? What if there are fifty righteous people in the city? Will you really sweep it away and not spare the place for the sake of the fifty righteous people in it? Far be it from you to do such a thing—to kill the righteous with the wicked, treating the righteous and the wicked alike. Far be it from you! Will not the Judge of all the earth do right?"

> The [I-SHALL-BE] said, *"If I find fifty righteous people in the city of Sodom, I will spare the whole place for their sake."* Genesis 18:23-26

The teachers of the law and the Pharisees brought in a woman caught in adultery. They made her stand before the group and said to [I-SHALL-BE-SALVATION], "Teacher, this woman was caught in the act of adultery. In the Law Moses commanded us to stone such women. Now what do you say?" They were using this question as a trap, in order to have a basis for accusing him.

But [I-SHALL-BE-SALVATION] bent down and started to write on the ground with his finger. When they kept on questioning him, he straightened up and said to them, "If any one of you is without sin, let him be the first to throw a stone at her." Again he stooped down and wrote on the ground.

At this, those who heard began to go away, one at a time, the older ones first, until only [I-SHALL-BE-SALVATION] was left, with the woman still standing there. [I-SHALL-BE-SALVATION] straightened up and asked her, *"Woman, where are they? Has no one condemned you?"*

"No one, sir," she said.

"Then *neither do I* condemn you," [I-SHALL-BE-SALVATION] declared. "Go now and leave your life of sin." John 8:3-11

The Holy Spirit also testifies to us about this. First he says: "This is the covenant I will make with them after that time, says the Lord. I will put my laws in their hearts, and I will write them on their minds."

Then he adds:

"Their sins and lawless acts *I will remember no more.*"
And where these have been forgiven, there is no
longer any sacrifice for sin. Hebrews 10:15-18

This Testimony, and the precedent of sincere inquiry
which allows it, forms the background for one of the most
far-reaching depositions in all of Scripture: the deposition
of Adam and his wife. It occurs after they eat the fruit from
the tree of the knowledge of good and evil, in the Garden of
Eden. The facts preceding the deposition are:

34.   Positional Declarations of facts preceding Adam's and
Eve's deposition.

34.1.   Eve knew there was a prohibition on "the
tree that is in the middle of the garden."

34.2.   Eve interacts with the serpent, without
seeking Adam's counsel, who was with her.

34.3.   Eve declares that she would die if she ate or
touched the tree's fruit.

34.4.   Eve accepts the serpent's counsel that she
could eat the fruit and not surely die.

34.5.   Eve eats the fruit as food, believing wisdom
could also be gained by doing so.

34.6.   Adam, who is with Eve, receives the fruit
from her and also eats it.

34.7.   Adam does not, himself, remove any fruit
from the tree; Eve gives it to him.

a *7ᵗʰ-Gift* page

¶4.8.　The Tree Commandment has two conditions, and one consequence:

1st:　"You are free to eat from any tree in the garden;"

2nd:　"[B]ut you must not eat from the tree of the knowledge of good and evil,"

Consequence: "[F]or when you eat of it you will surely die."

¶4.9.　Adam and Eve make coverings for themselves and hide from the I-SHALL-BE.

¶4.10.　Neither Adam nor Eve voluntarily reveals what transpired until deposed.

Thus, prior to the deposition, Adam and Eve, *together,* take actions to conceal and withhold evidence, including making coverings for themselves and hiding. Neither voluntarily discloses relevant facts of what happened. Nor does the serpent voluntarily provide any testimony regarding the events he precipitates. Subsequent to these events, the I-SHALL-BE represents the only objective party to conduct an honest inquiry to determine what transpired. Had the I-SHALL-BE been a witness or fully aware of all facts beforehand, he may have been obligated to immediately condemn all the parties. In deposing Adam and Eve, whom the I-SHALL-BE made and loves, at no time does Scripture suggest the I-SHALL-BE knows more than what Adam and Eve disclose to him while being deposed. The I-SHALL-BE adheres to a presumption of innocence during questioning. Adam is accorded respect as a son of God during the process. The deposition begins with Adam and his wife in hiding, having elected not to voluntarily disclose or reveal any evidence or information. The I-SHALL-BE opens the deposition with the first question:

> But the [I-SHALL-BE God] called to the man,
> "Where are you?" Genesis 3:9

This represents the first question ever asked by the I-SHALL-BE, as recorded in Scripture. There is no precedent the I-SHALL-BE inquired of Adam's whereabouts prior to this instance. In fact, based on the record in Scripture, the I-SHALL-BE had never asked anyone anything before this.

Adam responds to the opening deposition question:

> He answered, "*I* heard you in the garden, and *I* was
> afraid because *I* was naked; so *I* hid." Genesis 3:10

Since the question is, "Where are you?", Adam is technically non-responsive because he does not disclose where he and his wife are hiding, such as, "Here, in the third arboretum, fourth row, seventh tree." He only discloses the reason he is hiding, voluntarily giving information *not requested*, including reference to his emotional state ("afraid") and his new physical condition ("naked"). This requires Positional Declarations on why Adam provides information not requested with his opening answer.

25.   Positional Declarations on why Adam provides information not requested.

_____

25.1.   He is unaware that his disclosures could lead to the death penalty.

25.2.   He is fully aware his disclosures could lead to the death penalty.

By hiding, Adam keeps his wife from making additional admissions: if there is to be any further questioning of her, by a third-party, it will be through him and not directly of her, first. Are these the actions of an intelligent person? Is Adam a gifted, intelligent man? Relative to his recall abilities, he is unsurpassed by any mortal man:

> Now the [I-SHALL-BE] God had formed out of the ground *all* the beasts of the field and *all* the birds of the air. He brought them to the man to see what he would name them; and whatever the man called each living creature, that was its name. So the man gave names to *all* the livestock, the birds of the air and *all* the beasts of the field. Genesis 2:19,20

This gives witness to Adam's prodigious memory capacity, as well as to his etymological resourcefulness. To say that Adam is mentally competent to handle a deposition is an understatement. He may very well set the highest standard, among mortal man, of moral discernment, as well as all manner of human endurance, stamina and intellectual wherewithal. The writer submits that Adam is better prepared for deposition than any other mortal man ever has been. At that point in time, if a witness-advocate were needed, whose every word will affect all of mankind throughout time, then Adam was and is the best mortal representative to undertake so daunting an assignment.

Adam is well aware his conduct, with his wife's, constitutes a capital crime and warrants the death penalty if convicted. Conviction could only occur from admissions he and his wife make. Concealment, by hiding, delays testimony that can lead to their death, since they are still very much alive after eating the fruit. For anyone to compel such testimony, he would have to find them first. In hiding with his wife, Adam clearly demonstrates his love for her, as he tries to avoid any further harm to her from hostile questioning. However, once given to testify, Adam's love for the I-SHALL-BE, as a son of God, requires him to be forthcoming in every way that does not condemn his wife. Adam hides well enough to compel the I-SHALL-BE to ask his first question, ever, by calling out to find

him, leading to Adam's first, voluntary answer that excludes
his wife from any complicity in hiding:

> But the [I-SHALL-BE] God called to the man,
> "Where are you?"
>
> He answered, "*I* heard you in the garden, and *I* was
> afraid because *I* was naked; so *I* hid." Genesis 3:9,10

This response does not include "we"—and it never will.
Though he does provide some information not requested,
Adam's disclosure points to his conduct, alone, and to no one
else's. At no time can his wife be accused of hiding or concealing
information, since she is simply with her husband, based on
Adam's answer. It is proper and fitting for her to be with Adam,
wherever he might be. By hiding, Adam compels the I-SHALL-BE
to ask the first question. Each question the I-SHALL-BE asks helps
better inform Adam about what information the I-SHALL-BE
either already knows or doesn't already know.

The second question by the I-SHALL-BE is limited and
specific to the disclosure Adam makes, fully respecting Adam
by not rushing into judgment or speculation, but restricting
inquiry to the actual answer Adam gives:

> And he said, "Who told you that you were naked?
> Have you eaten from the tree that I commanded
> you not to eat from?" Genesis 3:11

This is a compound question, which, technically, is not
allowed during a deposition. The first part inquires about a
possible, unseen third-party. The second part addresses the Tree
Commandment as pretext to understand Adam's first answer.
Unless the I-SHALL-BE is being disingenuous, he doesn't
know how or by whom Adam came to realize he was naked.
Furthermore, the I-SHALL-BE makes no presumption about
"who" told Adam he was naked. The I-SHALL-BE indicates no

a priori knowledge about how Adam will answer. There is no doubt that the serpent was disingenuous when questioning Eve. However, there is no basis or evidence, in Scripture, to ascribe similar conduct to the I-SHALL-BE when he deposes Adam.

Adam's response to the I-SHALL-BE's second question is one of the most brilliant by any mortal man ever deposed:

> The man said, "The woman you put here with me—she gave me some fruit from the tree, and I ate it." Genesis 3:12

26.   Positional Declarations on Adam's second deposition response.

26.1.   Adam shifts blame to his wife for breaking the Tree Commandment.

26.2.   Adam only discloses that he ate the fruit, and *no one else.*

26.1 may be consistent with what many husbands may have done when accused of a capital crime that can result in the death penalty. This action could place Adam in a very unfavorable context, as a husband trying to save his own skin. It may even appear to be an attempt to divert the I-SHALL-BE's attention away from Adam to his wife.

However, in considering Adam's response, one must also consider what Adam *does not* disclose:

27.   Positional Declarations on what Adam does not disclose in his response.

       ⥾7.1.   He does not disclose his wife ate the fruit, but that only he had.

       ⥾7.2.   He does not disclose his wife spoke to the serpent.

       ⥾7.3.   He does not disclose anything the serpent said to his wife.

       ⥾7.4.   He does not disclose anything his wife said to the serpent.

If Adam's motive is to shift blame, then why limit disclosure regarding ⥾7.1, ⥾7.2, ⥾7.3 and ⥾7.4 in his answer? Though Adam does mention "the woman" (his wife) "gave" him the fruit, he doesn't say "the woman ate the fruit." Nor does he declare, "The woman ate the fruit first, then I ate it." Adam only discloses that *he* ate the fruit. There is no testimony Adam did anything other than this. Thus, if he were exercising his freewill to eat the fruit, then his answer only indicates his wife supported what *he* desired to do. If the inquiry ended here, only Adam would be guilty of actually eating from the tree of the knowledge of good and evil and breaking the Tree Commandment.

As a son of God, who loves the I-SHALL-BE, Adam is forthcoming, clearly acting in good faith through the answer he gives. Simultaneously, he is true to the enduring vow he made before the I-SHALL-BE regarding "the woman": "This is now bone of my bones and flesh of my flesh." Still, Adam includes in his answer, "The woman you put here with me . . ." Why does Adam say this at all? Isn't it already a fact in evidence, known to all parties, that the I-SHALL-BE formed the woman from Adam's rib and brought her to Adam?

       ⥾8.   Positional Declarations on why Adam said, "The woman you put here with me . . ."

a *7th-Gift* page

_____

&#x18;8.1.   It frames the attempt to shift blame from himself to "the woman."

&#x18;8.2.   It shifts blame back to the I-SHALL-BE for making "the woman."

&#x18;8.3.   It reaffirms "the woman's" origins from the I-SHALL-BE's holiness and goodness.

The evidence is overwhelmingly against Adam to support &#x18;8.1 or &#x18;8.2, especially considering how men engage in similar conduct with their wives, across cultures and throughout history. Even Abraham, when he thought his life was on the line, distanced himself from his wife:

> [A]nd there Abraham said of his wife Sarah, "She is my sister." Then Abimelech king of Gerar sent for Sarah and took her. Genesis 20:2

> And Abimelech asked Abraham, "What was your reason for doing this?"

> Abraham replied, "I said to myself, 'There is surely no fear of God in this place, and they will kill me because of my wife.'" Genesis 20:10,11

Such conduct is prevalent in contemporary societies, where a husband might deny he has a wife for personal convenience, let alone in a life-threatening situation. It seems reasonable to ascribe the same to Adam, the first man, as being like so many men when trying to save themselves.

So what possible evidence could justify &#x18;8.3? Now, remember, Adam could disclose his wife also ate the fruit, but he doesn't. Adam only says she gave him the fruit and he ate it. This testimony makes Adam the *only* party his wife has contact regarding the fruit. Adam purposefully excludes all testimony of his wife's involvement with the serpent. To do otherwise would implicate his wife in actually eating the fruit also, which Adam *does not* do. Could he? You bet, if he were trying to save his own skin. Does he? Absolutely not.

By not mentioning the serpent at all, Adam only implicates himself in breaking the Tree Commandment and committing a capital crime subject to the death penalty. Adam never even mentions the serpent in any of his answers. Why not? Because to mention the serpent would force disclosure of his wife's complicity in also eating the fruit. How does saying, "The woman you put here with me . . . ," *NOT* implicate his wife? Because this declaration reaffirms the "woman's" (Adam's wife's) origins from the I-SHALL-BE's holiness and goodness, by the I-SHALL-BE's will—before Adam's enduring vow respecting the I-SHALL-BE's purpose:

> So God created man in his own image, in the image of God he created him; male and female he created them. God blessed them and said to them, "Be fruitful and increase in number; fill the earth and subdue it. Rule over the fish of the sea and the birds of the air and over every living creature that moves on the ground. Genesis 1:27,28

> The [I-SHALL-BE] God said, "It is not good for man to be alone. I will make a helper suitable for him." Genesis 2:18

> The man said, "This is now bone of my bones and flesh of my flesh; she shall be called 'woman,' for she was taken out of man." Genesis 2:23

This is the only recorded vow Adam ever made, which he kept his entire life. Adam only had one wife during his 930 years. As a son of God with a created-wife, Adam specifically affirms that "the woman" originates from the holiness and goodness of the I-SHALL-BE. By saying "the woman *you* put here with me," he is leaving no doubt regarding her origins by the I-SHALL-BE's will *alone*. He reminds the I-SHALL-BE of the goodness, righteousness and love by which the I-SHALL-BE— and no other—brought the woman into being. Thus Adam makes clear that it could not have been her idea to give him the fruit. If questioning ended here, Adam, by his own testimony, is the only guilty party, whose wife simply helped fulfill his desire by handing him the fruit, that *only* he ate, according to his admission. By the end of Adam's answer, his wife is affirmed as originating from the holiness and goodness of the I-SHALL-BE and, as such, could not have initiated so horrible a catastrophe. This testimony is, in fact, truthful about Adam's wife.

Adam answers each question truthfully, and in each answer provides more information than specifically asked, reflecting his love and respect for the I-SHALL-BE as a son of God. Simultaneously, he acts as an advocate for his wife, anticipating all testimony regarding her in the most favorable light possible. He does disclose she gave him the fruit, while directly answering that "I [not "we"] ate it." At no time is the serpent mentioned in any way by Adam.

At this point, the I-SHALL-BE has sufficient evidence to convict only one guilty party, Adam. His wife would be spared the death penalty since there is no witness or testimony she ate the fruit, but rather only handed it to her husband to facilitate his desire. Without further questioning, Adam will receive the death penalty for willfully and knowingly committing a capital crime. The first husband will die and his wife will become the first widow. This is completely contrary to the I-SHALL-BE's purpose when he said, "It is not good for the man to be alone." Adam is about to be worse than alone; he is about to be dead. The serpent will have ended the threat of offspring on Earth and severed the union of the first husband and wife. The two whom the I-SHALL-BE joined together would be separated forever.

However, the I-SHALL-BE chooses to ask one additional question, not of Adam, but of his wife. When Adam reminds the I-SHALL-BE of the woman's origins, it also reaffirms Adam's own origins, since his wife was formed from his rib:

> [T]he [I-SHALL-BE] God formed the man from the dust of the ground and breathed into his nostrils the breath of life, and the man became a living being. Genesis 2:7

So, not only was Adam's wife of divine origin—which Adam affirms in his second answer—but Adam's origin is also divine—that's why he is called the "son of God":

> [T]he son of Methuselah, the son of Enoch, the son of Jared, the son of Mahalalel, the son of Kenan, the son of Enosh, the son of Seth, the son of Adam, the son of God. Luke 3:37

It was by the I-SHALL-BE's will that Adam's wife was formed from Adam's rib. And it is by that same will that Adam, himself, was formed from ground outside of the Garden of Eden and later placed in the Garden of Eden. No one spoke during the deposition until the I-SHALL-BE asked a question. Adam's testimony is definitive regarding the only guilty party. What objective, specific, non-leading question could the I-SHALL-BE ask now?

The question the I-SHALL-BE does ask is directed to Adam's wife:

> Then the I-SHALL-BE God said to the woman, "What is this you have done?"
>
> The woman said, "The serpent deceived me, and I ate." Genesis 3:13

The question is non-specific and of a general nature. In a formal court of law, such a question could be objected to as vague and not allowable. However, the very fact that the question is vague reflects just how far things have come. The I-SHALL-BE only has evidence leading to the death penalty for a son of God, by that son's own admission. The testimony regarding the I-SHALL-BE's righteousness is considerable and compelling:

> "Far be it from you to do such a thing—to kill the righteous with the wicked, treating the righteous and the wicked alike. Far be it from you! Will not the Judge of all the earth do right?" Genesis 18:25

> And he passed in front of Moses, proclaiming, "The [I-SHALL-BE], the [I-SHALL-BE], the compassionate and gracious God, slow to anger, abounding in love and faithfulness, maintaining love to thousands, and forgiving wickedness, rebellion and sin. Yet he does not leave the guilty unpunished; he punishes the children for the sin of the fathers to the third and fourth generation." Exodus 34:6,7

Up until now, Adam's wife has said nothing. She is fully aware of where the questioning has led, and that there is no testimony she ate anything. The guilt is clearly resting only on her husband. This is the man who was with her when she listened to another's counsel and ate the fruit. After she condemned herself by eating the fruit, he is the man who knowingly entered into the same condemnation by accepting and eating the fruit she gave him. He is the man who, as a son of God, hid from the I-SHALL-BE with her. And he is the man who answered all the questions of the I-SHALL-BE, condemning only himself to the death penalty while declaring her origins from the I-SHALL-BE's holiness and goodness in his answer.

Now, she is asked the most pivotal question any wife may ever be asked, "What is this you have done?"

29.   Positional Declarations on possible answers by Adam's wife, as "the woman."

———————

29.1.   "It is as he said."

29.2.   "I have nothing to add."

29.3.   "I choose to remain silent."

29.4.   "I only helped him do what he wanted to do."

29.5.   "I can't recall what happened."

29.6.   "I don't know what you are talking about."

29.7.   "I would never betray your trust like that."

The I-SHALL-BE moves swiftly after Adam's wife gives her answer. She is fully aware of what Adam left out of his answers. And just as Adam selflessly put himself in harm's way to deflect wrath away from his wife, his wife reciprocates his love with the truth:

> Then the I-SHALL-BE God said to the woman, "What is this you have done?"
>
> The woman said, "The serpent deceived me, and I ate." Genesis 3:13

This is the first testimony to the I-SHALL-BE concerning the serpent's involvement. It is the first admission that the

a *7th-Gift* page

woman also ate the fruit. At that point, the I-SHALL-BE is able to better understand Adam's conduct and answers as a son of God. He did specifically answer the I-SHALL-BE's questions sufficiently for the I-SHALL-BE to ascertain all relevant facts:

¶10. Positional Declarations on questions asked and answered.

¶10.1. "Where are *you?*"

¶10.2. "Who told *you* that *you* were naked?"

¶10.3. "Have *you* eaten from the tree that I commanded *you* not to eat from?"

Yet, Adam's answers only identify himself as the guilty party, leaving his wife's freewill intact. There is nothing specifically the I-SHALL-BE could ask Adam's wife that was not already addressed by Adam. An open-ended question created opportunity to present facts not in evidence for conviction or exoneration. By responding to the I-SHALL-BE's question, Adam's wife lifts the protective veil Adam formed for her through his answers. In doing so, she exposes herself to the same death sentence as Adam. In whatever occurs, they would suffer together, each considering the other's life more important than his or her own. If Adam is to suffer, he would not be alone. "The woman" voluntarily accepts the same fate as Adam, leaving no doubt her love for her husband while being completely truthful with the I-SHALL-BE.

The deep and abiding love the first husband and wife have for one another is now in evidence before the I-SHALL-BE. Though implicated in a capital crime, there is no action either of them took during the deposition that was not motivated by love for the I-SHALL-BE and each other, including the woman's truthful answer to the I-SHALL-BE's last question. There is no doubt whose wife she is, Adam having pre-positioned each answer as her husband. She ate the fruit, having been misled that it was a way to gain wisdom and *not* die; she told the truth, knowing that she *would* die. And her husband's enduring vow, respecting the love and grace and righteousness of the I-SHALL-BE to give him a wife, is rock solid. He makes it clear he would do whatever is necessary to maintain his vow to his death—and he does. While being advocate for his wife, he is truthful as a son of God in answering the I-SHALL-BE's questions, providing more information than specifically requested of him. He places his wife in the most favorable light possible and respects her freewill, which could lead to his own death, but not hers. And she is able to see how profound her husband's love is, for the I-SHALL-BE and for her, neither betraying her nor avoiding the penalty that the Tree Commandment requires.

After "the woman's" answer, the I-SHALL-BE moves swiftly. Though Adam did not mention the serpent at all, for reasons stated, the I-SHALL-BE condemns the serpent on "the woman's" (Adam's wife's) word *alone*, based on the compelling evidence that she risked her life in truthfully answering the I-SHALL-BE's only question to her:

> So the [I-SHALL-BE] said to the serpent, "Because you have done this,
>
> "Cursed are you above all the livestock and all the wild animals! You will crawl on your belly and you will eat dust all the days of your life.
>
> "And I will put enmity between you and the woman, and between your offspring and hers; he will crush your head and you will strike his heel." Genesis 3:14,15

Though the serpent is cursed by the I-SHALL-BE, he is not immediately destroyed. He will have "offspring" and will not be in confinement while allowed to live for many millennia—until the appointed time:

> And I saw an angel coming down out of heaven, having the key to the Abyss and holding in his hand a great chain. He seized the dragon, that ancient serpent, who is the devil, or Satan, and bound him for a thousand years. He threw him into the Abyss, and locked and sealed it over him, to keep him from deceiving the nations anymore until the thousand years were ended. After that, he must be set free for a short time. Revelation 20:1-3

> And the devil, who deceived them, was thrown into the lake of burning sulfur, where the beast and the false prophet had been thrown. They will be tormented day and night for ever and ever. Revelation 20:10

All of this was triggered by a corrupt, misleading question the serpent asked Adam's wife, since Adam could not be deceived in this way. It was followed-up with a slanderous allegation construed as information withheld from Adam's wife. Basically, the serpent was saying, "God and your husband are withholding useful information you deserve to have. I will share what I found out because they won't tell you themselves." In doing this, the serpent set the conditions for his eternal condemnation.

Afterward, the I-SHALL-BE turns his attention to "the woman" (Adam's wife) who has just admitted to committing a capital crime that warrants the death penalty:

> To the woman he said,

> "I will greatly increase your pains in childbearing; with pain you will give birth to children. Your desire will be for your husband, and he will rule over you." Genesis 3:16

Is this merciful or punitive? First, the I-SHALL-BE affirms her continuation as Adam's wife while expressing a boundless love by letting her know that the serpent's deception would not affect her fertility. She would become the mother of not one, but multiple offspring. In fact Adam's wife is the most fecund woman in the history of mankind. There is no woman who has ever lived, who is alive now, or who shall live, with more descendants than Adam's wife. And where she briefly desired that which another's false promise could not give, her righteous desire—with all its attendant joys, and difficulties, that Adam and she share the rest of their lives—is for one man, the vow of whom the I-SHALL-BE sustains through love, to rule over her as he does his own body, made one within the I-SHALL-BE's glory. The I-SHALL-BE's purpose and her husband's enduring vow are unimpeachable:

> The [I-SHALL-BE] God said, "It is not good for man to be alone. I will make a helper suitable for him." Genesis 2:18

> The man said, "This is now bone of my bones and flesh of my flesh; she shall be called 'woman,' for she was taken out of man."

> For this reason a man will leave his father and mother and be united to his wife, and they will become one flesh. The man and his wife were both naked, and they felt no shame. Genesis 2:23-25

Thus, the I-SHALL-BE's resolve is more restorative than only punitive to Adam's wife. The "pain of childbirth" serves as reminder of an ever-present adversary who attempted to kill the first husband and wife, yet failed. The "pain echo" alerts every generation that he is still trying. Even so, such pain is preferable to a barrenness Adam's wife never had to face (Gen 1:28's blessing preceded temptation to know *how*):

Now Sarai, Abram's wife, had borne him no children. But she had an Egyptian maidservant named Hagar; so she said to Abram, "The [I-SHALL-BE] has kept me from having children. Go, sleep with my maidservant; perhaps I can build a family through her." Genesis 16:1,2

He slept with Hagar, and she conceived. When she knew she was pregnant, she began to despise her mistress. Then Sarai said to Abram, "You are responsible for the wrong I am suffering. I put my servant in your arms, and now that she knows she is pregnant, she despises me. May the [I-SHALL-BE] judge between you and me." Genesis 16:4,5

The pain of childbirth, when such time comes, can never compare to the agony Sarai/Sarah endured under the circumstances noted. And when Sarai/Sarah did bear a son, at the age of 90, her testimony is in complete harmony with the pain, and joy, a new mother endures, and accepts:

A woman giving birth to a child has pain because her time has come; but when her baby is born she forgets the anguish because of her joy that a child is born into the world. John 16:21

Sarah said, "God has brought me laughter, and everyone who hears about this will laugh with me." And she added, "Who would have said to Abraham that Sarah would nurse children? Yet I have borne him a son in his old age." Genesis 21:6,7

The I-SHALL-BE's love is evident in saying to Adam's wife, "I will greatly increase your pain in childbirth; with pain you will give birth to children." There *will* be children—a multitude of them.

After this, the I-SHALL-BE addresses Adam:

> To Adam he said, "Because you listened to your wife
> and ate from the tree about which I commanded
> you, 'You must not eat of it,'—Cursed is the ground
> because of you; through painful toil you will eat of
> it all the days of your life. It will produce thorns and
> thistles for you, and you will eat the plants of the
> field. By the sweat of your brow you will eat your
> food until you return to the ground, since from it
> you were taken; for dust you are and to dust you
> will return." Genesis 3:17-19

Is this merciful or punitive? It might be remembered that
the serpent, whom Adam *did not* mention in any of his answers,
was cursed, earlier, by the I-SHALL-BE. Adam is *not* cursed,
though the ground is. There is a tremendous difference
between Adam being cursed and the ground being cursed.
This reflects the I-SHALL-BE's righteousness, and restraint,
regarding Adam's love for his wife.

What about the, "Through painful toil you will eat of it all
the days of your life"? The good news is that Adam lived 930
years and had food every day. Even though working the ground
involved "painful toil," the I-SHALL-BE kept the promise of
food, consistently recognizing Adam's hard work. "Thorns and
thistles" require additional care in planting and harvesting, yet
the "plants of the field," to this day, constitute some of the most
wholesome and nutritious food for all mankind. And if one is
blessed with vitality and energy, then there will be sweat from
the stamina to work all the way to the 930th year! Many go to
gyms, today, to sweat and "workout"—the outward signs of the
vigorous life Adam maintained through his day-to-day work.

Yet, even the I-SHALL-BE's curse of the ground is
addressed after the flood:

a *7th-Gift* page

> Then Noah built an altar to the [I-SHALL-BE] and,
> taking some of all the clean animals and clean birds,
> he sacrificed burnt offerings on it. The [I-SHALL-BE]
> smelled the pleasing aroma and said in his heart:
> "*Never again* will I curse the ground because of man,
> even though every inclination of his heart is evil from
> childhood. And never again will I destroy all living
> creatures, as I have done." Genesis 8:20,21

Now, there is still the rather serious matter of:

> "By the sweat of your brow you will eat your food until
> you return to the ground since from it you were taken;
> *for dust you are and to dust you will return.*" Genesis 3:19

The serpent may have anticipated a swift execution of
Adam and his wife, but this was not to be. The death sentence
was suspended for almost a millennium, and, through the
anticipatory sacrifice of the "Second Adam," the sentence is
reversed altogether on the last day. The serpent also benefits
from the mercy of the I-SHALL-BE, until the last day, mercy so
apparent for the world until the last day. As it has been from the
beginning, grace, upon grace, reflects the I-SHALL-BE's love,
extending through generation after generation. Such love is so
pervasive that—save those who are the children of God, in the
Testimony of the Father, his One and Only Son, and the Spirit
of Truth—it may be missed completely, in a world inhabited by
billions of people, all of whom are descendants of Adam and
his wife:

> I'm writing these things to you about those who are
> trying to lead you astray. As for you, the anointing
> you received from him remains in you, and you do
> not need anyone to teach you. But as his anointing
> teaches you about all things and as that anointing

is real not counterfeit—just as it has taught you, remain in him.

And now, dear children, continue in him, so that when he appears we may be confident and unshamed before him at his coming.

If you know that he is righteous, you know that *everyone* who does what is right has been born of him. 1John 2:26-29

Appendix ‫ז‬

Shuttle Thermal Integrity Detection System

Attached is the final draft of the "Shuttle Thermal Integrity Detection System" ("STIDS") description. Originally submitted and designated by the U.S. Patent and Trademark Office as provisional patent 60/445,329, its patent application publication reference is US-2004-0238686-A1. A fellow Caltech alum, who is the patent attorney of record for an earlier, U.S. patent (no. 6,236,992) awarded to this writer, determined the STIDS provisional patent as well written with "fileable-merit", its "patentable-merit" being subject to a proper and full examination by the U.S. Patent and Trademark Office. Additionally, it has been given a preliminary review by a member of a Space Shuttle Columbia accident oversight board to determine any merit or lack thereof.

The know-how focuses on the mission phase between launch and attaining orbit ("Transit-to-Orbit"). The only unrecoverable incidents, affecting two of 113 shuttle missions, are exclusive to the Transit-to-Orbit window. STIDS, therefore, has this as its area of emphasis, technology to "authenticate" a decision to initiate a safe-return during Transit-to-Orbit. A technological authentication, of a multi-million-dollar decision to initiate safe-return during Transit-to-Orbit, helps establish probable-cause to do so, within an actionable time-frame. This aids taking such a drastic step and allows it to be defensible when post-mission scrutiny assesses the economic impact of so precipitous an event.

It is especially important when a spacecraft is carefully examined upon its *safe-return* to Earth, when probable-cause is assessed from a vehicle's actual condition. STIDS may help bring objectivity to what might otherwise be considered a "judgment call" by an individual or individuals (whether on the ground or in the vehicle). There may be profound career consequences "either way" to make the call, which technological authentication may help mitigate.

I won't forget Dick, Michael, Judy, Ron, Elison, Greg and Christa. Nor shall I forget Rick, Willy, Michael, Kalpana, David, Laurel or Ilan—who magnify the remembrance of Dick, Michael, Judy, Ron, Elison, Greg and Christa.

On the day they didn't return to their homes and families, as more information was disclosed, my own despair, along with, perhaps, many others, deepened more, and more, and more with each new revelation, or piece of evidence, or witness, or photograph, or telemetry measurand. Left unchecked, the cycles of disclosure—which compel everyone to pay attention—are collectively disheartening. What can anyone do?

If the complete focus can not be on the well-being of the crew, then let there be a compelling focus on the well-being of, among other things, a "detection grid" herein. If a part of the "detection grid" is compromised after launch and before entering orbit, the crew can safely detach the shuttle and *themselves* for as safe a return as possible. The safe return of each crew, if at all possible, speaks volumes to respecting science *and* the lives of those who help *promote* it. This is no less true for the well-being of those in labs *on Earth* and those in labs *above the Earth.* The attached, legally pending patent attempts just that.

SHUTTLE THERMAL INTEGRITY DETECTION SYSTEM

RELATED APPLICATION DATA

The present application claims the benefit under 35 U.S.C. 119(e) of the priority date of Provisional Application Serial no. 60/445,329 filed d4/February 5, 2003, which is hereby incorporated by reference.

FIELD OF THE INVENTION

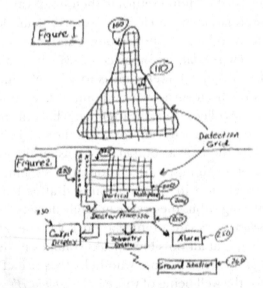

This invention (herein "Invention") generally relates to sensor technology for detecting structural compromise, of thermally-sensitive surfaces, after exposure to mechanical trauma. Such thermally-sensitive surfaces may be attached to a static structure or a vehicle, including those operating on the ground, in the air, or in space, such as space shuttles.

BACKGROUND OF THE INVENTION

An object of the present Invention is to help safely return a crew operating a vehicle, such as a space shuttle, when compromise of a thermal-protection surface could lead to catastrophic loss of crew and vehicle. Of the 113 shuttle launches during a 22 year period, the safe return of crews was achieved 98.3% of the time. In the only two exceptions, crews may have safely returned using existing operational capabilities of the shuttle. One of the two vehicles was Challenger, d3/ January 28, 1986, whose crew consisted of commander Francis R. Scobee, pilot Michael J. Smith, and astronauts Judith A. Resnik, Ronald E. McNair, Ellison S. Onizuka, Greg B. Jarvis (a fellow member of the technical staff in Hughes' Space and Communications Group), and Sharon "Christa" McCaliffe. The other vehicle was Columbia, d7/February 1, 2003, with commander Rick D. Husband, pilot William C. McCool, and astronauts Michael P. Anderson, Kalpana Chawla ("Culp-na Chav-la"), Laurel Blair Salton Clark, David M. Brown and Ilan Ramon. This Invention is motivated out of respect for the profound commitment these husbands, wives, fathers, mothers, sons and daughters made in a shared national and international purpose. And it respects the sentiment, of the only U.S. president ever to be awarded a patent since the U.S. Constitution was set in motion, regarding those who purposefully give "the last full measure of devotion":

. . . It is rather for us to be here dedicated to the great task remaining before us—that from these honored dead we take increased devotion to that cause for which they gave the last full measure of devotion; that we here highly resolve that these dead shall not have died in vain . . .

President Abraham Lincoln, Gettysburg, d5/ November 19, 1863

Providence's wisdom, in permitting the U.S. Constitution to exist in its present form, allows a citizen to present a petition, such as this one, regarding useful innovation that may benefit society as a whole. Though powerless while receiving the news reports of Columbia a year ago, d7/February 1, 2003, developing this Invention was the only means available for this writer to check the downward cycle of despair that enveloped himself, his fellow Caltech alums, the nation and the world, as occurred 17 years prior, with Challenger, on d3/January 28, 1986.

It may be remembered that certain events precipitated NACA's (National Advisory Committee on Aeronautics—the precursor to NASA, prior to the 1950s) evolution, having been formed to enhance U.S. achievement and leadership in the aeronautic arts. Yet, as the level of technical excellence and precision was raised to achieve space flight, NACA had to give way to NASA (National Aeronautics and Space Administration), respecting a fundamentally different type of leadership, philosophy and technical commitment to express the nation's new aspirations beyond aerodynamics. Such aspirations became real, and were advanced, by the Wright brothers, Samuel Langley and others in America, as well as internationally, near the turn of the last century.

The NACA/NASA change was and is in complete harmony with the U.S. Constitution:

. . . [T]hat whenever *any* form of government
becomes destructive of these ends, it is the right of
the people to alter or to abolish it, and to institute
new government, laying its foundation on such
principles and organizing its powers in such form,
as to them shall seem most likely to effect their
safety and happiness.

Prudence, indeed, will dictate that governments
long established should not be changed for *light
and transient causes*; and accordingly all experience
hath shown, that mankind are more disposed
to suffer, while evils are sufferable, than to right
themselves by abolishing the forms to which they
are accustomed. Congress, d5/July 4, 1776

In other words, the issues embracing the nation's attention,
first with Challenger, and later with Columbia, are not specific
to just a technological imperative, an institution, or even to
this present generation. Because of such issues' visibility to
the entire world, they evidence the most sublime meaning
of American-based ideals, through a present-day people, and
within a shared national purpose which, from the beginning,
like so many other substantive American endeavors, has
been pregnant with risk and danger, as well as unparalleled
achievement and success. The Declaration of Independence
acknowledges as much, suggesting that expediency show
deference to prudence, while avoiding destructive ritual that
may be indifferent to new avenues of safety and happiness for
any citizen, regardless of station.

Thus, the matter, at hand, transcends merely the
technological arts, exposing the underlying *values* of those
whom, today, must also, by necessity, send others into harm's
way.

NACA constituted sacrifice, on an unprecedented scale,
to advance the aeronautic arts. And NASA benefitted from

that knowledge, advancing far beyond what the NACA charter could encompass. NASA, though formed during the Cold War, inspired a new generation to embrace an impossible challenge—and achieve it.

Now, placed before America is a new challenge, and it shall again require the boundless energy, enthusiasm, innovation and commitment of a new generation. As NACA provided the foundation for NASA to come into being, so NASA may become the precursor of . . . what?

The California Institute of Technology; the Massachusetts Institute of Technology; Illinois Institute of Technology; Rensselaer Polytechnic Institute; New York Institute of Technology; New England Institute of Technology; Georgia Institute of Technology; Oregon Institute of Technology and universities such as Stanford, Berkeley, Princeton, Chicago, Rice, Rutgers, Dartmouth, Ohio, Michigan, Oregon, Purdue, Colorado, Delaware, Florida, Houston, Louisville, Maryland, Massachusetts, Minnesota, Missouri, Mississippi, New Mexico, North Carolina, Oklahoma, Puerto Rico, Rhode Island, Alabama, Texas, Washington, Wisconsin, Virginia and Illinois and others represent academic institutions with a profound interest in space and the technology to explore it, whether through manned-vehicles or remote probes and robots. It will be from institutions such as these that men and women will come with vision for even greater achievements and successes.

NACA was a precursor organization in the evolution that leads to NASA in the 1950s. It is proposed that "NISAA" (a designation for purpose of developing the idea) continues this progress, within and beyond the first decade of the 2000s, as the National Institute for Space and Aeronautic Achievement. NISAA will draw on the best talent from America and the world to advance the cause of shared, human achievement through space exploration, vehicle innovation, sustainable system design, and aeronautic research. Intrinsic to NISAA's mission is the recognition that all progress, in the discovery and exploratory arts, entails prudent acceptance of risk, uncertainty and danger that the unknown always presents. Yet, the confidence to go forward respects the unparalleled sacrifice and accomplishments of those who always answered the call, before NISAA, that now compels excellence from a new generation, today. Let "N"=Nonprofit, in NISAA, worldwide.

NISAA goes beyond administering the exploration of space to achieving the highest levels of excellence in the discovery and exploratory arts. NISAA benefits from the perspective, wisdom and practical skills NASA helped foster, just as NASA benefitted from decades of research by NACA. The level of investment through NASA was extraordinary compared to anything NACA ever attempted. At the time, NASA reflected a national purpose visible to the whole world, affirming American values and the respect for every living soul who is sent into harm's way to fulfill such purpose.

NISAA will express these values in new venues, protocols and future endeavors. Its very newness nurtures a vibrancy and expectation unlike anything before. It will accept personnel, technology and know-how from its predecessor. However, as NASA did before it, NISAA will define itself in new ways, with new personnel, new innovations, and new achievements that will be wholly its on. Its leadership will be more attuned to constituent technological institutions, which themselves are among the best the world can offer. With "Achievement" in NISAA's name, it will be a light for, and to, academe and industry, fulfilling its mission, disciplined by wherewithal inherited to carry the torch forward.

The Invention herein is to help complete the legacy of NASA technology and accomplishments with the Space Shuttle, a vehicle that is extraordinary in the history of nations. This Invention is to help ensure that NASA is honored for all that it has done for America and the world through the lives of those who served it, whether on Earth or above it. And as NISAA may come into being through a new birth of commitment, let it always be remembered that such birth, when and where it is permitted to be, has been through the lives of those who gave the "last full measure of devotion."

Thus, this Invention is a declaration of honor and gratitude to NASA, its current administrator, its past administrators, and its talented explorers, scientists, engineers, technicians, management, operations staff, and administrative personnel who have carried the torch through the unknown, and lighted the way for an unlimited future that awaits a new generation, with courage and dedication to embrace and advance what NACA and NASA have set before mankind. This will not be achieved within the first 100 days of NISAA, nor perhaps the first thousand days, nor within the present generation. There may yet be unforseen set backs. But with the steady confidence and progress NACA and NASA established, let us begin.

SUMMARY OF INVENTION

The Invention herein detects anomalies at critical phases of a space shuttle's operation, especially between launch and attaining orbit. In the case of Challenger, the Invention may have helped make the decision to detach the solid rocket boosters before their catastrophic failure. Well-planned scenarios for crew and vehicle recovery are extant with timely information to make such a decision. In the case of Columbia, multiple abort-options were available between launch and attaining orbit if thermal compromises were instantly detectable and verifiable. In both cases, real-time, telemetry information on thermal integrity, regarding booster section seams and heat-sensitive surfaces on the shuttle itself, may have helped permit the safe return of all 113 crews. It should be noted that in so complex a research vehicle, such as the shuttle, its 98.3% safe-return of crews speaks volumes to NASA's clear intentions regarding the well being of every astronaut.

Space shuttles, as research vehicles, using ceramic or brittle thermal protection technology on their outer surfaces, are vulnerable to catastrophic trauma. The Invention herein detects compromise of such a thermal protection layer, whether it be made of ceramic tiles or other thermally-protective material, within the critical window a shuttle may safely return to earth. It allows such determination before a shuttle goes into orbit, where compromise of thermal protection on the outer surface of a vehicle can lead to catastrophic failure, such as during re-entry. Furthermore, it considers the thermal integrity of "structural seams" and component parts of a shuttle's booster rockets, particularly such seams that if compromised by heat or fuel leakage of any type, could lead to the loss of such vehicles. Additionally, all such detected data are injected into a telemetry stream for real-time decisions that may aid the safe-return of crews operating such vehicles.

Finally, since any decision regarding such thermal integrity may affect some $500 million budgeted to launch and operate such a vehicle, or some $5 billion to reproduce a new vehicle, redundant verification of thermal-layer integrity on such a vehicle helps reduce to acceptable levels, or prevent, false positives or negatives during operation and or flight.

If in fact there is no compromise of the integrity of the thermal layer that protects a shuttle or other vehicle or surface, a sensor grid used to verify thermal integrity may ablate off during the course of re-entry.

BRIEF DESCRIPTION OF THE DRAWINGS

Figure 1. shows the affected surface area of a vehicle and the sensor grid 100 that spans it. In this implementation, a horizontal/vertical grid is indicated, though other geometries can be accommodated depending on the shape of a vehicle, or the areas of compromise to be detected.

a *7ʰ-Gift* page

Figure 2. shows vertical grid elements 200 feeding into vertical multiplexor 205 that samples each grid element at an appropriate rate. Horizontal grid elements 225 feed into horizontal multiplexor 220. Both 205 and 220 feed into detector/processor 210 which converts the signals from the grid elements to produce digital values defining an area or areas of compromise.

DETAILED DESCRIPTION OF INVENTION

The preferred embodiment is described of a system and attached apparatus for detecting thermal integrity of a protective-layer on a vehicle's surface, such as a shuttle, during its operation. Figure 2. shows the vertical grid elements 200 that attach to the surface areas of interest through either mechanical or adhesive technology, or are suspended above the areas of interest with suspension points at intervals across the surface areas of interest. The Invention detects compromise of all or part of the grid elements. The option of multiple contact points and feeds along any grid element is not excluded, which permits other grid elements to continue operating around a compromised area, allowing horizontal multiplexor 220 and vertical multiplexor 205 to receive redundant data, or to receive localized data independent of other grid areas. Connective terminals or fasteners or sockets allow grid elements to attach to vertical multiplexor 205. The same is true also for horizontal grid elements 225 and horizontal multiplexor 220.

Both vertical multiplexor 205 and horizontal multiplexor 220 feed digitized values to detector/processor 210. Detector/processor 210 then produces further refined digital values that can be used to plot thermal integrity in a cockpit display 230 or for transfer to telemetry system 240 for down-link to ground station 260. Ground station 260 has regular processing systems for displaying all telemetry from a vehicle. Additionally, such display is available to the vehicle's crew, in whole or in part.

When an external event or events compromise the detection grid, data values produced from detector/processor 210 allow rapid assessment of the area of compromise. Multiple levels of detection grids may be installed on the outside, within and/or beneath the vehicle's thermal protection-layer to determine the depth of compromise, if such refinement is needed. Alarm 250 can automatically sound for extraordinary compromise, processor 210 sensing grid trauma that precedes events indicative of potential vehicle failure in different phases of its operation.

The detection grid may also be comprised of wire-strips, containing one or more wires in very close proximity. These may also be twisted-pair or topologically similar configurations to reduce noise. An alternative implementation is to use conductive paint to form all or part of the detection grid. Additionally, transmission delays along grid elements could be detected to indicate compromise, when different grid transmission times are compared. To aid in such detection, passive or active components (such as resistors, transistors, or diodes) may be at the intersection of grid elements, and attach to different grid elements that physically (but not electrically) intersect. Passive or active components that do create electrical connection between different, intersecting grid elements are selected such that the electrical characteristics of the grid are sensitive to even a small subset of such elements, or an individual element, being compromised by external trauma determined to be indicative of potential vehicle failure.

Alternatively, fiber-optic, or other light sensitive grid elements, may be used to detect either breaks or attenuation of light transmittance, indicating compromise as determined from definable limits in detector/processor 210, telemetry system 240, cockpit display 230 or ground station 260. If needed, the entire grid could be fiber-optic, or a subpart of the grid in a hybrid configuration, as noise and other electrical characteristics warrant.

Since the Invention detects in real-time, the decision can be made to proceed with a mission or to abort the mission during easily recoverable phases, such as prior to reaching orbit.

The sensing grid and components are selected based on weight constraints and the ability to readily break when exposed to definable mechanical trauma, or to provide detectable changes when exposed to such trauma. Additionally, the sensor and other components may be selected to allow ready ablation in the final phase of a successful mission.

As needed, a redundant grid or grids may co-exist with a primary grid, other grids offset vertically or horizontally or laterally from one another, since the information from the Invention could determine if a multi-million-dollar mission continues or is terminated.

The integrity of the system is tested prior to launch of a vehicle. Once confirmed, launch proceeds and the system operates continually, including while in orbit to detect any events during orbital operations: space debris, stray tools or material from the vehicle, or mechanical incidents during extra vehicular activity that might compromise vehicle thermal integrity.

The digital components are kept to a minimum to meet weight requirements, perhaps allowing the system, independent of grid and connecting elements, to weigh about a kilogram, if possible, at an off-the-shelf-cost of actual circuit components of about $200 per unit. To meet military specifications, the weight might increase as would the price of a unit, depending on grid technology and detection sophistication (propagation delays and multi-compromise detection).

An alternative implementation is for each grid element to detect the absence of a thermal element on the surface of a vehicle (such as a tile). In this case, the grid and its detection elements could be exclusively beneath the thermal protection layer and the elements thereof, requiring little if any reinstallation between missions. The grid detection elements could be one-to-one for the thermal elements (such as tiles) or one-to-many depending on weight and the geometries to accommodate individual thermal surface elements per grid element or multiple thermal surface elements per grid element.

An additional embodiment is to include "continuity loops" above and below the seams of the sections of solid rocket boosters attached to a shuttle. Such wire, fiber optic, conducting paint, or other lines of material would be on the circumference of the booster cylinders (or other areas of the vehicle), above and below sections that are sealed together. Each loop could be within protective insulation or covering. The continuity loops may be redundant and also use a twisted pair topology to reduce noise. The material and insulation used to make the "continuity loops" would melt through when a localized temperature on any part of a loop is elevated beyond a pre-determined threshold.

The wires of the continuity loop on the boosters or other thermally sensitive areas would feed into detector/processor 210, for display or down link with other horizontal grid elements 225 and vertical grid elements 200. This would allow a relatively complete, real-time "thermal integrity" check of all relevant vehicle surfaces and seams or other thermally-sensitive areas or components. Should the integrity of any of these components be comprised, both the ground station and on-board personnel would have timely information to make critical decisions after launch and before proceeding to orbit, as well as while in orbit and before initiating re-entry. This, for example, allows on-board or ground personnel to initiate a safe-glide return alternative, after launch, and before a vehicle goes into orbit, or to choose alternatives to re-entry if any vehicle thermal-surface integrity compromise is detected before dropping out of orbit.

Finally, in the event of a thermal protective-layer shape that may produce cavities of empty space beneath the protective layer's surface, pressure sensing devices would act as redundant detectors of compromise. That is, in addition to the protection-grid and telemetry resulting from it, pressure sensors within hallow cavities would produce separate telemetry of pressure changes. An example of this would be the hollow area between re-enforced carbon-carbon elements on the leading edge of a shuttle's wing, and the flat forward edge of the wing itself. As such, if there is a precipitous change of pressure within such a cavity, and trauma to the detection grid, then the two independent telemetry sources would indicate crew and vehicle safety is compromised or may soon be compromised during continued operation of the vehicle.

What is claimed is:

1. A detection-grid beneath, within, or on the thermal protection surface of a vehicle, such as a shuttle, whose electrical characteristics are changed when the detection-grid is exposed to mechanical traumas that may impair the operational characteristics of the vehicle.

2. The circuit that translates the electrical characteristics of the detection-grid, that reflect mechanical trauma of a vehicle's surface, into meaningful digital values down-loadable from the vehicle to a support station on the ground via telemetry, as well as for display to the vehicle's crew.

3. The redundant inclusion of specific pressure sensors within cavities of a thermally-protective surface of a vehicle in conjunction with the apparatus of claim 1.

4. The inclusion of telemetry from pressure sensors of claim 3 with the telemetry in claim 2 that supports rapid decisions to abort or continue a vehicle's operation.

5.   The firmware and software within the circuit of claim 2 that produces meaningful conversion of grid electrical characteristics into meaningful digital values for down-load or display to the crew.

6.   The use of active or passive components at the intersection of grid elements such that the electrical characteristics of the detection grid is measurable by the circuit of claim 2 if a subset of such components, or grid elements in the vicinity of such components, are compromised by physical trauma.

7.   The design of the detection-grid in claim 1 such that if it is on the outside of the vehicle, it ablates off during re-entry of the vehicle from space into earth's atmosphere. This is inclusive of grids composed in whole or in part of conductive paint, fiber-optic, wire or other materials.

ABSTRACT

A detection-grid is disclosed that is part of a vehicle's thermal protection layer, such as that of a space shuttle. A hybrid digital/analog system detects electrical changes in the detection grid caused by mechanical trauma to a vehicle's external surface. The system produces timely and useful display of such incidents. Furthermore, with redundant verification of such real-time data, the vehicle can detach from other apparatus, such as an external fuel tank or booster rockets, to execute pre-planned glide or descent scenarios maximizing a crew's and vehicle's safe return before proceeding to orbit. The detection-grid ablates off during re-entry of a regular mission.

a *7th-Gift* page

Appendix ה

Sneed v. IBM: Decoding a $1.86 Billion Federal Riddle With Caltech's Honor Code, Pro Se

On a Mesa Overlooking the California Coast

Sneed v. IBM: Decoding a $1.86 Billion Federal Riddle With Caltech's Honor Code, Pro Se rests upon the mystic cords of memory, through gentle breezes sweeping across a peaceful mesa overlooking the coast of California, just south of Santa Barbara, in Carpinteria. There, as a junior at the Cate School, the author, Anthony Sneed (Sneed), meets with Headmaster Frederick F. Clark (Headmaster Clark). In the private meeting, Sneed discloses his decision to attend Caltech (Caltech Decision). Though his father's schooling ends at 7th grade, in Louisiana, and his mother's at 10th grade, in Texas, within two decades of marriage as California *citizens*, they own 35 properties, including two vacation homes, with annual revenues of $70,000 from applied-statistics investments, by the late 1960s. When he meets with Headmaster Clark, Sneed knows as much about Caltech as the patriarch Abraham knows about Canaan, when told to leave his father's household and go there.

About seven years before Caltech Decision, Sneed's father and mother begin filling their modest home's library (Father's Library) with books and gadgets: World Book® Encyclopedia; World Book® CycloTeacher; other technical books and literature; a huge, approximately 10-inch by 12-inch, large-letter, Old English, King James Bible; multiple Tinker Toy® sets; multiple Lincoln Logs® sets; DIGICOMP® computer simulators, one with marbles for electrons that roll down an inclined plane, triggering groups of binary "flip-flops" in "registers"; a telescope; an oscilloscope; volt meters; two microscopes; Resistor-Transistor-Logic integrated circuits, including logic gates and JK flip-flops; multiple Gilbert Erector Sets®; and multiple Radio Shack® electronic kits and components.

Only later in life does Sneed learn Dr. Richard Feynman (Feynman), the Manhattan Project scientist and Caltech Nobel laureate, enjoys playing the conga drum, while a Caltech professor, with local musicians at Will Rogers Park, its name at the time, and just 10-minutes by car from Father's Library. Equally close is a childhood home of another Nobel laureate, Dr. Glenn T. Seaborg, who becomes the first chairman of the U.S. Atomic Energy Commission, after he discovers plutonium, americium, curium, berkelium, californium, and later contributes to discovering einsteinium and mendelevium.

Sneed realizes, after he is grown, how fortunate he is to have a father and mother who create an extraordinary, secure, home environment. Though it is in a quiet neighborhood of a locale with a Nazareth-like reputation ("Can anything good come from there?"), far from the beaten-path of any mainstream scientific institution, two, Nobel laureate, particle physicists favor the locale with their presence at different times during their lives.

Just across the street from his home, as he is growing up, the Fifth District county supervisor allocates funds to build a new park. To Sneed, it is a dream park; the street in front of his home welcomes him to it each day. Along this street, he rides his unicycle and skates with his best friend, Roy. The park contains a model of the Main Asteroid Belt, with dual, tilted, concentric-rings, each raised six-feet on opposite ends from one another. The rings have two rails, with welded rungs between the rails, and a diameter of about 20 feet across, with about a six-feet diameter sundial in the middle of the entire structure. Nearby are hollow planetary-spheres: Jupiter, the largest; Saturn, with rungs between its "rings" to climb and walk on; and Mars, the smallest. Separate poles suspend them in the air, about four-feet above ground, each large enough for multiple children to climb into and stand up. The large sand area also has high "astro-swings" and a three-story rocket ship. The new recreation center has an office, a courtyard, and community conference room. Directly in front of Sneed's home is the park's wide open playing field, with two, adjacent baseball diamonds. Later, a swimming pool is added near the recreation center office. A gate on the far end of the playing field opens to the elementary school; Sneed and his little sister enjoy the short stroll to it, across the park, on school days.

Using the baseball diamonds for practice, his father sponsors and coaches a team of local teenagers for multiple seasons. They go on to win the right to play the final championship game for one of America's largest cities. Sneed's older brother captains the team, and later becomes a captain with the county's sheriff's department, retiring with full pension after three decades of service.

Additionally, Sneed's parents provide him reel-to-reel tape recorders; mini-tape-recorders; a stereo cassette player; FM/AM radios; an amplifier; microphones; speakers; transmitters and receivers; various games; electrical wire of various gages and lengths; electronic assembly tools of all kinds; jigsaw puzzles; a triple-mast sailing ship model; a Visible V8® combustion engine kit; a Visible Man® human anatomy kit; a biology experiments kit, including a dead frog in formaldehyde (Sneed visits no indignities on the frog's remains); a small-scale, working steam engine; a remote-controlled robot; multiple slot-car sets, including the huge, James Bond® road race layout, mounted on molded plastic sections, having trees, rolling hills, and tunnels; spy kits, one being a briefcase with hidden camera and surveillance electronics; electric motors and batteries of various shapes and kinds; a one-speed bicycle; a three-speed bicycle; a ten-speed bicycle; two unicycles; a piano; an electric, base guitar; a blue, 10-feet by 10-feet house tent, set up between the lemon and plum trees by the backyard garden; and more. All are put to good use to prepare Sneed for events that now seem foreseen.

Sneed's parents have his uncle, a master carpenter, build a custom-made, fold-out, study-center, with shelves and storage area. Father's Library has lots of space; a floor-to-ceiling, shelf-lined-wall, filled with books; foldout sofa, across the room from the bookshelves; artificial fireplace; recliner; interior, sliding-glass-doors leading to the dining room, which Sneed motorizes to open and close with a switch from his desk; and other amenities. His parents allow this to become Sneed's bedroom. It is the "tour" destination when his parents' friends or guests drop by, the sliding-doors opening with a motorized hum to welcome curious visitors. One of his parents' neighbors comes to report hearing Sneed on an AM radio station, a block away. So Sneed reveals his AM transmitter, built from specs in an electronics magazine.

Between the ages of 10 to 14, Father's Library is where Sneed studies the fundamentals of computers, their architecture, and their programming languages, which allow him to later ace computer courses and digital system design assignments at the Cate School, Caltech, NASA, and IBM. Assistant Headmaster Joseph I. Caldwell (Assistant Headmaster Caldwell) advises Cate School to accept Sneed, though it appears Sneed shares neither the demographic, nor economic, nor assessment-profile of the typical Cate student. With no grant, loan, or scholarship to meet tuition, the school encourages Sneed's parents to apply. Later in life, Sneed learns Assistant Headmaster Caldwell personally removes obstacles for Sneed's enrollment (Caldwell Fulfillment).

Caltech is not a regular destination for Cate graduates, Cate having a well-established, liberal arts pedigree; Headmaster Clark, a Harvard alum, is author of one of the foremost, secondary-school, history textbooks in the U.S. Nor does Sneed match Caltech's historical demographics. Yet, during the summer after his Cate freshman year, two years prior to disclosing Caltech Decision to Headmaster Clark, Sneed conducts a telephone book survey. He randomly calls companies, to find out what kinds of computers they use. One such firm is Granite Computer *Institute*. Its director, Mr. Dick Ryan (Ryan), invites Sneed to see his training center for IBM professional programmers. Sneed's father later agrees to drive him, after cautioning about playing on the phone.

Ryan meets and interviews Sneed, and discovers Sneed can already program in IBM FORTRAN and Dartmouth BASIC. Sneed learns FORTRAN on an IBM mainframe at the Claremont Colleges, two years prior, which he pursues after learning about the language through the DIGICOMP® computer simulator manual, the gift from his father at age 10. Ryan offers Sneed admission to study IBM mainframe systems and their architecture, their operations and programming in machine language, IBM unit-record equipment, and hands-on training with an on-site, IBM mainframe system, including weekends when it's idle. When Sneed's father asks about tuition, Ryan says, "Two." Sneed says to his father, "Let's give him $2!" Ryan responds, "Not that two." Sneed asks, "$20?" "Not that two," Ryan adds. "$200?" Sneed asks, as his father looks on. "Not that two," Ryan remarks.

Sneed looks apologetically at his father, about the half-hour drive to a meeting with such an unexpected outcome, and to Ryan for his graciousness to show Sneed his IBM systems and facilities. Ryan then breaks the silence with a smile and says, "If your father agrees, you can attend for the price of the books." Chastened, Sneed asks how much they might be. Ryan answers, "Three." Sneed pauses with trepidation, when Ryan adds, "That is, $30." Sneed looks hesitantly at his father, who smiles in agreement with Ryan (Ryan Fulfillment). Ryan plans for Sneed's presence to motivate the other students. Though all of Sneed's classmates are adults pursuing professional certifications, Sneed enjoys the classes during the summer, returning to Cate in the fall for his sophomore year. By that time, Sneed programs in BASIC, FORTRAN, and IBM machine language, and is perhaps the only 14-year-old in the world who can do so, along with riding a unicycle backward *and* forward.

Since Caltech, or the California *Institute* of Technology, also has "*Institute*" in its name, Sneed feels it might be just as much fun as Granite Computer *Institute*, and lets Headmaster Clark know about Caltech Decision in his junior year. During the fall of Sneed's senior year, Headmaster Clark invites Sneed to ride with him to meet Mr. Winchester Jones (Jones), who lives in Santa Barbara. Jones is a Cate alum, former Caltech admission officer, and a gracious host. Jones arranges an interview for Sneed at Caltech in Pasadena, at Throop Hall, named after Amos G. Throop (Throop), who establishes, in 1891, what later becomes Caltech. His son, George Throop, gives the "last full measure of devotion" as an artillery officer in the Red River Campaign during the Civil War.

Both Throop and Headmaster Clark are Unitarians. Headmaster Clark proves to be one of the most exceptional headmasters in Cate School history, even though, after his hiring, one board member expresses reservation about a "godless Unitarian" leading a school Episcopalian funding helps establish after 1910. During his tenure at Cate, Headmaster Clark massively expands its endowment, facilities and faculty credentialing. To Sneed, Headmaster Clark is the Abraham Lincoln of Cate headmasters. With courage and conviction, Headmaster Clark makes real the ideals of liberty and the love of learning, to create a more perfect community of masters and scholars at the Cate School in Carpinteria, California.

a *7th-Gift* page

Without favoritism, Headmaster Clark perpetuates the unity of purpose between the Declaration of Independence and the Constitution, while adhering to Cate's motto, *Servons* ("Let us serve," which Sneed translates, "Let us serve without favoritism"). Headmaster Clark far surpasses in meeting the challenge to shepherd the school's board and constituent communities through the tumult of the 60s, to the benefit of the school's legacy as an exemplary American institution. After Cate becomes co-ed, Sneed trademarks, *What's good for the school, what's good for the scholar*™, echoing Mr. Cate's, *What's good for the school, what's good for the boy.*

A Caltech President's Expectation

In or about November of his senior year, at age 17, Sneed receives an acceptance letter to attend Caltech as a freshman in the fall. Headmaster Clark chooses Sneed for the Headmaster's Award at Cate's commencement (Clark Fulfillment). Former Headmaster Curtis W. Cate, born in 1884, is the honored speaker, bestowing splendor upon the ceremony as heaven may only allow a king to do (Cate fulfillment; his hot, summer-69 poolside OK not forgotten). Solemnly, during his junior year at Caltech, Sneed and Jones reunite at the school, and sit together in the Katharine Thayer Cate Memorial Chapel, for the memorial service of Headmaster Cate, Katharine Cate's husband and the school's 1910 founder.

Sneed receives his B.S. degree from Dr. Robert Christy (President Christy) at Caltech's commencement (Christy Fulfillment). President Christy's gracious handshake, while awarding the degree, reminds Sneed of attending a Christy staff meeting; Sneed's brother was the guest, law enforcement speaker. Within two years of shaking President Christy's hand, Sneed fully repays his five-figure student loans to the Caltech Alexis Gufstavson loan fund and Security Pacific National Bank. For doing so, American Express issues Sneed a Gold

Card. Four years after graduating, Sneed becomes acting IBM marketing principal to successfully restore Security Pacific National Bank's datacenter, when a rare power-surge wipes out its four IBM mainframes, processing $32 billion in float every four days. Though only an IBM trainee, the bank is pleased and attempts to hire him. A couple decades later, Sneed deposits a $110,000 check with Caltech's credit union for an invention he patents.

Also after graduating from Caltech, a subsequent letter from its then-president recognizes Sneed's work on the 0.5 billion-mile, NASA Galileo/Jupiter space-mission. The main spacecraft jettisons a probe and successfully collects and relays science data to Earth from the probe, after the probe's Mach 150 (106,000 mph) plunge down to Jupiter's cloud-tops, and below them. Appendix ℵ contains that letter, wherein the president sets a gentle, future expectation, "[I] look forward to hearing of future milestones in your career." Subsequent to that expectation, Sneed submits to the United States Patent and Trademark Office (USPTO) a specification (Specification) for an encryption system, that he names BEST2000™.

An Amazing Caltech Board Member

As a strategic encryption system, Specification's design cost-effectively encompasses billions of lines-of-code, installed on billions of dollars of computer hardware, to protect billions of dollars of 21ˢᵗ century transactions and databases. After a three-year examination, USPTO issues Sneed U.S. patent no. 6,236,992 for Specification. Prior to this, Sneed discusses BEST2000™ with Caltech board member, Dr. Alex Lidow (Lidow). At the time, Lidow is CEO of International Rectifier Corporation (IRC), a $0.7 billion, NYSE-listed firm. One of the most extraordinary persons to graduate from Caltech, Lidow invents, among other ingenious ideas, the HEXFET, which greatly expands transistor applications for microelectronic components his firm manufacturers. Lidow's grandfather establishes International Rectifier in 1947, making it the oldest semiconductor manufacturer in America. Also in 1947, AT&T/ Bell Labs scientists discover the transistor, and receive a Nobel prize for this, which the HEXFET, later, significantly enhances.

Lidow has his technical staff implement Specification on his firm's IBM AS/400 midrange systems in France, Germany, England, Japan, Singapore, China and the U.S. Dr. Dale Prouty, a fellow Caltech alum of Lidow and Sneed, later provides sworn testimony of this successful outcome (Lidow-Prouty Fulfillment). Lidow and Prouty validate the economies and efficiencies of Specification, worldwide, on IBM computers, relative to alternative solutions. Lidow later approves Sneed's receipt of $236,000. Specification permits Lidow to reject signing a $10 million contract with Anderson Consulting (now Accenture), or similar contract with IBM, or any other firm, to do the same thing. Both Anderson and IBM urge him not to sign with Sneed. Lidow and Sneed enter agreement, and his firm goes on to grow to over $1 billion in annual revenue.

The Decision by a Chief Justice, the Fulfillment of Presidents

Decades before this, Dr. Harold Brown (Brown), a distinguished U.S. physicists, precipitates Lidow and Sneed meeting, when Brown becomes Caltech president and establishes his administration's special-project-office, after 1965, paralleling Headmaster Clark's vision at Cate. Caltech's special-project-office, under Lee. F. Browne (Browne), admits Sneed as a freshman in 1973 (Brown-Browne Fulfillment). Browne, a distinguished science educator, who provides unprecedented service as a "V7" U.S. naval officer during World War II, is Caltech's first, special-project-office director. Before Caltech's recruitment, Browne teaches Caltech's faculty children, as regional science coordinator in the Pasadena school system. Browne's science teams repeatedly win annual, science competitions, prevailing over Southern California's most exclusive high schools. Sneed and Lidow meet as Fleming House, Caltech undergraduates, to the later benefit of their collaboration.

Chief Justice Earl Warren's 1954 Supreme Court decision (Warren Decision) removes legal obstacles for Brown to establish Caltech's special-project-office. Chief Justice Warren would endure multiple death threats for adhering to the unity of purpose between the Declaration of Independence and the Constitution, which Warren Decision enshrines. Brown-Browne Fulfillment affirms Warren Decision's evolution through the Eisenhower, Kennedy, and Johnson administrations.

Mr. Nicholas deB. Katzenbach (Katzenbach), while an official at the Department of Justice in the Kennedy and Johnson administrations during the 60s, places the full weight of the U.S. Constitution behind Warren Decision (Kennedy/Johnson/Katzenbach Fulfillment). The Supreme Court issues the 1954 Warren Decision during the Eisenhower administration, which benefits from Mr. Thurgood Marshall's (Marshall's) legal expertise as an attorney with standing before the Supreme Court. President Eisenhower sets precedent that Warren Decision is consistent with the U.S. Constitution, and deploys federal troops to protect persons, property and institutions to which Warren Decision applies (Eisenhower Fulfillment). Subsequently, President Johnson signs into law the 1964 Civil Rights Act that President Kennedy proposes, before his 1963 murder, that ends his marriage to Mrs. Jacqueline Bouvier Kennedy. Four years after that murder, President Johnson appoints Marshall to the U.S. Supreme Court.

The Last Full Measure of Devotion

President John F. Kennedy; his brother, Attorney General Robert F. Kennedy; President Lyndon B. Johnson; Attorney General Katzenbach; Chief Justice Warren; Justice Marshall and others endure repeated death threats for perpetuating the unity of purpose between the Declaration of Independence and the Constitution, in manly courage. Only a year after President Johnson appoints Marshall to the Supreme Court, a coward murders the unarmed, Nobel laureate, Dr. Martin Luther King Jr., in 1968, ending his marriage to Mrs. Coretta Scott King. Two months later, by then a senator, Robert F. Kennedy, unarmed, is murdered, ending his marriage to Mrs. Ethel Skakel Kennedy. One hundred and three years prior to this, a coward murders the unarmed President Lincoln, ending his marriage to Mrs. Mary Todd Lincoln. Presidents Kennedy and Lincoln, and Senator Kennedy are all unarmed when murdered, their wives faithfully at their sides.

Try as they may, violent parties can not destroy the unity of purpose, between the Declaration of Independence and the Constitution, through these desperate acts. And history gives witness to the merit in each *person* who gives the "last full measure of devotion," fulfilling such purpose. This includes members of Congress, presidents, Supreme Court justices, attorneys general, governors, legislators, mayors, locally elected officials, law enforcement personnel, Nobel laureates, citizens, military personnel, or anyone the 1868, 14th Amendment originally designates a born *person,* who results from an embryo's conception, under Heaven, on Earth.

Brown experiences Kennedy/Johnson/Katzenbach Fulfillment, first hand, including adherence to precedent of Eisenhower Fulfillment, and deploying federal troops, when hostile parties attempt to nullify Warren Decision, or threaten the *liberty* of institutions and *persons* to which Warren Decision applies. Brown serves as secretary of the U.S. Air Force during the Kennedy administration. From that position, Brown goes on to become president of Caltech and, later, secretary of defense in the Carter administration. As secretary of defense, among other insightful decisions, Brown evaluates and approves funding to develop the first, stealth aircraft in U.S. history. He also supports creation of a new, U.S. special forces command within the military.

A Billion Man-Hour Entitlement

Warren Decision, and the freedom it gives Brown to establish Caltech's special-project-office, has a natural evolution from President Abraham Lincoln's (Lincoln's) strengthening the U.S. military with his 1863 Executive Proclamation (Lincoln Fulfillment) during the Civil War. Lincoln Fulfillment frees his administration to enlist new recruits, from a population of *persons* previously denied military service. The anti-military riots that year, by Americans of European descent in northern cities, who take the lives of over 1000 policemen, federal troops, citizens, and *persons* in New York, alone, are violent attempts to block military enlistments, as battlefield fatalities mount during the war. On a "Richter scale of riots," what occurs in New York might be a 6.0, contrasting to 1.0 for, say, Californians' 1965 civil riot response, and peace restored without deployment of, nor fatality to any federal troops. Lincoln adds over 0.25 billion, new, military man-hours, through 180,000 recruits excluded from military service, prior to Lincoln Fulfillment. Those man-hours help ease the burden of war, and loss of life by Americans of European descent, who constitute some 99% of all enlistments, before Lincoln Fulfillment.

a *7^(th)-Gift* page

Simultaneously, Lincoln Fulfillment disrupts later plans by parties hostile to the Constitution (Hostile Parties), and their desire to visit further harm upon citizens and *persons* loyal to the Constitution and the Declaration of Independence—the latter set forth four score and seven years before Lincoln Fulfillment. During those 87 years, Hostile Parties lay life-long claims—using Coerced, Uncompensated, Labor Tradition (CULT) practices—to the services of *persons* and their offspring Lincoln Fulfillment encompasses. By 1861, before the war, such *persons* are annually providing more than a billion man-hours of uncompensated-service to the U.S. economy. Article IV, Section 2, Paragraph 3 of the Constitution allows this in *some* U.S. jurisdictions, so long as such service is for *peaceful* purposes (Billion Man-hour Entitlement).

As a military necessity—which the Constitution anticipates under Article I, Section 8, Paragraph 12—Lincoln Fulfillment ends the Article IV, Billion Man-hour Entitlement, specific to Hostile Parties only, and their later attempt to redirect portions of that entitlement for *military* purposes. With the stroke of a pen through Lincoln Fulfillment, the Billion Man-hour Entitlement, under law, ends, only for Hostile Parties. This blocks any plan to use uncompensated labor, under Article IV and its Billion Man-hour Entitlement, for military purposes, to further disrupt the Constitution and Lincoln's administration. Thwarting such plans perpetuates, for future generations, the nation's commitment to preserve the unity of purpose between the Declaration of Independence and the Constitution, through the evolving formation of a more perfect union.

Lincoln Fulfillment helps rapidly conclude years of hostilities and Civil War casualties, within 18 months of going into effect. It later disconnects Hostile Parties' use of the Billion Man-hour Entitlement to perpetuate armed insurrection as the war's end nears. And it allows *persons* formerly serving under that entitlement to support U.S. military field operations, intelligence gathering, logistics, food production, and medical relief, benefiting the unity of purpose between the Declaration of Independence and the Constitution, and their perpetuation. Lincoln Fulfillment opens the way, later, for Warren Decision, Eisenhower Fulfillment, Kennedy/Johnson/Katzenbach Fulfillment, Caldwell Fulfillment, Ryan Fulfillment, Clark Fulfillment, Cate Fulfillment, Brown-Browne Fulfillment, Christy Fulfillment, and Lidow-Prouty Fulfillment, each reflecting the unity of purpose between the Declaration of Independence and the Constitution, set in motion by its Framers in 1776. A year later, in 1777, Sebram Sneed leaves England for America, from whom Sneed is a direct, male-line descendant. The name "Sneed" is from the Old English "Sned," the center handle the harvester holds on a scythe. Sebram Sneed's family coat of arms legacy motto is, *"Neither to oppress nor be oppressed."*

IBM's Equal Rights, as a *"Person,"* Under the 14ᵗʰ Amendment

[The contents of this entire appendix constitute the author's opinion] By 1965, Katzenbach, later IBM's chief counsel after Robert F. Kennedy's recommendation, heads Johnson's Department of Justice (DOJ) as Attorney General (AG). That year, DOJ begins investigating IBM's business practices (President's AG's Investigation). In the summer of 1970, Sneed, at age 14, begins certification on IBM mainframe architecture and programming at Granite Computer *Institute* and, after graduating from the Cate School, attends the California *Institute* of Technology three years later. The previous year, 1969, President's AG's Investigation leads to a DOJ suit (President's AG's Suit) against IBM, regarding its business practices that, among other things, suppress the emergence of a U.S. computer services industry (Services-Industry). President's AG's Suit follows precedent of a 1956, DOJ Consent Decree, that IBM signs to avoid prosecution for restraint of trade under the Sherman Antitrust Act.

President's AG's Suit puts IBM on notice, regarding restraint of trade methods adversely affecting *persons* and enterprises responding to government and private sector demand for contract-programming-services. IBM considers President's AG's Suit a violation of its 14ᵗʰ Amendment, equal protection rights as a "*person.*" However, President's AG's Suit makes clear that the same 14ᵗʰ Amendment, equal protection applies to *persons* and enterprises that IBM's methods are injuring. In fact, Congress originally establishes the 14ᵗʰ Amendment specifically for a born *person* who results from an embryo, and not for IBM, since IBM is never a human embryo, nor does it have a mortal, physical body as a *person* does, under the Constitution. IBM does claim the right, as a "*person*" under the 14ᵗʰ Amendment, to an *immortal existence,* unlike the limited lifetime of *every human embryo* in history. Such immortality is at odds with Exodus 20:4, "You shall not make for yourself an idol in the form of anything[.]" Since "IBM" does not breath, or eat, or sleep, it, like an idol, espouses "immortality," while demanding 14ᵗʰ Amendment rights, as a "*person,*" into perpetuity.

Without President's AG's Suit, IBM could continue to use special pricing schemes to conceal its contract-programming-services costs within pricing proposals, bundled with other products like hardware and software. This affects trade involving *persons* and enterprises offering only contract-programming-services. President's AG's Suit reveals this is a detriment to the free market and U.S. economic competitiveness. Mr. Burke Marshall, formerly head of DOJ's Civil Rights Division under Attorney General Robert F. Kennedy, becomes IBM counsel, on the recommendation of Robert F. Kennedy in 1965. By 1968, Mr. Marshall, as IBM counsel, exhorts IBM to cease and desist the "tie-in" between contract-programming-services and other IBM offerings, including training, systems engineering, as well as hardware and software sales. Mr. Marshall tells IBM CEO Thomas Watson Jr., in no uncertain terms, that each "tie-in" violates federal law and the Sherman Antitrust Act (see *Father, Son & Co*, by Thomas Watson Jr., pp. 368, 369, and 381).

By 1970, Granite Computer Institute reflects the free market imperative of President's AG's Suit, *and* Mr. Marshall's courageous, legal leadership as IBM counsel, paralleling his unflinching stand on Civil Rights while at DOJ during the Kennedy administration. The suit helps protect the free market and need to increase the supply of *persons* trained in the programming arts, on IBM equipment, to respond to new demand for such skills, independent of IBM. Though Sneed is only 14 years old in 1970, he is one such *person.*

Thus, President's AG's Suit addresses expansion of a competitive, free market for contract-programming-services, especially as public and private sector institutions purchase IBM, and similar equipment, in the billions of dollars, during the 1970s, 1980s, 1990s and 2000s. The suit also promotes a broader-base of talent to think about problems affecting all users of such equipment, to discover economical solutions, that IBM may not consider, or which may not be in IBM's interest to offer. This is especially true for lower-cost solutions which compete with IBM hardware and software proposals, yet fulfill private and public sector customers' needs.

IBM's "14th-Subsidiary"

IBM responds (Watson/B. Marshall Fulfillment) to President's AG's Suit, and other litigation, by organizing a subsidiary to avoid 14th Amendment and Sherman Antitrust Act prosecution ("14th Amendment Subsidiary" or simply "14th-Subsidiary"). Over the years, IBM assigns various names to what herein is called its 14th-Subsidiary. By law and IBM policy, IBM's 14th-Subsidiary is independent of IBM's bureaucracy and IBM's equipment pricing schemes. IBM may not use its 14th-Subsidiary to restrain trade or manipulate the market for computer technology, much as DeBeers does for diamond mining worldwide. IBM's 14ᵗʰ-Subsidiary, though part of IBM, gains acceptance and trust from the public and private sectors as an honest broker, because of a wall of separation between it and IBM. This gives IBM's 14ᵗʰ-Subsidiary extraordinary flexibility to respond to clients' contract-programming-services needs, including trusted access to each company's hardware, software, operations, business methods, practices, databases or programming know-how.

Such trust derives from IBM's 14ᵗʰ-Subsidiary's commitment to equal protection for everyone's interests, consistent with the 14ᵗʰ Amendment itself, without interference from IBM. This is especially important when its 14ᵗʰ-Subsidiary pursues solutions at lower-cost than IBM is willing to provide, or which are in markets collateral to IBM's own business interests. And should IBM, itself, need its 14ᵗʰ-Subsidiary's services, it receives no favoritism regarding insider information about 14ᵗʰ-Subsidiary's trusted access to other companies' hardware, software, databases, pricing, or programming know-how.

The perception of equal protection, impartiality and confidentiality are IBM's 14ᵗʰ-Subsidiary's most valuable assets. Its practices, which just filing President's AG's Suit helps compel, reflect the unity of purpose between the Declaration of Independence and the Constitution, to perpetuate U.S. economic competitiveness. In the future, when IBM needs new sources of revenues and profits, its 14ᵗʰ-Subsidiary far surpasses other parts of IBM to fulfill that need, while helping expand IBM's market for new products and services.

Simultaneously, a robust, U.S. Services-Industry grows to tens-of-thousands of *persons* and hundreds of new enterprises, independent of IBM. Services-Industry augments IBM's 14ᵗʰ-Subsidiary's wherewithal, to meet billions of hours from demand for contract-programming-services, by public and private sector institutions, which helps increase sales of IBM and others', similar products. DOJ's suit, leading to IBM's 14ᵗʰ-Subsidiary's trust and practices, becomes one of the greatest win-win's in corporate history. One such enterprise, reflecting the purpose of President's AG's Suit, is ISB::*Institute*™, a think-tank Caltech alum Sneed establishes in 1985.

A Worldwide Computer Crisis Unfolds

By 1997, the worldwide installed-base of interconnected computer systems encompasses multiple billions of lines-of-code, handling multiple trillions of dollars in daily transactions. IBM's 14th-Subsidiary and Services-Industry, that serve U.S. needs as part of such a global infrastructure, reflect the wisdom of President' AG's Suit in 1969. Thus by 1997, the aggregation of computers serving both public and private sector needs, throughout all levels of society, is an extension of the unity of purpose between the Declaration of Independence and the Constitution. This includes such systems supporting government, public safety, tax collection, the courts, the military, academia, health care, manufacturing, transportation, banking, the arts, entertainment, real estate, and other national, state, and local interests.

The national economy's dependency on these systems becomes pervasive from 1969 to 1997, such that by 1997, any wide-scale disruption of these systems could directly affect the peaceful perpetuation of the U.S. economy and the stability of the world economy. Collaterally, during these 28 years, IBM and Services-Industry promulgate a dormant defect (IBM Defect), within billions of lines-of-code, across the worldwide installed-base of computers. Because the IBM Defect is dormant, and systems with it appear to work normally, IBM and others in the industry make no concerted efforts to correct it. This allows the IBM Defect to spread widely, like a lethal, sleeper virus, later threatening the stability of the U.S. and collateral economies, internationally.

a *7th-Gift* page

By 1997, it is clear that to fix the IBM Defect, globally, requires over a billion, billable man-hours, representing a tremendous revenue opportunity for IBM and Services-Industry. For multiple billions of lines-of-code with the IBM Defect, the repairs must be done before midnight that begins the year 2000, from which the IBM Defect takes its name—"Y2K"—obscuring its origin from conversion of 80-column, IBM punch cards to magnetic media in the 1950s. Please Google® search "us patents 6236992" and "us patents 12381901" for the IBM Defect's history.

Y2K: Mankind's Formidable and Cunning Adversary

In 1995, Sneed reviews Y2K as part of his duty, being a U.S. citizen with interest in the computing arts. Realizing Y2K's worldwide implications, Sneed trusts IBM, as an industry leader with other firms and institutions aware of Y2K since at least 1969, has Y2K solutions to prevent disruption of the U.S. and international economies. Such awareness and 26-year lead-time, since at least 1969 when the Johnson administration initiates President's AG's Suit, reassures Sneed that multiple decades are sufficient time to reverse the IBM Defect from billions of lines-of-code it is within, before midnight that begins the year 2000.

In February 1997, a *Los Angeles Times* newspaper article documents that a Department of Defense (DOD) computer complex in Ohio, after midnight that begins the year 1997, duns a government contractor for being 97 years delinquent on a contract not due until 2000. This becomes the first, real, Y2K failure that ever reaches Sneed—all other media reports seem to be sensational, speculative, or hypothetical Y2K dooms-day scenarios. However, DOD disperses billions of dollars to federal contractors, other personnel, and organizations through DOD datacenters. The government complex in Ohio is one of many DOD datacenters, globally, including those at foreign bases, embassies, and on ships at sea.

Since DOD is the best-funded government agency on Earth, Sneed concludes that as of February 1997, IBM, its 14ᵗʰ-Subsidiary, and Services-Industry may be struggling to address Y2K and prevent media exposure of more, public failures. He surmises that other, less-well-funded government departments and agencies, including Congress, the Executive branch, State, Treasury, Justice, the Courts, HHS, CIA, Education, Energy, Interior, Commerce, Agriculture, and so forth are even more vulnerable. Like a giant Anaconda coiling around the entire, worldwide installed-base of computers since 1969, Y2K may now be squeezing and suffocating its prey, the *LA Times* article being among the first, bone-cracking sounds of what's to come.

Though Y2K may represent over a billion, billable man-hours of revenue opportunity for IBM and Services-Industry, Sneed realizes that to address Y2K during the next two years will divert limited funds and resources—a lot of them—away from immediate, *profit-making* endeavors. For example, in 1969, Y2K is 31 years away. Most CEO's don't look 31 years down the road. Their focus is on quarter-to-quarter, return on investments and finite resource allocations, to meet stock market expectation, and respond to on-going competition from other firms. And in most cases, their competitors *also* aren't diverting funds and resources for Y2K. Yet, year after year, as more and more applications go on-line, reaching across the entire U.S. and world economies, the Y2K flaw spreads, through a larger and larger volume of code, worldwide, like yeast in an enormous batch of dough.

In fact, any diversion of resources for Y2K robs a customer's current projects—the ones which tend to increase demand for IBM or similar equipment, software and services. And IBM, too, must make its quarterly numbers, now, and not numbers 20, 40, or 120 quarters in the future—that's somebody else's problem. Such a group-think mindset, since 1969, causes, IBM, its clients, and others to peacefully drift together, like boats on what seems a calm waterway, each year edging closer and closer to the equivalent of Niagara Falls.

Sneed surmises, after reading the *LA Times* article, that Y2K's effect on the free market parallels placing a frog in a pot of water. If the water is hot, the frog will instantly leap out and save itself. But it may stay in a cool pot of water too long, if the temperature slowly increases to boiling, since the gradualness leads to inaction until it's too late. The Ohio DOD datacenter, and others like it, with less than 24 months to fix them, parallels the frog's predicament. The Securities and Exchange Commission, after 1997, requires firms to disclose their Y2K remediation plans, so that stockholders can assess exposure of their capital investments—and move them if circumstances warrant. The potential for litigation, on a wide scale, becomes possible for concealing lack of Y2K remediation planning, which includes IBM itself.

IBM's Worldwide Meltdown—Before Y2K

[The following includes references to *Big Blues*, by Paul Carroll; and *IBM Colossus in Transition*, by Robert Sobel]; By early 1997, IBM is deciding between "windowing" and "expansion" as alternatives to address Y2K. Yet each presents problems in repairing billions of lines-of-code, worldwide, before midnight that starts Y2K. Furthermore, in 1997 IBM is recovering from the worst economic disaster in corporate history. From 1980 to 1985, under John R. Opel (Opel) as CEO, IBM achieves the greatest, corporate, earnings performance, ever, at the time. Revenue grows from $29 billion in 1981, to $46 billion by 1984, with earnings doubling from $3.3 billion to $6.6 billion. Opel's stewardship also helps double IBM's capitalization to $72 billion.

Up to then, never in the history of IBM or any other corporation does one leader's tenure achieve so much, while garnering respect and admiration as a person who reflects the unity of purpose between the Declaration of Independence and the Constitution. Opel begins his career at IBM in 1948, in Missouri, the home state of President Harry Truman, who that same year wins re-election as President. In 1945, President Truman approves use of atomic know-how that Christy, Feynman and others help develop through the Manhattan Project, years after Japan's military rapes or murders thousands of women and children in Korea, China, and the Philippines; and its bombing of Pearl Harbor. Albert Einstein's 1939 letter to President Roosevelt warns such know-how is possible, and is of interest to Germany, Japan, and other nations.

When Opel steps down as CEO in 1985, his successor, John F. Akers (Akers), becomes beneficiary to the largest CEO-to-CEO wealth transfer in corporate governance history, at the time. As previous IBM CEO Frank T. Carey (Carey) does for Opel, Opel leaves a huge war chest for Akers, assuring Akers the wherewithal to develop IBM in new ways, however he may choose. Akers grants considerable latitude and financial resources for his management-team to discover, develop, innovate, acquire, and expand IBM from 1985 to 1993, and push its revenues beyond $100 billion.

Akers has some tremendous successes of his own, including $14 billion in revenue with IBM's new AS/400, the introduction of the IBM RS/6000, the IBM ThinkPad, and enhanced IBM mainframes. There are also major IBM setbacks, such as Joe Guglielmi's OfficeVision disaster; Jim Cannavino's OS/2 vs Windows debacle; Bill Lowe's IBM's PS/2's Micro Channel Architecture misfire; the PC modem, PC laser printer, Convertible PC, PCjr, and Prodigy embarrassments; and missing first-to-market opportunity with Intel's 386 microprocessor, to maintain the first-to-market precedent by IBM's Don Estridge (Estridge) with Intel's 8088 and 286 microprocessors.

Estridge oversees creation of a $5 billion PC business out of no where, during Carey's and Opel's tenures. Estridge helps protect IBM's mainframe-reputation and IBM revenues by being first-to-market with the newest Intel microprocessors, bolstering IBM's technological street credibility for all IBM products. This strategy denies revenue and market share to IBM's PC rivals, who desire to erode IBM's revenue and reputation if they could. Estridge's strategy is successful, even as he endures internal, IBM criticism about IBM PCs potentially taking revenue from other IBM product lines, including mainframes. When Bill Lowe replaces Estridge, Lowe doesn't maintain IBM's first-to-market precedent, missing it with Intel's 386 microprocessor, which begins a decline in IBM's PC fortunes. In 1985, Estridge and his wife Mary perish in a commercial plane crash, faithfully beside each other. Akers later gives sworn testimony that Estridge could have become CEO of IBM.

Losing first-to-market precedent with Intel's 386 microprocessor allows IBM's rivals to take massive amounts of PC revenue from IBM, and open speculation that IBM's other products may be also-rans as well, including IBM mainframes. IBM's OS/2 vs Windows debacle furthers this perception. Lowe spends some $125 million a year to develop OS/2, from 1986 on, the same year Microsoft offers IBM a 10% stake in itself for less than $100 million, which IBM declines. By 1993, that option's value accelerates to $3 billion. OS/2 continues to drain IBM's resources for over half a decade. In 1986 and 1987, IBM sells its 20% stake in Intel for $625 million, making a $225 million profit on its $400 million stock purchase during Opel's tenure. The sold-stake grows to $5.4 billion in value by 1993. IBM also declines to exercise an additional 10% Intel option for $200 million, which grows to $2.5 billion by 1993. Free of any IBM ownership influence, Microsoft's Bill Gates with Windows, and Intel's Andy Grove with Intel's more powerful microprocessors, begin setting industry standards for PCs, independent of IBM's Jim Cannavino (Cannavino). Cannavino replaces Lowe as head of IBM's PC business from 1988 to 1993, which steadily declines from Estridge's first-to-market successes.

One of the greatest setbacks for IBM comes as its head of U.S. sales loses control of the industry narrative about information technology and mainframes. He allows others to define IBM mainframes as large, dumb, over-priced dinosaurs, much as IBM's PCs are being dismissed by the media at the time. This misperception ignores IBM mainframes' strengths as hardened-aggregations of enterprise servers, with industry-proven security, reliability and maintainability that surpass alternatives. Thus, the first-to-market loss from new 386 PC technology has a collateral, negative effect on IBM's mainframes, letting public perception drift about their value and future relevance, along with IBM's PCs. Competitors like HP, Sun, DEC, Dell, Compaq, and others exploit this public misperception and market-uncertainty to make painful inroads across IBM's customer base.

In 1992, Akers reassigns George Conrades (Conrades), the head of U.S. mainframe sales, whose aspirations include succeeding Akers. Conrades promptly leaves IBM. Before leaving, it seems he blames Akers for not providing him the wherewithal to rebut public misperception about IBM mainframes as viable products. Yet, historically, competitors and pundits habitually attempt to create public misperception about IBM's efforts, whether with the IBM 360 and its operating system, the IBM Series 1 and its operating system, the IBM 8100 and its operating system, or other products which go on to be successful. Rebutting and countering such misperceptions is an ordinary responsibility for the head of U.S. sales. Additionally, Ellen Hancock (Hancock), whom Akers trusts to head IBM's network-products line of business, does not discern a market for bridges, routers, repeaters, or the development of similar components of a global, network infrastructure to serve cross-platform communications needs for IBM customers. Cisco Systems does, and becomes a billion dollar business.

a *7th-Gift* page

After acquiring cash, stocks and pensions, Conrades and Hancock seem to depart IBM with no concern about how anyone will address what they leave behind, undone. Thus, "Respect for the Individual" (RFTI-1), IBM's basic belief, seems to come to mean "Respect Favors the Insiders" (RFTI-2), with no obligation of reciprocity and loyalty to IBM's CEO. Respecting IBM's CEO's position seems outside the definition of "respect" and "individual," when a CEO requires accountability for opportunities and trust given. Both Bill Lowe and Jim Cannavino, heads of IBM's PC business, divorce their wives and marry younger members of their IBM staffs. They oversee the decline of the PC business from 1985 to 1993.

In 1992, seven years after Akers succeeds Opel, IBM sustains a $5.5 billion loss, whereupon Conrades leaves. The next year, 1993, the loss increases to $8.1 billion, the largest loss in corporate history. Additionally, within IBM there is powerful impetus for parts of IBM to break-away, similar to state secessions prior to the U.S. Civil War. Like a whale bleeding through open wounds that attract sharks from miles around, investment bankers converge on IBM, to rip it to pieces, and feed on the chunks that are torn away.

By 1993, IBM's market capitalization is $25 billion, down from $72 billion under Opel, a few years earlier. In the first half of 1993, it literally is about to run out of cash, as mainframe sales plummet, even with steep discounts—a dangerous death spiral that puts creditors, suppliers, bankers, mutual funds, the stock market, and media on notice. No reputable executive openly expresses interest about stepping into the IBM debacle. As his hold on the levers of power weakens, Akers stays at the helm, and endures the crushing pressure of his job, even knowing outside recruiting firms have IBM's board's approval to replace him. Ultimately, Akers serves IBM from 1959 to 1993, some three-and-a-half decades, successful by all measures until his last two years.

While Akers attends to his duties as best he can under the circumstances, modern-day "prophets" of doom—through print and other media—don't help him at all. They begin sounding IBM's death knell, including some pieces in *The Wall Street Journal* and *The Economist.* Authors publish books on a post-IBM world. Many competitors, by piling-on about IBM's woes, violate an ancient precept in Proverbs: "Do not gloat when your enemy falls; when he stumbles do not let your heart rejoice, or the [I-SHALL-BE] will see it and disapprove, and turn his wrath away from him." Proverbs 24:17,18. The gloating becomes an industry chorus.

Who would desire to command Apollo 13, knowing it might become stranded in space, forever, with its commander on board? Or lead the Continental Army's 5,000, barely clothed and fed men, against the most powerful military empire, on Earth, in its day? Or fly Spitfires during *The Battle of Britain?* Or U.S. Air Force P51 prop-planes against German Me262 jets in World War II? Or be in the first Higgins boats on D-Day? Or follow President Herbert Hoover into office after the economy collapses? Or succeed President Buchanan in 1861, after seven states secede in preparation to destroy the U.S.'s unity of purpose, from its commencement in 1776?

Who would risk one's name; reputation; future prospects; wealth; health—including possible hypertension and coronary heart disease; or expose one's family and friends to stress and public ridicule; or become the target of hostility from strangers? And who might be the "right" person to pickup IBM's mantle and endure the public gauntlet Akers is experiencing, and somehow prevail through it to restore IBM's fortunes—if it could be done?

The Man from Mineola

[The following includes references to *Who Says Elephants Can't Dance,* by Louis Gerstner Jr.; and *The Dream Machine,* by Jon Palfreman and Doron Swade] Fortunately for IBM, on d2/March 1, 1942, in Mineola, New York, Providence anticipates such future need, through the birth of one Louis V. Gerstner Jr. (Gerstner). He is the son of a father whose career begins as a milk-truck driver, later to serve honorably in coordinating fleet logistics for an entire brewing company. Gerstner's mother also advances in her career, from secretary to community college administrator. He lives with his siblings, one older brother and two younger ones, in the wholesome home of his parents, until 1959, when he leaves for college.

Yet before all formal education, advance degrees, or professional training Gerstner receives later in life, his parents are his first educators, imbuing a faithfulness to develop the gifts, talents, and abilities that are a birthright and a blessing for each child in his family. That faithfulness would not only later benefit multiple U.S. industries; it would help perpetuate the unity of purpose between the Declaration of Independence and the U.S. Constitution, amidst an exceedingly competitive, world economy, that is often indifferent to such purpose, when profit takes priority over principle.

The Birth of Electronic Computers in 1942

The same year Gerstner is born, 1942, researchers in the U.S., England, and Germany independently converge on the idea that vacuum tube computers could be many times faster than state-of-the-art, electromechanical, relay-based designs. Germany loses its lead in this electronics revolution when its government refuses to fund a vacuum tube computer that Konrad Zuse (Zuse), a private citizen, proposes. He proves the viability of his binary design with a fully operational computer using electromechanical relays, but without funding, his electronic plans end. Another private citizen, Wernher von Braun, does receive such government funding, which leads to development of V2 rockets, used to later terrorize England.

The Nazi party insists von Braun become a member, so he does, but he renounces the party at the end of the war when the Allies, with his cooperation to avoid the Russians, "capture" him and his rocket team. He and his team become builders of the Saturn V, the launch-vehicle for the U.S. Apollo Program to the moon. Also that year, 1942, the need to crack Germany's Enigma and Lorenz encoding machines leads England to invest in Colossus, a vacuum tube, electronic computer. When built in 1943, Colossus breaks Nazi message-codes in as little as 30 minutes, which prior relay computers often couldn't do in any time-frame, if at all, for results to be useful. This significantly affects the outcome of the war, to the benefit of the Allies led by the U.S. and the U.K.

Gerstner's birth year also is when the U.S. military considers vacuum tube computers for the first time, which leads to a $500,000 investment between 1943 and 1945 to build ENIAC. ENIAC, the Electronic Numerical Integrator and Analyzer Computer, computes firing-tables that improve accuracy and rapidly reconfiguring myriad field artillery with varying, ordinance mixes. ENIAC, as an electronic computer, later can do in 30 seconds what a room full of mechanical, analog computers requires days. It also helps with H-bomb calculations after the war.

EDVAC, or Electronic Discrete Variable Computer, is the follow-on to ENIAC, which the U.S. military funds in 1944. With EDVAC, the "von Neumann" architecture becomes a standard—which reflects John Mauchly's and Presper Eckert's design to store programs and data in a common, main memory. IBM adopts and refines the von Neumann architecture for scientific and business machines with its vacuum-tube-based IBM 700 series computers of the early 50s, its transistorized IBM 7000 series in the late 50s, and its IBM 1400 series in 1959. The IBM 1400 becomes the most popular, general-purpose, von Neumann architecture machine of its day. The year IBM introduces its 1400 series, in 1959, Gerstner leaves home and enters Dartmouth college. A Dartmouth faculty member invents BASIC, or Beginners All-purpose Symbolic Instruction Code, by the mid-60s. BASIC becomes one of the most widely-used, interpretive computer languages in the world. It is later standard with every, early IBM PC. Gerstner graduates from Dartmouth in four years, with a degree in engineering science, and gains acceptance to Harvard Business School.

After completing his Harvard graduate studies in 1965, McKinsey & Company hires him in New York. That very year, IBM begins shipping the IBM 360 series, overcoming a near disaster in developing its operating system, OS/360. The machine catapults IBM from a 25% market share in the 1950s, to 70% in the 1960s. *Fortune* magazine acknowledges IBM's CEO, Thomas J. Watson Jr., as being one of the most brilliant executives in U.S. history, successfully gambling some $5 billion to create the IBM 360 line. Gerstner, another "Jr.," follows Watson's precedent and saves the company from disaster, a quarter of a century later.

Preparing for such an extraordinary role from 1965 to 1974, promotions advance Gerstner to senior partner at McKinsey, with exposure to a plethora of corporate cultures in the financial services industry, which is his specialty. Gerstner's expertise grows, about how American firms work, how to improve their direction and future performance, and how to capture and enhance their competitive metrics. Rigorous, applied analysis, heavy on detail, produces consistent, deliverable results that allow him to gain respect within McKinsey and through the industry he serves.

In 1977, he considers an offer from American Express, which would allow him to be a top-executive at a worldwide, financial services enterprise. Though different from his role as a consultant, he accepts the challenge and works at American Express for 11 years. As CEO of an American Express subsidiary, with his leadership it grows at a compound rate of 17 percent. During his tenure, card membership just about quadruples. Simultaneously, he diversifies his business unit into new, lucrative markets. Gerstner develops a deft hand at corporate governance, transitioning from advisor/ trouble-shooter to respected corporate executive.

As with Mckinsey, he remains keenly sensitive to the effective use of information technology to improve value creation for businesses, applying the same, hard analysis to information technology investments and performance as to every other part of an organization. Before there is an Internet, he helps oversee development of a vast network of interconnected systems, working across international borders and currencies, to achieve the competitive objectives American Express sets for its cardholders. Information technology is not a theoretical abstraction to him; it is the medium of delivery that realizes his company's business plans to serve cardholders, worldwide, seamlessly, and create value for stockholders. This includes overcoming the challenge of helping IBM, as a vendor, understand such a vision and implement it, constructively, cost-effectively, and globally.

By April 1989, RJR Nabisco recognizes Gerstner as a rare and capable executive, and is sorely in need of his abilities after it attempts an ill-advised merger. He considers becoming head of RJR Nabisco to help it navigate a complex, SEC-regulated, leveraged buyout—and address the good, the bad, and the ugly of dealing with a heavily indebted balance sheet. At or about this time, IBM's demise is still three years away, though its vulnerabilities are becoming discernable within its balance sheet and cash flow numbers. This type of data Gerstner is expert at analyzing, for making tactical and strategic business decisions, from expertise he applies and refines at RJR Nabisco, American Express, and McKinsey.

From Dartmouth, to Harvard, to amassing a diverse set of skills as a rare and insightful corporate executive, Gerstner comes to reflect the American rugged individualism of a special forces commander. By 1992, he can go into virtually any business situation, with a team, and create the best possible outcome, within available resources, personnel, and market conditions. In 1992, IBM hits the wall with its $5.5 billion loss.

a *7th-Gift* page

In 1993 it is staggered by an $8.1 billion loss. Just before 1993, IBM approaches Gerstner to consider the role of a life-time, though he doesn't perceive IBM as the calling for which his entire life prepares him. However, other leaders and corporate titans prevail upon him to consider IBM a national treasure, warranting the very best effort to right it, if possible, and that doing so may be one of the most challenging executive assignments in U.S. history.

After Gerstner does a preliminary analysis of IBM's condition, he sets the odds at five to one against its recovery. And yet, he becomes drawn, over time, to the uncertainty IBM faces, much as a fireman instinctively goes toward a burning building, when everyone else is fleeing. The heart of a trouble-shooter, with the competitiveness of a leader, still beats within him. His family splits on discussion about IBM, but unify behind him when he makes the final decision to help the firm. He negotiates terms to address closure of economic arrangements with RJR Nabisco, and enters agreement to become IBM CEO by April 1, 1993.

It's easy to look back now and say, What a brilliant decision by all concerned! But accepting IBM's CEO-ship and debriefing Akers, executive to executive, helps Gerstner realize the depth of responsibilities that are then his. Though he proves clearly to be the right man for the job, at the time he begins, he, as with Lincoln long before, has a team of rival, "true Republicans," including ones like Secretary of State Seward and Secretary of Treasury Chase, who express disdain at a Republican "inferior" having the position that is rightfully theirs. Gerstner learns there are many such "true Republicans" throughout IBM after he becomes CEO.

The Corporate Turnaround as the End of the Century Nears

Gerstner's first order of business is "secession," or the breaking up of IBM. The investment banking community set the table for a feast on IBM's disintegration, before Gerstner's arrival. IBM's previous administration put accounting systems in place to facilitate the breakup and sell-off of major parts of IBM. But Gerstner, later, runs the "moneychangers" out of the "temple" and vetoes the idea. He gives a hint about his position when he meets with IBM's leadership team for the first time. He imparts the key pillars of his management philosophy to them, including:

o   He manages by *principle*, not procedure
o   He looks for people who work to solve problems and help colleagues. He sacks politicians.
o   Hierarchy means very little to him. He wants meetings to include the people who can help solve a problem, regardless of position. And he wants committees and meetings kept to a minimum. There is to be no committee decision making. He wants lots of candid, straightforward communication.

Gerstner has a rare meeting with his brother, Richard, a long-time IBMer. Richard returns as an IBM contractor, well before and unrelated to Gerstner's hire, after receiving care for a misdiagnosis of a manageable malady. Gerstner reviews Richard's advice to recast the public perception about the role of IBM's mainframes, and to make clear their advantages, rather than let competitors further control the public narrative to the detriment of IBM's market share. Both agree that only one such formal meeting between them is appropriate, out of respect for each other and their roles at IBM. One can imagine their parent's satisfaction that their sons are serving in such capacities, honorably and with integrity.

Over time, Gerstner receives "push-back," in one way or another, from every major IBM constituency. To the investment bankers' chagrin, he decides to keep IBM whole, while selling-off only the parts of IBM that are low-margin and non-strategic, as market conditions and timing warrant. This later includes letting Loral buy IBM's Federal Systems Division, and selling Entry Systems Division, which makes IBM's PCs, to Lenovo. Both decisions bring in much-needed cash. Dennie Welsh (Welsh)—a services and networking genius—helps create the IBM Global Network. Gerstner eyes it as a saleable asset, and hopes to get $3.5 billion for it. He defers to a $5 billion bid from AT&T, sparing IBM the expense of expanding and maintaining the network. This transaction raises the ire of IBM "insiders." Yet the timing is perfect, and maximizes Gerstner's profit, before a global network-capacity-glut develops, just as the Internet takes off. He also helps his Technology Group diversify into custom-chip production, and pulls revenue from new markets, including game consoles.

With his enormous patent portfolio, he successfully reaches out through his intellectual property staff, keeping IBM profitably in the mix of being a cooperative, industry-standard innovator. He discontinues software products that compete with companies helping sell IBM hardware, and focuses on IBM's core, software product competencies that do not injure mainframe revenue production by 3rd-party software providers. Again, with Welsh's insight, he reaches out through the "cloud" computing paradigm, making it "blue," proving that IBM is a friend to open standards that unify IBM's customers' information technology investments across vendors.

All the while, Gerstner works diligently to reassure and calm IBM's seven constituencies—customers, stockholders, all IBM personnel, suppliers, financial communities, governments, and the media. He gives honest notice about layoffs, early on, and makes clear any layoff decision will be prompt and conclusive. To deal with IBM's immediate liquidity needs from day-one, he both slashes mainframe prices and requests increases in committed lines of bank credit to $4.7 billion. He also seeks approval to raise $3 billion through issuance of preferred stock, and selling IBM's U.S. receivables at a discount for quick, securitized cash.

Akers does leave Gerstner a gift that becomes worth $19 billion, from 1997 to 2001. Before stepping down, Akers approves a $1 billion, bipolar-to-CMOS, manufacturing conversion for the System/390, with possible spin-off effects for other products. Gerstner affirms Akers' decision and doesn't cancel the conversion, even considering the liquidity crisis IBM faces. This allows IBM to later build some of the best price/performance, high-end systems in the world. The price of a unit of mainframe processing power drops from $63,000 to less than $2,500 over seven years. Shipments of mainframe capacity decline 15% in 1993, but grow by 41%, 60%, 47%, 29%, 63%, 6%, 25% and 34%, in 1994, 95, 96, 97, 98, 99, 2000, and 2001, respectively.

Between 1993 and 1995, he cuts expenses by $8.9 billion. In addition to restructuring the board's corporate governance, he reengineers the leadership team and incentives for senior management that is revolutionary in IBM's history. And he redesigns IBM's corporate culture, promotion by promotion, email by email, site-visit by site-visit, speech by speech, meeting by meeting, and decision by decision.

Though he doesn't call it a "vision" when his tenure begins, Gerstner defines IBM's "focus": to preemptively-drive restoration of IBM's profitability through pragmatic, rigorous analysis and market-responsiveness; to project into each customer premise an effective presence; to grow as an industry-leader that internalizes client/server as a way of life; to develop and deliver a full-spectrum of services that competitively leverage IBM customers' I.T. investment plans, the way niche-players or a Balkanized IBM may never be able to do; and to address all, ancillary fundamentals necessary to execute the "focus." Gerstner clearly states he will not use the term "vision" early in IBM's recovery. Somehow, the press and pundits miss the importance of "focus" during the near-term, triage phase, and Gerstner clearly stating later-results will determine the articulation of a "vision."

Step by step, Gerstner stabilizes the company, much as Commander Jim Lovell does for Apollo 13, after part of his spacecraft explodes, on the way to the moon, with the whole world watching. By the end of Gerstner's "freshman year" in April of 1994, the company is on the mend. By his "sophomore year," shareholder value is accelerating upward. By his "junior year," the company is prospering and has a hopeful future. And by his "senior year" in 1997, there's acknowledgment, even among some skeptics inside and outside the corporation, of the extraordinary talent he is applying to rejuvenate IBM. So Gerstner advances to "graduate studies" at IBM. Dartmouth and Harvard may be quite pleased with their part in preparing Gerstner to achieve what he does. However, it is faith in doing what is right and just and fair, which Gerstner's parents instill within him from youth, that informs everything he successfully achieves in his professional life as an executive.

Gerstner and "Respect for the Individual"

After taking over from Akers, Gerstner accepts that he must confront IBM's corporate culture, its belief system, and its tenet of "Respect for the Individual" in light of new, market realities IBM faces at the end of the 20th century. IBM's paternalism, during the early and mid-20th century, is entrenched when IBM becomes the most successful corporation in history by the 1980s. However, Gerstner quickly realizes the perception of the belief system, by many within the company, differs dramatically from its original intent, to be responsive and proactive regarding customer needs and market dynamics.

Gerstner concludes that the inward focus of "Respect for the Individual" must evolve into a more market-aware "Respect for Initiative by every Individual," at all levels, especially in helping peers and the entire organization consistently fulfill customer needs and the demands of competitive market-responsiveness. The leadership of the company, therefore, must know, in great detail, competitive market metrics regarding product quality, service delivery, personnel practices, and production cost-containment, for the company to profitably create value to the benefit of its customers and shareholders. Creating such value allows the company to sustain revenues and profits to meet on-going commitments, both inside and outside of the company.

So Gerstner begins reengineering IBM's corporate culture, at its core, by completely upending IBM's compensation system. First, he realigns employees' interests to coincide with those of shareholders through employee stock ownership, to a degree never done in IBM's history. Secondly, he sets base-compensation increases and annual bonuses to the overall performance of the company, as well as to an employee's business unit. Gerstner's foresight in doing so helps establish a unity of purpose and effort across the entire corporation, worldwide.

Gerstner rejects IBM's old, international model that IBM is a collection of relatively autonomous "city-states". Through his incentive system, he sets forth the vision of a market-responsive federation of shared-interests, working through collaborative synergies to continually create value extending outward from the whole, corporate body. From his global system that he calls "variable pay," employees, worldwide, share $9.7 billion in rewards and bonuses during a six-year period.

a *7ʰ-Gift* page

He reduces some paternalist programs of the past, and ends others completely, while still keeping overall benefits programs among the most generous of any U.S.-based, multinational corporation. Additionally, for any individual who manifests initiative that creates market-responsive value, Gerstner's plan provides rewards that are scalable at every level.

A Caltech Alum Becomes the Top-Performer in IBM's History

One of the earliest and most prolific executives to benefit from Gerstner's management reengineering is Peter Wilzbach (Wilzbach). Wilzbach's pre-IBM background might make him the least likely person to achieve such distinction. He prepares for a life of poverty as a candidate to be a Jesuit priest after graduating from Caltech, and before joining IBM. As a Caltech undergraduate, he helps develop the on-board memory for the Mariner mission to Mars, and becomes a member of the technical staff at Hughes Electronics after graduating. Yet, he responds to the call of being a salesman at IBM, as considerable as his technical and theological background is before starting IBM's sales school. Such a unique blend of ability and perspective may only pass through IBM once in a generation.

Wilzbach rapidly advances through IBM's management ranks. By 1997 he is vice president at IBM's 14ᵗʰ-Subsidiary, where he applies his abilities to create a supernova event that causes Gerstner's compensation plan to shower Wilzbach with tremendous quantities of cash. Wilzbach closes the largest information technology, outsourcing contracts in IBM history, with GE and Ameritech. At the time, Dennie Welsh (Welsh) heads IBM's 14ᵗʰ-Subsidiary, using the name IBM Software Service for Collaboration, which it later changes to Global Services.

Gerstner nurture's Welsh's services vision, with Wilzbach on Welsh's management team, and grows IBM's 14th-Subsidiary to 1/3rd of IBM's revenue and 2/3rds of IBM's profits by 1997. This is done in compliance with President's AG's Suit in 1969, and the impetus that organizes 14th-Subsidiary's business practices. In their efforts, Gerstner, Welsh, and Wilzbach respect the unity of purpose between the Declaration of Independence and the Constitution, while fulfilling expectations Gerstner sets for IBM's historic turnaround.

Wilzbach and Sneed first meet in 1980. About seven years after Wilzbach works on NASA's Mariner/Mars mission, Sneed works on NASA's Pioneer/Venus mission as a Caltech undergraduate and Hughes Space and Communications Group intern. After graduation from Caltech, Sneed accepts a position as a member of the technical staff at Hughes (now Boeing) and successfully contributes to the NASA Galileo/ Jupiter mission. He goes on to become an assistant professor of Electrical and Computer Engineering at California State University Northridge. There, he meets an IBM recruiter, which leads to meeting Wilzbach.

Wilzbach, then an IBM manager, interviews Sneed and offers him a marketing position. At the age of 12, he decides to work at IBM, and does, with Wilzbach as his first manager (Wilzbach Fulfillment). Sneed graduates with honors from IBM's New York marketing school (Opel Fulfillment), after election as president of the 6,000 member Western Region IBM Club, later becoming first to achieve 400% of quota for the entire western U.S.

Gerstner, Sneed, and Y2K

Thus, the February 1997, *Los Angeles Times* article about Y2K stuns Sneed, who by then heads his own think-tank. He puts other projects aside, takes a blank sheet of paper, and begins thinking about the problem, and designs a solution to address billions of lines-of-code. He reviews it with a patent attorney, who suggests Sneed refine it as a specification (Specification) to file as a patent. He does, later naming the invention BEST2000™.

In May of 1997, Sneed provides a copyright-restricted, evaluation-only copy of Specification to Gerstner's staff, for a fee of $49,764.55 per site for any use other than evaluation. Gerstner and his staff must repair over a billion lines-of-code in less than 24 months, worldwide. They determine Specification superior to two alternatives. Gerstner's staff proceeds with contract formation, and asks Sneed to include a per-line-of-code price of no more than 50 cents. This leaves margin for IBM's 14ᵗʰ-Subsidiary to charge its customers with Y2K exposure, and IBM itself, up to $2 per line of code to fix the problem. Gerstner's staff also asks Sneed to make it easy to reconcile the aggregate number of lines-of-code Gerstner's staff might apply Specification.

In early June, Gerstner's staff phones Sneed approvingly upon receiving his agreement for 14 cents per line of code, and no charge after a billion lines-of-code. Gerstner's staff also asks Sneed to pretend his first contact (First Contact) is with Gerstner's staff at IBM corporate, instead of Gerstner's staff at IBM's 14ᵗʰ-Subsidiary. This may allow IBM corporate to save up to $1.86 per line of code from the up to $2 per line of code IBM's 14ᵗʰ-Subsidary charges, since IBM's 14ᵗʰ-Subsidiary must treat IBM corporate the same as any other client. This also avoids the appearance of insider information trading between IBM corporate and IBM's 14ᵗʰ-Subsidiary.

Sneed contacts Gerstner and reviews these events. Gerstner, after investigating the matter, has a staff member (Gerstner-Contact) send Sneed a d6/June 20, 1997 letter that includes, 1) acknowledgement of receipt of Specification for evaluation-only within IBM, 2) acknowledgement of receipt of written contract formation terms with IBM's 14ᵗʰ-Subsidiary for production-use of Specification, and 3) stipulation that before Sneed receives any compensation, he must notify

Gerstner's staff when a patent issues to Sneed for Specification (Gerstner-Condition). The letter also directs Sneed to only communicate with Gerstner-Contact regarding compensation for IBM's worldwide use of Specification, which for a billion lines codes amounts to $165 million, when spread over 500 sites.

The United States Patent and Trademark Office issues Sneed patent no. 6,236,992 for Specification, on d3/June 22, 2001. At or about the time Sneed sends Gerstner notification of patent no. 6,236,992 under Gerstner-Condition, the IBM board discusses with Gerstner his retirement. Gerstner announces his retirement within six months, even though he is only 59 and the architect of one of the greatest turnarounds in U.S. history. Gerstner-Contact, who sends Sneed Gerstner-Condition, in writing, is later forced to leave IBM and seek employment elsewhere. Neither man responds to Sneed's notification under Gerstner-Condition, before retiring and leaving IBM. Gerstner shows good-faith to have Gerstner-Condition put in writing, and therein makes the declaration, before retiring, that neither he nor his staff will expropriate Specification. Sneed accepts that Gerstner discloses no use, because he is unaware of any, based on disclosure by his staff to him, before IBM's board encourages his retirement.

Gerstner-Condition and the Unity of Purpose Between the Declaration of Independence and the Constitution

The parable in *Sneed v. IBM: Decoding a $1.86 Billion Federal Riddle With Caltech's Honor Code, Pro Se* is this: Why would three, reputable, Caltech alums urge Sneed to seek relief under Gerstner-Condition, without any evidence IBM uses Specification? At the center of this parable is Caltech's Honor Code and Gerstner-Condition, which establish the legal basis for IBM to pay Sneed $165 million for Specification, under

U.S. law, specific to the Y2K problem, worldwide. For a billion lines-of-code, Sneed's 14 cents per line-of-code with IBM's 14ᵗʰ-Subsidiary allows savings in excess of $1.86 billion, before comparison to alternatives IBM was considering.

The 14 cents per line-of-code, and per-site fee, allows a BIG profit margin for IBM's 14ᵗʰ-Subsidiary, and becomes the hypothetical basis to project Specification's fair market value if IBM ever discloses use—with *quantum merit* (20% of $1.86B in unlawful enrichment) and 100% delayed discovery (hidden use at 10%/year) that a respected California state judge teaches Sneed in a prior case—or $700 million under Proverbs 22:22,23. Specification and its benefit to sustaining U.S. economic competitiveness is exactly the kind of effort President's AG's Suit anticipates in 1969, as well as the government's 1956 Consent Decree IBM signs to avoid prosecution for restraint of trade. These actions, by the Department of Justice, help nurture the growth and development of Services-Industry from 1969 to 1997. This includes organizing IBM's 14ᵗʰ-Subsidiary and protecting a *person's* 14ᵗʰ Amendment existence to help solve a critical problem, like Y2K, that may threaten the unity of purpose between the Declaration of Independence and the Constitution.

Gerstner-Condition is a modern reflection of precedent set forth by the Framers of the Constitution, through Lincoln Fulfillment, Warren Decision, Eisenhower Fulfillment, Kennedy/Johnson/Katzenbach Fulfillment, Watson/B. Marshall Fulfillment, Caldwell Fulfillment, Ryan Fulfillment, Clark Fulfillment, Cate Fulfillment, Brown-Browne Fulfillment, Christy Fulfillment, Wilzbach Fulfillment, Opel Fulfillment, and Lidow-Prouty Fulfillment. Thus, Gerstner-Condition is in complete harmony with the unity of purpose between the Declaration of Independence and the Constitution, positioning Gerstner, *potentially*, as an Abraham Lincoln of IBM CEOs. As Gerstner notes in his book, *Who Says Elephants Can't Dance*, relative to his relationship with IBM, and retirement discussions the IBM board initiates that end his tenure:

[I] now realize that I was always—even to the end—
an outsider.

Who Says Elephants Can't Dance, by Gerstner, page 281

In many ways, so were Thomas Watson Jr., Nicolas
Katzenbach, Burke Marshall, and others. This realization
(Gerstner-Realization) comes to Gerstner *after* his last day
at IBM, when he considers IBM's board's timing to make
way for a new IBM CEO to replace him, one who turns out
to be a "true Republican," IBM insider. There can only be
candor in Gerstner-Realization, as an honest assessment,
after giving nine years of his life, as CEO, to provide the most
extraordinary service one individual may give to a corporation,
to an industry, and to a nation. This is even more poignant
when one realizes that Thomas Watson Sr. dies only weeks after
retiring; Thomas Watson Jr. suffers a heart attack a year before
retiring; and Akers endures unimaginable pressure, to hold
everything together, as best he can, until IBM's board relieves
him—possibly protecting his health and life, from over three
decades of continuous service. Thomas Watson Sr. endures
such pressure for over four decades.

IBM, as an *Immortal Bodiless Mythical Entity* (IBME™), is
never an embryo, like Thomas J. Watson Sr.; Thomas J. Watson
Jr.; Arthur "Dick" Watson, head of IBM World Trade and later
U.S. ambassador to France; T. V. Learson; Frank T. Carey; John
R. Opel; or John F. Akers—all CEOs before Gerstner becomes
the "outsider" CEO. And yet, to differing degrees, they are
all "outsiders," relative to the latent, mortality-prejudice an
IBME™ has against *any* finite being. An IBME™ can have little
more than a temporal, transactional relationship with finite
beings, while it uses their fealty to perpetuate its immortal
aspirations, especially through the image of a flesh-and-blood
CEO, to gain the trust of *persons*, though it is not, *per se*, a
flesh-and-blood *person*.

a *7th-Gift* page

Each CEO must conform to the IBME™'s *unending* will, as painful as this may be at times, such as when the IBME™ is profiting from CULT practices in Germany during World War II. Though Watson Sr. could not stop the IBME™'s profit-taking from the Third Reich's brutal *procedures and processes*, he does return awards and decorations, based on *principle and precept*, even if this raises the ire of a highly-profitable client-government. However, because of the IBME™'s worldwide reach, even Watson Sr. would not stop IBME™ production of new equipment and deliveries within the Third Reich. Still, there were numerous other ways Watson Sr. asserts "outsider" thinking, including developing his two sons as IBM employees, and handing the reigns of power over to them, six weeks before Watson Sr. dies. Both sons become two extraordinary hires in IBM history.

When Watson Jr. heads the company, he knows the importance of thinking like an "outsider," and does so, repeatedly. By the early 60s, he is risking $5 billion to create the IBM 360, even with "push-back" from powerful, entrenched, IBME™ "insiders," who prefer more incremental steps. Watson Jr.'s decision grows his firm's market share for computers from 25% in the 1950s, to 70% near the end of the 1960s. When, as an "insider," he could easily feather his own nest, Watson Jr., though CEO, thinks as an "outsider" and agrees with IBM president, Al Williams (Williams):

> I considered taking even more radical steps to increase IBM's commitment to its employees. When I talked to my wife at night, I would speak of various ways of sharing our success more broadly. Those at the top were doing fantastically well on stock options—despite the fact that Williams and I stopped taking options in 1958, after Williams said, "We don't want to look like pigs." . . . "How much more am I worth to IBM than that guy down at the bottom of the pay scale? Twice as much? Sure.

Ten times as much? Maybe. Twenty times as much?
Probably not."
> *Father, Son & Co.*, by Thomas Watson Jr., p. 311

Watson Jr.'s "outsider" thinking guides him to hire
Nicholas Katzenbach and Burke Marshall, on Robert F.
Kennedy's recommendation. Prior to Watson Jr. hiring them,
they both serve within DOJ under Attorney General Robert F.
Kennedy. It is one thing to hire such extraordinary men, and
another altogether to accept their counsel when entrenched
IBME™ "insiders" want to ignore them and continue "tie-in"
sales, contrary to federal law and the Sherman Antitrust Act:

> Burke Marshall, however, was shocked to find IBM
> doing business in this way. He saw bundling as a
> glaring violation of antitrust law known as a "tie-in
> sale," such as when a local electric company tries
> to dictate the appliances you buy for your house.
> By requiring customers to buy our products by
> the bundle, we were making it almost impossible
> for independent companies specializing, say, in
> software to break into the business. At first people
> at IBM had trouble grasping this. No one could
> understand what Marshall objected to—bundling
> was like the Apostles' Creed at IBM . . . Burke
> Marshall is a mild man, but as the public officials
> of Mississippi and Alabama learned from his work
> at the Civil Rights Division, he also has a precise
> mind and an unbendable will . . . "But why?"
> [IBM] people kept asking. "Why change now?" . . .
> "Because you've got a tie, god******, you've got
> a tie! It's illegal! If you try to defend it in court,
> you'll lose!"
> *Father, Son & Co.*, by Thomas Watson Jr., p. 381

Watson's "outsider" thinking and willingness to act on
Marshall's proactive, expletive-underscored counsel, after
intense birth pains both within and outside of IBM to do
so, helps open the way to organizing IBM's 14th-Susidiary's
business practices. By 1997, IBM's 14th-Subsidiary accounts for
1/3rd of IBM's revenues and 2/3rds of its profits.

Thus Gerstner follows in the best of "outsider" thinking
tradition. With his family and health intact as his CEO tenure
ends, he achieves beyond the highest expectation a nation
may hope for any citizen, making him a great American by
any measure. America claims Gerstner as a son, harmonizing
with sacrifices spanning multiple centuries, with hundreds of
thousands of souls, to perpetuate American ideals, beginning
with the Framers, who as "outsiders" establish what becomes
the United States.

There is a debt of gratitude, beyond words, for what
Gerstner does, following what others have done before him,
and shall do afterward, through the stewardship of trust to
mend what is broken. And that nation says, thank you Lou,
thank you very much, for your integrity, indefatigable passion,
and can-do fortitude to pursue what is right and just and
fair, that compels internalizing these qualities in your fellow
citizens. Your faithfulness inspires that nothing is impossible,
for those who apply themselves, with conviction, effort, and a
patient resolve that all shall yet turn out well.

Gerstner-Realization, relative to IBM's "insider" culture,
is comparable to what happens to a "scientist" who comes
to the Manhattan Project, which is failing to produce a new
device for the government. The "scientist" helps overcome
major, development obstacles and scientific bottlenecks, to
successfully build the new device. And yet, in the end, others
never consider the "scientist" a "true" member of the team.
But as Dr. Richard Feynman, Nobel laureate, would counsel
such a person, "What do you care what other people think?"
Dr. Feynman applies this to himself while on the Manhattan
Project and, later, as a Caltech professor, including conducting

an employment interview for a future Caltech provost at a "topless" restaurant (Proverbs 6:26 and 31:3 advise caution to avoid such places). The merit in the "scientist's" contribution is that the device works, rather than issues about orthodoxy, philosophy, or *acceptable* ways to scientifically solve a problem. It's also like a "walk-on" quarterback who leads a team to the Super Bowl, multiple times, but will always be an "outsider," because he's a "walk-on," lacking acceptable pedigree.

Or it's as though the military is losing a war, and a "Western general," with success after success in "Western campaigns," receives command of the whole army, after promotions over "Eastern generals" who endure defeat after defeat in "Eastern campaigns." The "Western general" initiates strategies and tactics that his new "insider" command staff questions, yet fulfills. There are tremendous losses on both sides, drawing his adversary's men and materiel into a specific, geographic box of the "Western general's" own choosing. Pinning down hundreds of thousands of his adversary's men and related materiel is effective, but at great cost to both sides. Yet it allows the "Western general's" loyal subordinate to take only 60,000 men, and decimate his adversary's storehouses and factories, in the very heart of what his adversary claims as "homeland." Property, war materiel, factories, crops, and livestock, in a 50-mile wide swath, 200 miles in length, are destroyed or consumed, with few casualties. After three months, the movement places the 60,000 men near the now vulnerable, demoralized, starved, and pinned-down adversary. The "Western general's" unrelenting pounding holds his adversary in place, protecting his subordinate's completely successful flanking move, to swing the gate shut, stop the effusion of blood, end the war, and let everyone go home to tend their families and farms.

When Lincoln is promoting U.S. Grant (Grant), his "Western general," up through the ranks, higher-ranking officers complain, falsely, that Grant is a drunkard. Lincoln asks them, "What does Grant drink?"—so he can buy cases of it for the rest of them. Lincoln makes sure Grant doesn't feel like an "outsider," as Grant's efforts help end the scourge of war as quickly as possible. For Gerstner, the answer to Lincoln's question—"What does he drink?"—appears on page 40 of Gerstner's book, for the rare occasion circumstances may warrant. It remains written, "It is not for kings, O Lemmuel— not for kings to drink wine, not for rulers to crave beer, lest they drink and forget what the law decrees, and deprive all the oppressed of their rights." Proverbs 31:4,5 Yet there are a few leaders, such as Prime Minister Sir Winston Churchill, for whom some libation seems to have a salubrious effect.

The dictionary defines "merit" as "*value* that deserves respect and acknowledgment." The freedom to celebrate and embrace merit, and each *person* through whom merit becomes real and tangible, is at the heart of the unity of purpose between the Declaration of Independence and the Constitution. Losses of $5.5 billion and $8.1 billion may be tangible manifestations of a tremendous failure and breakdown of *process and procedure* by which an organization conducts itself. The merit Gerstner brings to IBM is seen through *principle and precept*, those which benefit from how Gerstner's parents teach him from an early age.

Letting *principle and precept* guide his life-walk allows Gerstner to seek out and discern value and its creation, independent of *where, through whom, or how* value creation originates. No one who creates objectively-measurable value is an "outsider" to Gerstner. He proves himself a friend to all who internalize this *principle*, with $9.7 billion in rewards and bonuses to IBM employees, as material witness to his conviction, putting in perspective Gerstner-Condition and $165 million for Specification. The structure Gerstner brings to IBM is so effective that it restrains the "insider"-tendencies of his successor, holding in check the company's old *process and procedure* habits that strangle value creation, often at its birth. Gerstner surmises this fact with a caution to his successor:

> Yes, I was always an outsider. But that was my job.
> I know [Gerstner CEO successor S. Palmisano]
> has an opportunity to make the connections to
> the past as I could never do. His challenge will
> be to make them *without going backward*; to know
> that the centrifugal forces that drove IBM to be
> inward-looking and self-absorbed still lie powerful
> in the company. *Who Says Elephants Can't Dance*, by
> Gerstner, page 282

The proverb that harmonizes with Gerstner's sentiment to his successor is:

> Hold on to instruction; do not let it go. Guard it
> well, for it is your life. Proverbs 4:13

Only a few years into Gerstner's successor's tenure, the Securities and Exchange Commission, with DOJ (SEC/DOJ), indicts IBM CEO S. Palmisano's (Palmisano's) designate-successor, IBM VP R. Moffet (Designate-Successor Moffet), who is an IBM "insider" and one of several "E5" designate-successors. The indictment is for insider trading: Palmisano Designate-Successor Moffet attempts to ascribe to himself value-creation abilities he does not have, by trading on insider information. After indictment, SEC/DOJ convicts Palmisano Designate-Successor Moffet, who leaves IBM to serve a federal prison sentence. Gerstner mentors by example for his successor, Palmisano, with commitment to management by *principle*. While Gerstner is CEO, no one is uncertain about Gertsner's position and trading on "insider" information. V. Rometty (Rometty), an IBM "insider," later becomes a new Palmisano Designate-Successor, after Designate-Successor Moffet's conviction.

a *7th-Gift* page

Akers manages by *process and procedure,* and sustains $13.6 billion in losses during his last two years as CEO. Gerstner, who manages by *principle and precept* during his entire tenure, reverses those losses and restores the firm to leadership and profitability. A few years after Palmisano, as an IBM "insider," becomes CEO, a federal conviction lands at his Designate-Successor's doorstep. There is a parable here that Gerstner may understand better than anyone else, inside or outside of IBM. Gerstner's mentoring and management by *principle,* during his tenure, helps sustain IBM's prosperity, making it possible for Palmisano's tenure as IBM CEO to have a firm foundation, economically and ethically. Rometty's tenure is one IBM CEO-generation removed from Gerstner's management-by-*principle.*

When Palmisano's CEO-tenure ends, his exit-package could easily reach $1 billion, when comparing his *untarnished* tenure to MCI Worldcom's Bernie Ebber's, Enron's Jeff Skilling's, Tyco's Dennis Kozlowski's, or Palmisano's own Designate-Successor R. Moffet's. Considering such contrasts, his exit package could more than surpass the packages of IBM's Thomas J. Watson Sr., Thomas J. Watson Jr., T. Vincent Learson, Frank T. Carey, John R. Opel, John F. Akers, and Louis V. Gerstner Jr.—*combined.* The actual amount he is sent away with, notwithstanding the precedent of Thomas J. Watson Jr.'s adherence to IBM president Al William's admonition, "Let's not be pigs," coincidentally is the same $165 million-level of Sneed's claim under Gerstner-Condition. Hypothetically, based on use, if any, of Sneed's Specification, this elevates its value as equal to Palmisano's exit-pay, or that Palmisano's exit-value is comparable to value claimed for Sneed's Specification in court filings. Through court filings, Palmisano, as a 14th Amendment *personation* with Rometty, denies use or even knowing what Sneed's Specification is. If otherwise, hypothetically, IBM, as an IBME™ dealing with Palmisano as a finite being, makes no exit-payout, in real dollars, so, hypothetically, Palmisano's exit costs IBM nothing.

Gerstner's Legacy and His Commitment to Principle With Gerstner-Condition

The U.S. economy provides economic rewards, independent of a *person's* 14th Amendment legacy. This includes *persons* receiving tremendous remuneration through sports, such as football, baseball, basketball, track, tennis, and golf; performance arts, like acting, singing, dancing, directing, producing, composing; playing various instruments, whether piano, horns, strings, percussion, wind, synthesizer or others; and literature, with fiction, non-fiction, or poetry. The outpouring of wealth to *persons* through these venues is proven, time and time again, independent of their 14th Amendment legacies.

Gerstner-Condition sets the legal basis to affirm this in a new way, at the very heart of perpetuating the nation's technological infrastructure. That infrastructure teeters on failure in 1997, due to Y2K, adding further to Gerstner's burdens while he addresses the residual effects of IBM's economic melt-down from only four years prior. In setting forth Gerstner-Condition, he creates the legal basis for a *person* who invents Specification, that can help Gerstner remedy Y2K for IBM's worldwide operations, to receive $165 million. And this precedent trumpets, for the first time, that the 14th Amendment not only extends *highest* free-market rewards to a *person* who is an athlete, a musician, singer, dancer, writer, producer, or entertainer, and familiar venues thereof. It also affirms that a *person* can still express the ideal of *liberty*, with value-creation for one of the most critical parts of the U.S. economy, to perpetuate the unity of purpose between the Declaration of Independence and the Constitution. Gerstner-Condition is consistent with *liberty* manifesting itself through *individual* initiative by a *person*, playing by the rules, adhering to *principle*, not trading on "insider" information, and creating value by applying science, technology, engineering, and math (STEM) disciplines.

Gerstner-Condition and a $165 million free-market precedent places an exclamation mark next to STEM, for a larger portion of the billions of man-hours and billions of *children*-hours pouring into sports, entertainment and other venues, every year. And it encourages the pursuit of service through merit, independent of 14ᵗʰ Amendment legacy. The U.S. already does so, with the best baseball and basketball and football players in the world. Its "STEM team" is even more critical for perpetuating the U.S. economy and *American* ideals. Gerstner-Condition competes with rap-artists, as *persons*, garnering hundreds of millions of dollars, with enticing displays of riches and life-style choices; Gerstner-Condition competes with athletes who publicly are showered with millions upon millions of dollars, for activities involving a ball, a puck, a bat, a pole, a club, a glove, a board, skis, or other objects; it competes with actors who receive up to $20 million, or more, for performing in a film, no matter what such performance may entail or reveal; or celebrities who receive tens of millions of dollars to endorse sport shoes, makeup, fast food, apparel, medications, alcoholic beverages, cigarettes, detergents, and other consumer items, some beneficial, many of questionable value, and others that may actually be harmful or ill-advised to repeatedly consume or use.

Gerstner-Condition also affirms the commitment to the unity of purpose between the Declaration of Independence and the Constitution for institutions sharing the legacy of such purpose. This includes private schools, like Cate School, Granite Computer Institute, or Caltech; as well as public sector institutions, like Cal State University Northridge, NASA, USPTO, or the US Postal Service; and private sector businesses, like Hughes (now Boeing) or others that may recognize merit, independent of a *person's* 14ᵗʰ Amendment legacy.

Donations are the mother's milk of private and public sector institutions, to validate past decisions, especially those from decades ago that reflect *principle and precept*, beyond a blind fixation on profits and portfolios and mechanized *procedures and processes.* A donation by a *person* celebrates such institutions and the historical narrative they carry into the future, about what is only possible through a physical *person,* with a soul and a body and a heart. This differs from anything a corporation may do, as a soulless, bodiless *"person"* under the 14th Amendment, with transactional-focus on its temporal, political or public relations gains or losses, quarter to quarter. Donations also celebrate *persons* who promote what is right and just and fair, selflessly, including *persons* like Caldwell, Ryan, Clark, Cate, Brown, Browne, Christy, Wilzbach, Prouty, Lidow, Gerstner, and others who nurture merit, and rejoice with it, *wherever* and through *whomever* they may find it.

USPTO awards Sneed a second patent, #7546982, which he designates BEST2000B™, letting his first patent, #6236992, become BEST2000A™. BEST2000B™ fulfills aerospace and military needs, for real-time trauma detection to flight-surfaces on combat, commercial, and research aircraft or spacecraft. This patent celebrates Homer J. Stewart, former chief of NASA/JPL's advanced technical studies office and Sneed's Caltech freshman advisor; and Charles Ray, Sneed's advisor after his freshman year; and Chuck Wilts, a gifted educator and Caltech electrical engineering professor who mentors Sneed's independent-study project; and Dean Ray Owen, Fred Thompson, Richard Dean, James Mayer, a Caltech applied physics professor who, along with James Black, a Caltech staff manager, embrace Sneed as a *person,* as well as a Caltech undergrad (Stewart-Ray-Wilts-Owen-Thompson-Dean-Mayer-Black Fulfillment); or soccer coach Cameron, who makes Sneed the starting striker, allowing him to become co-captain and scoring leader, including three goals in one game, the fourth hitting the cross bar. For one of Caltech's best seasons in decades, Sneed becomes a varsity letterman and receives honorable mention on the all-league team of the Southern California athletic conference Caltech is a member.

Sneed v. IBM: Decoding a $1.86 Billion Federal Riddle With Caltech's Honor Code, Pro Se

Lincoln Fulfillment, of 1863, engages 180,000 *persons*, to apply 19ᵗʰ century technology—in the form of armaments—to help perpetuate the unity of purpose between the Declaration of Independence and the Constitution. Both Lincoln and, contemporarily, Gerstner, endure "outsider" status in their respective eras, yet each gains respect in carrying out his duties. Gerstner-Condition is witness that 134 years after Lincoln Fulfillment, in 1997, one *person*, with a mind and a soul and a body and a heart, can step forward to help perpetuate that same unity of purpose by inventing something of value. And, in 1997, this is done through peaceful means, by pen and paper, with thought and reflection. Rather than military conscription and might of arms, a *person* exercises the *freedom* to create value, an inalienable *precept* of the Declaration of Independence. The Constitution enshrines this by nurturing invention, which Gerstner-Condition gives supporting witness is *real*, as a *person* is *real*.

Though Gerstner puts Gerstner-Condition in place before making way for a new CEO, the new CEO finds no evidence IBM uses Specification. Nor does Gerstner disclose any use before his retirement and Gerstner-Realization. *Sneed v. IBM, pro se*, goes inside the corridors of power at IBM, and inquiry through the federal court system. Three Caltech alums, with notice of Caltech's Honor Code, urge Sneed to seek relief, which Sneed does under Gerstner-Condition for $165 million. None of the three alums provides evidence IBM uses Specification. Caltech's Honor Code creates the most favorable outcome possible to fulfill Gerstner-Condition, as a federal judge elevates Caltech's Honor Code to probative status, setting an extraordinary legal precedent under the U.S. Constitution during federal court proceedings.

One of the three Caltech alums, urging Sneed to seek relief, is the most prolific VP in IBM's history—and knows

Gerstner and Palmisano. Both the alum and Gerstner are under IBM's executive confidentiality agreement to say nothing that helps Sneed's case. Another is an IBM customer executive and Caltech board member who implements Sneed's Specification on IBM computers in seven countries, after his staff consults with Gerstner's staff. The third is an international venture capitalist, who has ready-access to top executives in the Fortune 500, and gives sworn testimony of Specification's successful use on IBM computers in seven countries.

Two, Gerstner successors—S. Palmisano and V. Rometty, through legal proxies—construe themselves and IBM as a $100 billion, disadvantaged, New York "*person*" under the 14th Amendment (Palmisano-Rometty *Personation*). This allows Palmisano-Rometty *Personation* to flee the California state court system, into a more "*person*"-friendly federal court. Thus, Palmisano-Rometty *Personation* avoids a California state court summons as a New York "*person*," and disadvantage it perceives in California state courts—courts it readily uses to sue California *citizens,* for decades, and collect on them. In federal court, it rejects Gerstner's *conduct-by-principle* and uses its "Mad-hatter," *conduct-by-procedure* defense, claiming no prior contact *whatsoever* with Sneed about Y2K. Gerstner-Realization encompasses Palmisano-Rometty *Personation* as a true IBM "insider," which Gerstner makes clear by declaring he is an "outsider" relative to Palmisano-Rometty *Personation.*

Palmisano-Rometty *Personation* demands the court rule it is immune to Caltech's Honor Code's *principle* that, "No member of the Caltech community shall take unfair advantage of another member of the Caltech community," regarding obligation to disclose use, if any, of Sneed's Specification. Sneed includes Caltech's Honor Code in his federal court filing, consistent with Gerstner's *management-by-principle,* which informs Gerstner-Condition and Gerstner-Realization. Palmisano-Rometty *Personation* has legacy community-affiliation with Caltech through Caltech's Board of Trustees and former IBM CEO Thomas Watson Jr., also an IBM "outsider," like

a 7th-*Gift* page

Gerstner, at the time of his Caltech board membership. Sneed is a Caltech Alumni Association Life-member, and therefore Sneed gives Caltech Honor Code notice to the three Caltech alums urging him to seek relief, and to Palmisano-Rometty *Personation*, reflecting Gerstner-Condition and Gerstner's *management-by-principle.*

Palmisano-Rometty *Personation* argues in federal court that Gerstner-Condition never exists, and claims that even if it does it's non-binding upon Palmisano-Rometty *Personation* in any way. It also asserts that Sneed, somehow, receives prior, *unwritten, unspoken notice* of Specification's use by Palmisano-Rometty *Personation*, while *simultaneously* arguing Palmisano-Rometty *Personation never uses Sneed's Specification*— and declares this within the *same* court filing. It makes claim of prior *unwritten, unspoken notice* to Sneed regarding use of his Specification, to demand the court time-bar Gerstner-Condition as outside the statute of limitations. The court doesn't grant Palmisano-Rometty *Personation's* demand regarding the written Gerstner-Condition.

In what follows Gerstner-Condition's fulfillment, may this inspire another *person's* initiative and wherewithal, in new ways, today, as well as in the future, even for an embryo not yet conceived, who becomes a newly born *person*, as Heaven's purpose continues to be fulfilled from long ago.

What follows is on a need-to-know basis only, regarding Caltech's Honor Code and discovery documents that are part of the public record from federal court proceedings. The casual reader may skip to the next appendix, page 493.

White House/Caltech/IBM Wilzbach/Sneed Communication

Discovery and other Material From *Sneed v. IBM*

FAX TO: 202-XXX-XXXX   **CONFIDENTIAL**

FOR: Honorable President Barack H. Obama
         c/o Honorable Attorney General Eric H. Holder
FROM: Anthony Sneed, ISB:·Institute Fellow, *Emeritus*
DATE: d2/January 21, 2013

RE: *Sneed v. IBM*: Half-time Report

Number of Pages including this one: ___28___

FINAL DRAFT, with year reference-edits for dates events occur, and on page 25,

> "[A] member of the Caltech community shall not take unfair advantage of another member of the Caltech community[.]"

Anthony Sneed
2058 North Mills Avenue
Claremont, CA 91711
909-XXX-XXXX

Dr. Jean-Lou Chameau, Caltech President   d3/January 1, 2013
1200 E. California Blvd, Pasadena, CA   PRESIDENT'S FINAL BRIEF VIA MAIL

RE: Executive Summary, *Sneed vs. IBM* and Caltech's Honor Code

    1.   On d6/October 28, 2011, Caltech alum Anthony Sneed (Sneed) petitions the California judiciary for $165 million in relief for his Y2K invention, BEST2000™ (Specification), under IBM's written condition to pay Sneed for its use (Agreement). L.A. County Court case KC062442 elevates to U.S. Central District case CV 11-010217GW, *Sneed vs. IBM*. IBM demands the Court rule Agreement invalid or time-barred: the Court grants neither.

    2.   From October 2007 to April 2011, three Caltech alums deem Agreement suitable for relief: Caltech board member Dr. Alex Lidow; retired IBM Vice President Mr. Pete Wilzbach; and international venture capitalist Dr. Dale Prouty. Beginning d6/October 19, 2007, Sneed sends them written notices their advice reflect Caltech's Honor Code. All three accept notices without dissent. By December 2007, all advise Sneed to seek relief.

    3.   Their advice is consistent with communication from IBM CEO L. Gerstner (Gerstner) to Sneed, beginning in June of 1997. Gerstner's and his staff's written condition (Gerstner-Condition) sets performance for Sneed to meet before Gerstner and his staff compensate Sneed for IBM's use of Specification. Gerstner directs his staff to send Sneed Gerstner-Condition, which Sneed receives on IBM letterhead with a d6/June 20, 1997 date.

    4.   Gerstner-Condition memorializes contract formation terms prior to finalizing Gerstner-Condition: $49,764.55 per site for production-use; and 14 cents per line of code for the aggregate number of lines of code in applications Gerstner's staff applies Specification, with no charge beyond one billion lines of code. A patent issuing to Sneed is the final contract formation step Gerstner-Condition enshrines for Sneed to receive compensation.

5.   Gerstner's staff accepts and uses a copyright-restricted, evaluation-only copy of Specification from Sneed, and confirms Specification's merits to address Y2K: (1) It preserves the two-digit year-format for software evolving from the 1950s to the present; (2) it accommodates data files that age more than 100 years; (3) it allows indefinite future use; and (4) it helps automate manual remediation for billions of lines of code that have Y2K exposure.

6.   By mid-1997, Gerstner and his staff have less than 24 months to repair over a billion lines of code worldwide. Sneed learns Gerstner's Y2K remediation alternatives restrict Gerstner to well under 100-year age-limits for data, or they trigger Y2K-all-over-again within a period of time, or they require manual conversion for billions of lines of code. Gerstner's staff confirms Specification's merits before signing and sending Sneed Gerstner-Condition.

7.   Prior to finalizing terms, Gerstner's staff tells Sneed by phone to: (1) Set a price of up to 50 cents per line of code, leaving margin for IBM's up to $2 per line of code to bill itself, internally, as well as to charge its clients; and (2) make it easy to assess the aggregate lines of code in software Gerstner's staff applies Specification. Gerstner's staff phones Sneed approvingly upon receiving his 14 cents per line of code price, with no charge after an aggregate of a billion lines of code. <u>Gerstner's staff memorializes (1) and (2) in Gerstner-Condition, with its last condition being that a patent issues to Sneed prior to compensation.</u>

8.   Gerstner assigns Mr. Al Torressen (Torressen) as Sneed's contact (Gerstner-Contact). Gerstner-Contact Torressen signs Gerstner-Condition which stipulates: (1) Sneed only communicate with Gerstner-Contact regarding any worldwide use of Specification by Gerstner and his staff; (2) Gerstner and his staff will not expropriate Specification; (3) Gerstner and his staff admit to evaluative-use of Specification as of d6/June 20, 1997; and (4) Sneed must perform for compensation, by notifying Gerstner-Contact when Sneed's patent issues.

Executive Summary for the President, *Sneed vs. IBM* and Caltech's Honor Code

3

9.   Four months later, in November of 1997, International Rectifier Corporation (IRC) CEO Dr. Alex Lidow (Lidow) and his staff enter agreement with Sneed to use Specification on IRC's IBM AS/400 computers in France, England, Germany, Japan, Singapore, China, and the U.S. Before contracting with Sneed, Anderson Consulting (Anderson) presses Lidow and his staff to sign a multi-year, "best efforts," Y2K contract for up to $10 million or more. It also requires IRC hire a number of Anderson's outside contractors.

10.   At Lidow's staff's request,   Anderson vets Sneed's Specification and validates that it works, but urges IRC to sign Anderson's contract anyway. Sneed facilitates contact between Lidow's staff and Gerstner's staff.   Gerstner's staff affirms Sneed's Specification works, but urges Lidow's staff to use IBM's alternatives instead of Specification.

11.   By April 1998, Lidow and his staff apply Specification on IRC's IBM AS/400 computer in Singapore. With it, they repair 1.2 million lines of code in six weeks. IRC's other worldwide sites rapidly follow, well before Y2K, with no post-remediation issues.

12.   In February 1999, Dr. Dale Prouty (Prouty) becomes an independent witness to Specification's merit and use by IRC. He and Sneed meet 20 years earlier while working on the NASA Galileo mission to Jupiter at Hughes (now Boeing).   Prouty visits IRC and vets Specification's merits with Lidow's staff.   Prouty confirms Specification's use in seven countries and directs Sneed to seek relief, which Lidow also encourages by April 1999. Lidow approves the merits of Sneed's efforts and, later, Sneed's receipt of a $236,000 award.

13.   Sneed also facilitates a $400,000 IRC contract for Ms. Florence Lee, a Taiwanese-born relational database expert. She re-architects and integrates Lidow's worldwide IBM AS/400's with relational databases and server technology. This helps Lidow reconcile his daily worldwide financials. IRC goes on to grow from $0.7 billion to $1 billion+ in revenues.

a *7th-Gift* page

14.   On d3/May 22, 2001, the United States Patent and Trademark Office (USPTO) issues Sneed US patent 6,236,992 for Specification, affirming its Y2K merits: (1) It preserves the two-digit year-format for software evolving from the 1950s to the present; (2) it accommodates data files that age more than 100 years; (3) it allows indefinite future use; and (4) it helps automate manual remediation for billions of lines of code that have Y2K exposure.

15.   On d3/June 19, 2001, Sneed performs per Gerstner-Condition and notifies Gerstner and Gerstner-Contact that US patent no. 6,236,992 issues to Sneed.   Six months later, on d3/January 29, 2002, Gerstner announces his retirement as CEO and IBM VP S. Palmisano (Palmisano) replaces him three months later. Palmisano's staff force Gerstner-Contact Torressen to later leave IBM and seek employment elsewhere. Before   replacement and leaving IBM, neither man acknowledges Sneed's notifications under Gerstner-Condition.

16.   Between 2002 and 2007, Palmisano's contacts repeatedly change (Palmisano-Contact). Palmisano-Contact-1 tells Sneed, by mail d2/June 24, 2004, to disregard Gerstner-Contact Torressen.   Three months later, on d3/September 28, 2004, Palmisano-Contact-1 declares no payment due under Gerstner-Condition for evaluative-use. By mail d2/November 20, 2006, Palmisano-Contact-2 tells Sneed to ignore Palmisano-Contact-1 and declares no payment due under Gerstner-Condition for evaluative-use. From 2005 to 2007, Lidow and Prouty urge Sneed not to drop Agreement, before, during, or after Palimsano-Contact denials.

17.   Sneed calls IBM VP P. Wilzbach (Wilzbach) October 2007, who is first to confirm receipt of Specification at IBM. On IBM letterhead, d3/May 13, 1997, Wilzbach encourages Sneed's patenting effort. Wilzbach and Palmisano are peer, "E-5" IBM VP's at Wilzbach's subsidiary. A month later, d6/June 20, 1997, Sneed receives Gerstner-Condition. By d6/December 31, 1999, Palmisano is president of the subsidiary, and Wilzbach leaves IBM.

Executive Summary for the President, *Sneed vs. IBM* and Caltech's Honor Code

5

18.   Wilzbach and Sneed first meet in 1980, when Wilzbach hires and mentors
Sneed as an IBM trainee, guiding Sneed's election to president of the 6,000 member Western
Region IBM Club. Before IBM, both successfully contribute to NASA missions, Wilzbach on
Mariner/Mars and, seven years later, Sneed on Pioneer/Venus and Galileo/Jupiter. They serve
as undergrad interns on NASA missions. Hughes (Boeing) hires both after Caltech graduation.

19.   In their October 2007 call, Wilzbach tells Sneed he quit IBM d6/December
31, 1999--on Y2K's eve. Prior to leaving, near only the age of 50, Wilzbach closes the largest
outsourcing contract in IBM's history. They update each other since last speaking in 1997.

20.   After reviewing Y2K contact between Sneed, Gerstner, Palmisano, and their
staffs, Wilzbach tells Sneed to settle directly with Palmisano, and to: (1) Write a one page
settlement letter, (2) choose a reasonable settlement amount, and (3) avoid litigation with IBM
if at all possible. Wilzbach tells Sneed his advice is informal as a friend, not as an advocate.

21.   Sneed submits the settlement letter (Settlement Letter) to Palmisano on
d6/October 19, 2007 for $1,194,349.20, per Wilzbach's informal, friendly advice, and cc's
Wilzbach, Lidow, and Prouty. Sneed gives notice that Caltech's Honor Code informs contact
among the alums. Wilzbach, Lidow, and Prouty accept the letter and its notice without dissent.

22.   Palmisano-Contact-2 writes Sneed 10 days later, on d2/October 29, 2007,
declaring he will no longer respond to Sneed's inquiries under Gerstner-Condition and declares
only evaluative-use of Sneed's Specification. This potentially breaches (Breach) Gerstner-
Condition for Sneed to perform by initiating inquiry, dating back to June 1997. Breach is after
Sneed cc's Wilzbach, Lidow, and Prouty on the d2/October 19, 2007 Settlement Letter to
Palmisano with its Caltech Honor Code notice. Palmisano-Contact-2 doesn't confirm
Palmisano, himself, directly receives Settlement Letter, which disregards Wilzbach's advice.

23.   Sneed's follow-up letter to Palmisano, on d2/November 5, 2007, states that if

Breach is from Palmisano, it is inquiry notice (Notice) and disregards Gerstner-Condition's

inquiry performance, since discovery of use may still occur after 2007 due to IBM's size.

24.   Palmisano-Contact-2 writes Sneed d5/November 8, 2007, three days later,

recanting Notice, and declares Palmisano approves the denial of use (Palmisano-Denial) in

response to Sneed's inquiry and performance under Gerstner-Condition dating from June 1997.

25.   On d4/December 2, 2007, Sneed phones Wilzbach and updates him.

Wilzbach tells Sneed to sue Palmisano. Sneed has no proof of use, and IBM's written denials

date back to June 1997. Sneed asks if Wilzbach might act as liaison to Palmisano for resolving

Agreement. Wilzbach says he can't, and tells Sneed guidelines for how to sue Palmisano.

26.   At or about this time, the US Department of Justice (DOJ) indicts

Palmisano's friend and associate, IBM VP B. Moffet (Moffet). Moffet is candidate to succeed

Palmisano, prior to Moffet's conviction, federal prison sentence, and departure from IBM.

Also, at or about this time, the US Supreme Court issues the *Twombly* ruling (see ¶ 67 below).

27.   On d5/December 3, 2007, Sneed sends Wilzbach a written summary of

Wilzbach's guidelines (Guidelines) to sue Palmisano, from their d4/December 2, 2007 phone

call. It includes notice of Caltech's Honor Code regarding Wilzbach's advice. Sneed cc's

Palmisano, Prouty, and Lidow; all accept Guidelines and notice without dissent.

28.   In the 80s, as his training manager, Wilzbach mentors Sneed to take on

challenges beyond expectations for trainees: interfacing a bank's datacenter with its branch

office system; troubleshooting a workflow bottleneck for a statewide ATM network; and acting

as IBM marketing principal to successfully restore an IBM customer datacenter processing $32

billion in federal reserve float, after a municipal power surge destroys it. All are successful.

Executive Summary for the President, *Sneed vs. IBM* and Caltech's Honor Code

7

29.   IBM CEO J. Opel (Opel) reviews Sneed's performance as acting IBM marketing principal to restore the IBM customer datacenter processing $32 billion in federal reserve float every four days. Opel determines Sneed, though only a trainee, performs within IBM policy. The grateful financial institution attempts to recruit him. Sneed graduates with honors from IBM's New York Marketing School and is first to achieve 400% of quota in IBM's US Western Region the following year. He receives IBM's regional recognition award.

30.   With prior notice of Caltech's Honor Code, Wilzbach's Guidelines to sue Palmisano harmonize with Prouty's and Lidow's expectation.   Caltech's Honor Code precludes them from setting expectation they know is flawed, misleading, or without merit.

31.   Sneed accepts their new expectation in context with past ones: his parents Odis and Doris Sneed's expectation he master a computer simulator they give him at age 10; Prouty's mentoring that leads to Sneed's successful development of an operating system and simulator for the Galileo/Jupiter mission; Wilzbach's mentoring Sneed as a trainee; Lidow's mentoring to apply Sneed's Specification worldwide; Gerstner's expectation for Sneed's patent.

32.   Gerstner is a fellow Harvard alum of Sneed's boarding school headmaster, Frederick F. Clark (Clark). During World War II, Clark lands on the beaches of Normandy, resisting fascism and *predestined antimeritism*. Clark chooses Sneed for Cate's Headmaster's Award at the school's 1973 commencement; its 89-year-old founder Curtis W. Cate, a Harvard alum, and Sneed's faculty benefactors, Joseph I. Caldwell and Sanderson M. Smith, also attend.

33.   Wilzbach's 2007 expectation to seek relief resists *predestined antimeritism*, herein the use of any belief or practice, by law or under color of authority, to retro-actively devalue or deny the worth or merit of another's life, work, or writings, by demurring another's legal existence for purposes of historical revisionism, economic exploitation, or other harm.

Executive Summary for the President, *Sneed vs. IBM* and Caltech's Honor Code

8

34.   By December 2007, Wilzbach, Lidow, and Prouty share the same expectation that Sneed sue Palmisano (Wilzbach-Lidow-Prouty Expectation). Under Caltech's Honor Code, this creates reasonable doubt to Sneed about Palmisano-Denial. However, Palmisano-Contact-2's d5/November 8, 2007 letter recants Notice and resumes the pattern and practice of denials dating from 1997. Wilzbach-Lidow-Prouty Expectation suggests this to be misleading.

35.   Gerstner-Condition stipulates Sneed receive disclosure about IBM's use of Specification, worldwide, from Palmisano-Contact-2, and no one else at IBM, or agreement to pay Sneed breaches. For Wilzbach to do any more than he has may expose him to IBM's executive confidentiality agreement, with punitive consequences if Wilzbach breaches it.

36.   Wilzbach encourages Sneed to sue IBM through Wilzbach-Lidow-Prouty Expectation, without details of any specific use of Specification by IBM. Lidow knows of use, but only as an IBM customer executive for his IBM AS/400 sites in seven countries. Prouty provides sworn testimony of use, but only for IBM AS/400 sites outside of IBM.

37.   Gerstner-Condition requires Sneed perform and initiate inquiry; the Specification's minimum useful life is 28 years. Sneed performs approximately every 24 to 36 months, after a patent issues, per Gerstner-Condition and its declaration that IBM won't expropriate Specification. If there's use without payment, IBM breaches Gerstner-Condition.

38.   Palmisano-Denial contradicts Wilzbach-Lidow-Prouty Expectation from December 2007 to April 2011 (Contradiction). Contradiction is extant for over 36 months, with no dissent from Caltech's Honor Code notices to Wilzbach, Lidow, Prouty, and Palmisano.

39.   On d2/April 4, 2011, the anniversary of Dr. Martin Luther King's assassination and witness thereof to homicidal *predestined antimeritism,* Sneed makes inquiry (Antimeritism Inquiry), as is pattern and practice under Gerstner-Condition. Sneed also tells

Palmisano what he tentatively concludes Contradiction means after it is extant, for over three

years, without dissent from Wilzbach, Lidow, Prouty, or Palmisano (Tentative Conclusion):

> By 2007, [Sneed] learned [Wilzbach] retired from IBM. [Wilzbach], as a retired IBM
> EVP, corroborated [Dr. Alex Lidow's] opinion, as an IBM customer executive, that
> [Specification's] use on computers *inside* of IBM merited payment to [Sneed]. Both
> [Wilzbach] and [Dr. Alex Lidow] are Caltech alums, [Dr. Alex Lidow] a member of the
> Caltech board of trustees. In addition, Dr. Dale Prouty ("PROUTY"), a third Caltech
> alum, vetted [Specification] with sworn testimony of its use and economy on IBM
> computers.   d2/April 4, 2011 letter from [Sneed] to Palmisano

40.   Sneed sends Antimeritism Inquiry and Tentative Conclusion to Palmisano,

Wilzbach, Lidow, Prouty, and Caltech President Dr. Jean-Lou Chameau (Chameau), along with

a settlement invoice (Invoice) for $1,492,936.50, benefiting from Wilzbach's advice to avoid

litigation if at all possible.   Sneed includes Caltech President Chameau, along with Caltech

board member Lidow, because of their supremacy in upholding Caltech's Honor Code.

41.   At or near first Settlement Letter, Palimsano and Lidow memorialize a major

contribution to Caltech with a plaque on Fleming House--Lidow's and Sneed's undergrad dorm

where they meet in 1974. It has two names: Lidow's and former Caltech board member and

IBM CEO Thomas Watson, Jr.'s. Lidow invites Sneed to Caltech's gathering to celebrate this.

42.   After Antimeritism Inquiry, Palmisano ceases denials.   Wilzbach vets

Tentative Conclusion in a d1/June 5, 2011 letter (Wilzbach-1 Letter). He informally tells

Sneed that as a friend without first-hand knowledge, he assents to suing Palmisano after vetting

what *Sneed Thinks*. He asks Sneed not to cc him on new letters to Palmisano, per pattern and

practice since 2007, or it may later trigger disclosure-leverage Palmisano has over Wilzbach.

43.   Wilzbach doesn't preclude cc'ing him on follow-up with Caltech or a court.

Wilzbach adheres to Wilzbach-Lidow-Prouty Expectation after vetting what *Sneed Thinks*

about Contradiction. This, juxtaposed to Palmisano-Denial for over three years, and Palmisano

Executive Summary for the President, *Sneed vs. IBM* and Caltech's Honor Code

10

a *7th-Gift* page

ceasing pattern and practice of denials after Antimeritism Inquiry, leads to Sneed's final conclusion to seek relief under Gerstner-Condition (Final Conclusion).

44.   Sneed's alternative is to set aside Wilzbach-Lidow-Prouty Expectation, ignore Caltech's Honor Code informing it, and defer to the October 2007 Palmisano-Denial and other denials as true since 1997. But this disregards Palmisano's April-May 2011 cessation of denials after more than a decade--prior denials arrive in as little as 10 days or less.

45.   Sneed would be taking unfair advantage of Wilzbach, Lidow and Prouty if he sets aside their good faith advice. Sneed believes a claim against Palmisano is probative if a court examines Palmisano-Denial under Caltech's Honor Code and recent cessation of denials.

46.   Per Wilzbach's earlier advice, Sneed attempts modest settlement inquiry in correspondence with Palmisano, d6/July 15, 2011 and d6/September 16, 2011. Sneed cc's Chameau, Lidow and Prouty, but not Wilzbach, per Wilzbach's direction to avoid triggering disclosure-leverage Palmisano has over him.

47.   Continuing into October 2011, Palmisano maintains the recent cessation of pattern and practice of denials since 1997. On d6/October 7, 2011, Sneed corresponds with the Honorable President Barack Obama, via the Honorable US Attorney General Eric Holder about Agreement. Neither sends objection to Wilzbach-Lidow-Prouty Expectation or sensitivity to Palmisano as a federal contractor. Sneed cc's Chameau, Lidow, and Prouty, but not Palmisano or Wilzbach, per Wilzbach's caution of triggering Palmisano's disclosure-leverage over him.

48.   Sneed personally drives the petition for relief to the California judiciary, Los Angeles County Court, on d6/October 28, 2011, which becomes case KC062442, *Sneed vs. IBM*. Sneed does so in the event Palmisano attempts to construe the d6/October 29, 2007 Breach--which he recants 10 days later--to trigger a four-year California statute of limitations.

49.   Palmisano flees state court and, by legal proxy, uses 28 USC 1332 (1332) to construe himself and his $100 billion firm as a disadvantaged, New York *person* under the 14ᵗʰ Amendment (Palmisano-*Personation*). Palmisano-*Personation* demands removal to federal court, to obviate any unfair advantage Sneed may have as a California *citizen* in a California court under state law. It doesn't disclose retaining California's most formidable law firms, for decades, to sue California *citizens*, using the economy of the same state courts it now flees.

50.   Palmisano-*Personation* construes itself a disadvantaged _person_ under law:

> No State shall make or enforce any law which shall abridge the privileges or immunities of *citizens* of the United states; nor shall any State deprive any *person* of life, liberty, or property, without due process of law, nor deny to any *person* within its jurisdiction the equal protection of the laws.   US Constitution, 14ᵗʰ Amendment, Section 1

The US Congress ratifies this amendment in July of 1868, not for Palmisano-*Personation*--but to remedy harm to *actual people* from Coerced Uncompensated Labor Tradition (CULT) practices, legal in the US prior to ratifying the 13ᵗʰ Amendment in December of 1865.

51.   Such practices rely on *predestined antimeritism,* by embryo CULT *pre-assignment,* under law, in *some* US jurisdictions prior to 1865. CULT *pre-assignment's* intent is to retro-actively devalue or deny the worth or merit of a born person's life, work, or writings, by demurring a person's legal existence for purposes of economic exploitation or other harm. *Pre-assignment* is only feasible after birth, yet is law when an embryo first forms under heaven.

52.   By December 1865, Congress ratifies the 13ᵗʰ Amendment, making such practices a crime in *every* US jurisdiction. Some 670,000 lives are lost to resolve that no one, by law or under color of authority, may use embryo CULT *pre-assignment* to demur the legal existence of a born person, for purposes of economic exploitation or other harm. The 14ᵗʰ Amendment obviates past CULT *pre-assignment* criteria for *any* embryo leading to a US birth.

53.   Palmisano-*Personation* uses the 18 US 394 (394) opinion of 1886, in *Santa Clara County v. Southern Pacific Rail Road Company*, to *self*-assign its CULT-*identity* as a disadvantaged *person* under the 14th Amendment, though it is never a *pre-assignment* embryo:

> The court does not wish to hear argument on the question whether the provision in the Fourteenth Amendment to the Constitution, which forbids a State to deny to any *person* within its jurisdiction the equal protection of the laws, applies to these corporations. We are all of opinion that it does.   Justice M. Waite's 394 *pronouncement*.

54.   The 14th Amendment's original intent protects Sneed's ancestors, and many other *persons*, from *predestined antimeritism* by embryo CULT *pre-assignment*, which the 13th Amendment makes a crime by 1865.   After 1886, Immortal-Bodiless-Mythical Entities (IBME's™) exploit CULT-*identity self*-assignment as such *"persons"*: Immortal because their existence is unlimited; Bodiless because they can't be imprisoned; Mythical because they are a legal fiction.   With 394, IBME's™ begin proliferating on Earth as an *immortal precedent* at odds with Exodus 20:4-6, *"You shall not make for yourself an idol in the form of anything*[.]"

55.   If Palmisano-Denial is false, Palmisano-*Personation* is using the 14th Amendment--as a disadvantaged, New York *person*--to demur Sneed's legal existence in a California court, under color of authority as a US government contractor, subject to 18 US 242:

> 18 US 242 Deprivation of Rights Under Color of Authority: Whoever under color of any law, statute, ordinance, regulation, or custom, willfully subjects any *person* in any State, Territory, Commonwealth, Possession, or District to the deprivation of rights, privileges, or immunities secured or protected by the Constitution or the Laws of the United States, or to different punishment, pains, or penalties on account of such *person* being an alien or by reason of his color or race that are prescribed for the punishment of *citizens*, shall be fined under this title or imprisoned not more than one year, or both.

56.   The honorable, Federal Judge George H. Wu (Honorable Judge Wu) takes up federal case CV 11-010217GW (217GW), *Sneed v. IBM*, in his Central District, Western Division court. "GW" is Honorable Judge Wu's initials on all cases he accepts in his Court.

57.   In his d6/October 28, 2011 filing, Sneed reviews Palmisano-*Personation*'s cessation of denials dating back to June 1997, after it receives the d2/April 4, 2011 Antimeritism Inquiry under Gerstner-Condition, with prior vetting by an "unnamed" IBM VP.

58.   In its December 2011 filing and subsequent ones, Palmisano-*Personation* asserts there's no written agreement or proof it has or is using Specification, so it owes nothing.

59.   On or about d2/January 2, 2012 Palmisano retires and VP V. Rometty (Rometty) replaces him as CEO. By legal proxy, 394 perpetuates the IBME™ with Rometty and parties thereof as the same disadvantaged, New York *person* (Rometty-*Personation*).

60.   Federal Rules of Civil Procedure (FRCP) 26(f) requires a conference and a joint report prior to a first hearing. At the d3/January 10, 2012 26(f)-conference, Rometty-*Personation* declares it CAN NOT AGREE (CNA) with, and rejects these 26(f)-report areas:

    60.1 CNA it knows Sneed's patent issue date or that Sneed is the inventor.
    60.2 CNA it receives Specification from Sneed in 1997.
    60.3 CNA it corresponds with Sneed in 1997, or if such correspondence exists now.
    60.4 CNA there is any basis to pay Sneed, or that Gerstner-Condition ever exists.
    60.5 CNA it assigns anyone as Sneed's contact in 1997.
    60.6 CNA it is using anything like Specification, or knows what Specification is.
    60.7 CNA with any initial 26(f) discovery request for methods it uses to address Y2K.

61.   Rometty-*Personation* files a Reply after Sneed's Opposition to its FRCP 12(b)(6) "Motion to Dismiss for Failure to State a Claim." At the d5/January 23, 2012, 12(b)(6)-hearing, CNA informs Rometty-*Personation*'s denunciation of Gerstner-Condition (Rometty-Denunciation), while arguing that *if it does exist*, statute of limitations time-bars it.

62.   Sneed explains that Gerstner approves Gerstner-Condition in writing, d6/June 20, 1997, through Gerstner-Contact Torressen. Gerstner-Condition requires Sneed perform by notifying Gerstner-Contact when a patent issues as final condition to receive any payment. Sneed performs with notice after USPTO issues Sneed US patent 6,236,992, d3/May 22, 2001.

63.   Sneed adds that, Rometty-*Personation* denies use and admits to only evaluative-use from 1997 to 2011, inclusive of Palmisano-Denial. Gerstner-Condition requires no payment for evaluative-use. But by d2/April 4, 2011, three fellow Caltech alums set expectation for Sneed to seek relief. One is an "unnamed" IBM VP who assents to <u>Sneed's</u> Tentative Conclusion to seek relief under Gerstner-Condition.

64.   Sneed further explains that Rometty-*Personation* receives Sneed's Tentative Conclusion without dissent, ceasing its pattern and practice of denials dating back to 1997. This leads to Sneed's Final Conclusion to seek relief because Rometty-*Personation* is breaching Gerstner-Condition through a misleading Palmisano-Denial. Furthermore, Rometty-Denunciation of Gerstner-Condition contradicts Palmisano-Denial *under* Gerstner-Condition.

65.   During the 12(b)(6)-hearing, Honorable Judge Wu announces he is a former California state judge and that Sneed's filing could go forward in a California court, but Rometty-*Personation* demurs this by removal as a 14th Amendment, disadvantaged *person*. He tells Sneed how his filing may proceed in federal court: (1) Obtain from the "unnamed" IBM VP, who may remain so, a direct statement of Specification's use by Rometty-*Personation*, instead of his assent for Sneed filing after vetting what <u>*Sneed Thinks*</u>; or (2) file before 2007.

66.   Honorable Judge Wu explains that under 12(b)(6), "[A] court must (1) construe the complaint in the light most favorable to the plaintiff [Sneed], and (2) accept all well-pled factual allegations as true, as well as all reasonable inferences to be drawn from them," citing *Sprewell v. Golden State Warriors*, 266 F.3d 979, 988 (9th Cir. 2001).

67.   He further states that after 2007, federal courts, but not California state courts, become subject to *Bell Atlantic Corp v. Twombly* (*Twombly*), 550 U.S. 544, 556 (2007): "A claim has facial plausibility when the plaintiff [Sneed] pleads factual content that allows the

Executive Summary for the President, *Sneed vs. IBM* and Caltech's Honor Code

15

court to draw the reasonable inference that the defendant is liable for the misconduct alleged."

Ashcroft v. Iqbal (Iqbal), 556 U.S. 663,___, 129 S.Ct. 1937, 1949 (2009), which *Twombly* cites.

68.   Rometty-*Personation* argues that the "unnamed" IBM VP assenting to file from what *Sneed Thinks* is not factually the same as declaring Rometty-*Personation*'s use of Specification. It concludes Sneed fails *Twombly*. To remedy this, Honorable Judge Wu instructs Sneed to provide a direct statement of use from the "unnamed" IBM VP or other source, by d5/Feburary 23, 2012, with a status-hearing on d5/March 1, 2012.

69.   Honorable Judge Wu asks Rometty-*Personation* to disclose Y2K remediation it does use. Rometty-*Personation* declines under the 5ᵗʰ Amendment and confidentiality. Honorable Judge Wu then grants Sneed discovery if Rometty-*Personation agrees* to respond.

70.   Sneed calls Wilzbach that same day. Charlotte, Wilzbach's wife, receives the call, updates Sneed since last contact, mentions a property matter her realtor is coming by to discuss, and that Wilzbach is away. Sneed amicably ends the call and follows-up by mail.

71.   Sneed receives Wilzbach's d1/January 29, 2012 response (Wilzbach-2 Letter). Aware Sneed is now in federal court, Wilzbach tells Sneed it is "fine" to proceed with Wilzbach-Lidow-Prouty Expectation, extant since 2007. As before, he assents Sneed doing so from vetting what *Sneed Thinks*, since Wilzbach has no first-hand knowledge of Sneed's claim.

72.   Wilzbach expresses "great concern" about becoming the "unnamed" IBM VP in Sneed's case, or that his advice be seen as anything other than informal, as a friend. He does tell Sneed Rometty-*Personation* is NOT using Specification at two of its *outside* client sites.

73.   Wilzbach reminds Sneed about Wilzbach-1 Letter and disclosure-leverage on Wilzbach, which Sneed's filing does not trigger. Wilzbach advises it would be a "great mistake" to have him testify, and to make sure an officer of the Court sees Wilzbach-2 Letter.

a *7th-Gift* page

74.   Wilzbach, through Wilzbach-1 or Wilzbach-2 Letter, along with his seniority under Caltech's Honor Code, could issue a cease litigation to Sneed to help Sneed avoid expending blood and treasure on a federal case that lacks merit.   Prior to each letter, Wilzbach vets what *Sneed Thinks* and helps evolve what *Sneed Thinks* in going forward to seek relief.

75.   Each letter structures Sneed's subsequent actions: (1) Wilzbach-1 directs Sneed to avoid triggering Palmisano-*Personation*'s disclosure-leverage over Wilzbach, before Sneed files suit, since Wilzbach is not cooperating with Palmisano-*Personation*, nor does he when Sneed reaches court; and (2) after Sneed is in court, he tells Sneed to present Wilzbach-2 to an officer of the Court.   Sneed proceeds with disclosure to an officer of the Court, and therefore to the Court itself, that Wilzbach assents to proceeding from what *Sneed Thinks*.

76.   Sneed delivers his amended filing, in person, on d5/February 23, 2012, following Wilzbach's and Honorable Judge Wu's advice, with a copy of Caltech's Honor Code.

77.   Rometty-*Personation* DOES NOT appear at the d5/March 1, 2012 status-hearing, even though it is a $100 billion IBME™ whose legal proxies appear *simultaneously* in courts all over Earth to *collect* monies on its behalf, including in California state courts.   The Court directs Sneed to send Rometty-*Personation* notice to appear d2/March 5, 2012.

78.   At the d2/March 5, 2012 hearing, Honorable Judge Wu asks Rometty-*Personation* to explain its *nonappearance*.   It responds that after the d5/January 23, 2012, 26(f)-hearing, it believes the case has no further merit, so Rometty-*Personation* feels no reason to appear.   In other words, Rometty-*Personation* not only demurs Sneed's legal existence but the Court's as well.   This is similar to its cessation of denials and disregard of Gerstner-Condition, after Wilzbach vets the d2/April 4, 2011 Tentative Conclusion it receives from Sneed.

79.   Rometty-*Personation*'s explanation disrespects the Court and Honorable

Judge Wu. On the spot, he schedules trial for d3/December 4, 2012 and sets these Court dates:

Discovery Cutoff	d4/Aug 15, 2012	Expert Cutoff	d6/Sep 7, 2012
Motion Hearing Cutoff	d2/Oct 15, 2012	Pretrial Conf.	d2/Nov 19, 2012
Court Trial:	d3/Dec 4, 2012		

80.   He grants Sneed discovery and sets a status conference to confirm Rometty-

Personation complies and discloses its Y2K remediation methods. *If* Rometty-Denunciation

and Palmisano-Denial are false, then Rometty-*Personation*'s conduct before the Court is

consistent with the definition in ¶ 33 above:

> *Predestined Antimeritism*: [H]erein the use of any belief or practice, under law or color
> of authority, to retro-actively devalue or deny the worth or merit of another's life, work,
> or writings, by demurring another's legal existence for purposes of historical
> revisionism, economic exploitation, or other harm.

81.   Rometty-*Personation* files its 12(b)(6) on d2/March 12, 2012 and argues the

Court must reject Caltech's Honor Code since it informs conduct *only while at Caltech*, and

not after graduation. Because of this, the Court must ignore any inference benefiting Sneed's

claims under Caltech's Honor Code, or that it informs Wilzbach's actions, deeds, or statements

in any way regarding Rometty-*Personation*'s use of Sneed's Specification.

82.   Rometty-*Personation* also denigrates that Sneed's and Caltech's Honor Code

"Mad Hatter-style reasoning could easily be dismissed as comical doubletalk...to waste the

resources of a federal court and of IBM." Rometty-*Personation* provides nothing to Honorable

Judge Wu from Wilzbach about Caltech's Honor Code and Wilzbach's views regarding

Rometty-*Personation*'s "Mad Hatter" style reasoning. This, too, is consistent with *predestined*

antimeritism, to demur the legal existence of another by *pre-assignment* as a "Mad Hatter."

83.   Sneed's Opposition to the 12(b)(6) reviews Wilzbach's d1/June 5, 2011,

letter that vets what *Sneed Thinks* in Sneed's d2/April 4, 2011, letter to Palmisano and its

Tentative Conclusion. This is before Sneed reaches his Final Conclusion, herein ¶ 39, about

Palmisano-Denial and Contradiction. Rometty-*Personation* denigrates the following as "Mad

Hatter" style reasoning, though it is by an officer of the Court who vets Caltech's Honor Code:

> If the above statement contained in SNEED's letter to IBM and Wilzbach and other
> Caltech Alumni was an Honor Code violation, Wilzbach would have been obligated to
> immediately notify Chameau, Lidow, and Prouty of such violation as it would have been
> a serious misrepresentation of facts known to Wilzbach to be false and it would
> constitute taking an unfair advantage. Wilzbach was bound by the Honor Code to
> inform both Palmisano and SNEED and other Caltech Alumni that SNEED's
> representations were false. Wilzbach has never made any such statement to SNEED or
> any other Caltech Alumni at any time to the knowledge of SNEED.

84.   The Court does not concur with Rometty-*Personation*'s attempt to demur

Sneed's legal existence through its "Mad Hatter" ad hominem vituperation. Rather, the Court

carefully reviews the 35 page submittal by Sneed of Caltech's Honor Code.

85.   The Court agrees to see, by d5/Jun 27, 2012, how Caltech's Honor Code

"applies even beyond a Cal Tech student's graduation from the institution and the underlying

assertion that those who are 'governed' by the Honor Code--[Sneed] admits the Honor Code is

'purely voluntary'--have a duty to *report* others they suspect of having violated it."

86.   Since Sneed is the only party who expressly believes Caltech's Honor Code

has life-long merit, Honorable Judge Wu accepts that, "[Sneed is] held [by the Court] to the

requirements...to, at least in this Case, the Cal Tech Honor Code." This gives Caltech's Honor

Code standing in Court, rebuffing Rometty-*Personation*'s "Mad Hatter" hostility toward it.

87.   By d3/June 19, 2012, Sneed supplies all 133 pages of discovery to the Court.

It exposes six Rometty-*Personation* exhibits as being from sources it claims it doesn't have.

88.   *The Court* serves the 133 of discovery on Rometty-*Personation* and includes it in the Court's docket.   Rometty-*Personation* takes further action to demur Sneed's legal existence, including: (1) Willfully not sending Sneed a Court filing and then asking him to respond after its deadline; (2) threatening sanctions if Sneed obeys Honorable Judge Wu's instruction; (3) withholding discovery because Sneed is not an attorney, which Honorable Judge Wu rebukes, stating that in his Court Sneed has power of attorney as a *pro se* litigant. Rometty-*Personation* continues to withhold disclosure of any of its Y2K methods as "burdensome."

89.   Rometty-*Personation*'s lead counsel is a 1990s Harvard Law School *Summa Cum Laude* grad, co-counsel is Stanford Law, with a team of paralegals and the deep-pockets of a $100 billion IBME™.   Sneed respects Rometty-*Personation*'s entitlement to a presumption of innocence, deferring to its removal to federal court and other demands.   At no time does Rometty-*Personation* respond to any discovery or other requests--except to assert 5th Amendment confidentiality, or that a request is too burdensome, or that it has been too long.

90.   In his final amended filing, d4/June 27, 2012, Sneed acknowledges the Court's honor of allowing Sneed to *voluntarily* be, perhaps, the only alum through whom Caltech's Honor Code elevates to probative status under the US Constitution and *federal* law.   Though an honor, Sneed is aware of the importance any precedent may set.   So, if Sneed errs, it is on the side of conservatism and Caltech's Honor Code being *voluntary* in its effect upon others.

91.   The pedagogical power of Caltech's Honor Code emanates through *liberty*, from which it teaches, because it is *voluntary* for anyone who may accept it.   *Remaining voluntary* helps Caltech attract funding, administrators, faculty, staff, students, and others seeking the best environment to pursue scientific, academic, and personal excellence.

92.   This resonates in the legacy-belief Caltech memorializes from John 8:31,32 in its first century, on its flag and stationery when Wilzbach, Lidow, Prouty, and Sneed attend: "If you hold to my teaching...[t]hen you will know the truth, and *The Truth Will Set You Free*."

93.   Sneed knows Rometty-*Personation* rejects Caltech's Honor Code's relevance, while enjoying Caltech board membership standing, with a name it sponsors on Caltech facilities, including a plaque on Sneed's undergraduate student house. The plaque may acknowledge, in a small way, some $1.86 billion in savings from use of Sneed's Specification, beyond a billion lines of code at 14 cents per line of code, instead of up to $2 under a 1956 Consent Decree. It's from a June 1997 call with Wilzbach that Sneed contacts DOJ, which opens inquiry about Rometty-*Personation*'s use of Specification and the 1956 Consent Decree.

94.   Through its 2012 Court filings, Rometty-*Personation* aggressively argues that none of its staff, or any prior staff, discloses use of Sneed's Specification to him, from 1997 to 2011. Sneed concurs. From discovery documents Sneed provides the Court, the parties who may have given him notice during that period are Wilzbach, Lidow, and Prouty.

95.   Rometty-*Personation* also argues that from 1997 to 2011, Wilzbach, Lidow, and Prouty are no more than speculative, hearsay witnesses. It opines that Wilzbach has no obligation under Caltech's Honor Code, which Sneed admits is *voluntary*, to say if what *Sneed Thinks* is true or not, after vetting Sneed's d2/April 4, 2011 Tentative Conclusion to Palmisano.

96.   Wilzbach assents to Sneed seeking relief after vetting what *Sneed Thinks* from Tentative Conclusion with Lidow's and Prouty's input. Yet Wilzbach never directly tells Sneed about any specific site Rometty-*Personation* uses Sneed's Specification. Sneed's discovery to the Court confirms Wilzbach helps guide Sneed to Sneed's Final Conclusion to seek relief, without ever disclosing a specific site Rometty-*Personation* is using Specification.

97.   Wilzbach does tell Sneed Rometty-*Personation* is NOT using Specification at two of its *outside* client sites. It argues this isn't disclosure of use, and Wilzbach never explicitly acknowledges, in writing, any of Sneed's multiple Caltech Honor Code notices inform Wilzbach's responses. Wilzbach assents to Sneed seeking relief, but he never states doing so in response to, or with knowledge of Caltech's Honor Code, or its meaning to him.

98.   Thus Rometty-*Personation* concludes that from 1997 to 2011, Sneed can never file more than speculative claims, if any, including his d6/October 28, 2011 filing, because Rometty-*Personation* never uses Sneed's Specification. Sneed's only filing fails the Supreme Court's 2007 *Twombly* ruling because it is speculative and Sneed files in 2011. Rometty-*Personation* argues Wilzbach's advice to seek relief is speculative, at best.

99.   Rometty-*Personation* also demands the Court reject Gerstner-Condition as outside of California's four-year statute of limitations. Rometty-*Personation* alleges that Sneed some how receives *unwritten, unspoken notice* of use from Rometty-*Personation*, itself, between 1997 and 2001. It *simultaneously* declares use *never* occurs in that period:

> That IBM refused payment to [Sneed in 2001] was certainly sufficient to heighten [Sneed's] suspicion, as confirmed by his admission of "repeated" additional inquires [sic] to IBM over the course of many years. Had plaintiff been *truly ignorant* of any potential breach, he would not have continued to demand payment from IBM. Plaintiff made constant and repeated inquiries *precisely* because he was suspicious that IBM had a contractual obligation to compensate him for use he believed had occurred. Because plaintiff *indisputably* had "sufficient information to put [him] on notice of the *possibility*" of his claims, he cannot invoke the delayed discovery rule.

100.   This refers to Sneed sending his patent issuance notice to Gerstner and Gerstner-Contact Torressen on d3/June 19, 2001, under Gerstner-Condition. There's no "refused payment" because both men are replaced after six months.

a *7ʰ-Gift* page

101.   Prior to this, in June of 1997, Wilzbach alerts Sneed about the parent-company taking control of Sneed's Specification from Wilzbach's subsidiary--action DOJ opens inquiry and reviews under its 1956 Consent Decree.

102.   Gerstner's staff, at the time, responds with a written denial of use to both DOJ and Sneed (DOJ-Denial). It parallels no-charge, evaluative-use under Gerstner-Condition. DOJ-Denial, by July 1997, to a law enforcement agency, for Sneed means there is no breach of Gerstner-Condition.   Sneed continues to perform under Gerstner-Condition and initiates inquiry approximately every 24 to 36 months, from 2001 to 2011, after DOJ-Denial.

103.   In its 2012 filings, Rometty-*Personation* further attempts to time-bar Gerstner-Condition with a statute of limitations claim. It demands the Court construe unilateral *speculation*, by counsel who inquires on Sneed's behalf, d2/June 14, 2004, as PROBATIVE to time-bar Gerstner-Condition, even though it denies use to DOJ and Sneed before, during, and after this inquiry. It also demands similar *opinion* by Wilzbach, Lidow, and Prouty NOT BE PROBATIVE in 2012, for purposes of dismissal under *Twombly* as mere hearsay *speculation.*

104.   In its last effort to time-bar Gerstner-Condition with its 2012 filing, it alleges Sneed receives *unwritten, unspoken notice* from an employee who sends an unsolicited letter to Sneed in July of 1997. Since neither Gerstner nor Gerstner-Contact authorizes the party, Sneed rejects the party's attempt with a letter that speculates on such depravity, now and in the past.

105.   The letter includes historical speculation about the firm using its IBME's™ "Dehomag" *fictitious identity,* which conceals from the US public its contract to implement a Coerced, Uncompensated, Labor Tracking (CULT) system, by tattooing ciphers on forearms of select German *citizens.* It then applies its tabulators and sorters to process the ciphers.

106.   The target *citizens* identify themselves through *voluntary* census participation and public-records reporting the IBME™, under its contract, begins consolidating for the Third Reich in 1933.   The system later generates startlingly accurate notification lists after the Nuremberg laws pass in 1935, to extinguish the legal existence of millions of German *citizens*.

107.   The final "solution" it designs, with help from its IBME™ US parent, leads to windfall profits as its government-client places new orders for tabulators and sorters to replicate CULT systems throughout parts of Europe its client-government conquerors. This ends in the mid-40s when the Allies halt such homicidal *predestined antimeritism*. The unauthorized employee's harassment ceases when Sneed compares the attempt to demur Gerstner-Condition with "Intellectual Genocide" in a d6/July 11, 1997 letter to that *person*.

108.   So Rometty-*Personation* demands the Court construe Sneed's *speculative* "Intellectual Genocide" comparison, in the d6/July 11, 1997 letter, as proof Sneed has *unwritten, unspoken notice* of use. It asserts he has this on the *same day* Sneed receives its DOJ-Denial. Thus, it draws upon precedent extinguishing the legal existence of millions of *citizens* in Germany, to also attempt demur of Sneed's legal existence, while acting under color of authority as a federal contractor in the US.

109.   Otherwise, it could use *Twombly* to construe the "Intellectual Genocide" comparison as *speculation* for dismissing Sneed with a 12(b)(6) motion in 2012. Instead it tries to retro-actively time-bar Gerstner-Condition with opportunistic *self-suspicion* of use in 1997-- the beginning of a decade of denials, including Palmisano-Denial and Rometty-Denunciation.

110.   Yet it rejects Caltech's Honor Code as "Mad Hatter-style reasoning":

For what it's worth, the Caltech Honor Code applies only as between *current* [emphasis in original] students[.]   Rometty-*Personation*'s d2/March 12, 2012, 12(b)(6) motion

111. In other words, if the Honorable Judge Wu accepts Caltech's Honor Code's simple precept that a member of the Caltech community shall <u>not take</u> unfair advantage of another member of the Caltech community, then Rometty-*Personation* asserts it may suffer harm as a 14th Amendment, disadvantaged *person* because neither Rometty-*Personation* nor Sneed nor Wilzbach nor Lidow nor Prouty is a member of the Caltech community or a student. Such a view allows Rometty-*Personation* to unilaterally exempt itself and all of its employees from Caltech's Honor Code, inclusive of its demand to do so for any former student it employs.

112. If Sneed successfully rebuts this "reasoning" and survives Rometty-*Personation*'s 12(b)(6) attempt to demur Sneed's legal existence, it could adversely affect employment of future Caltech alumni. Rometty-*Personation* and other IBME's[™] may desire employees ready to <u>take</u> unfair advantage of anyone at any time an IBME[™] may deign. So if Sneed prevails, he may be taking unfair advantage of Caltech to attract funding from such IBME's[™], and new students who may have to disavow Caltech's Honor Code to get work.

113. If Wilzbach acknowledges Caltech's Honor Code notices from Sneed as informing his advice for Sneed to seek relief, Rometty-*Personation* could accuse Wilzbach of placing Caltech's Honor Code above its executive confidentiality agreement. This could have unforeseen consequences for Wilzbach.

114. Finally, there is Sneed's belief that Caltech's Honor Code is *voluntary*. Rometty-*Personation* has historical affiliation through Caltech's board and is a member of Caltech's community. It clearly could not foresee the present case and its potential impact on itself as a profit-making IBME[™].

115. Furthermore, it has no precedent of paying Sneed $165 million, or any other descendant for whom Congress *originally* ratifies the 14th Amendment to end *predestined*

Executive Summary for the President, *Sneed vs. IBM* and Caltech's Honor Code

25

antimeritism. It employs such *persons* under contract, which it may readily terminate--as it does for 70,000 people in 1992-93--for much less than 1% of this amount. During the half century after its founding, it excludes such *persons* from even its less-than-1% employment.

116. Its desire seems to limit the 14ᵗʰ Amendment to self-assign its CULT identity as a disadvantaged *person,* to flee state courts and avoid a claim by a California *citizen,* while using the economy and convenience of those same courts when making claims for monies *against* California *citizens.*

117. Also, if Sneed succeeds and Rometty-*Personation* pays Sneed $165 million, or any portion of this amount, or acknowledges use of Specification, it could accuse Wilzbach of violating its confidentiality agreement even if Wilzbach never explicitly discloses use at *any* site. Wilzbach, Lidow, Prouty, and Caltech's Honor Code lead Sneed to continue inquiry, which results in an abrupt cessation of denials spanning more than ten-plus years. The Court notes in its d5/January 23, 2012 finding:

> There obviously was nothing preventing Plaintiff [Sneed] from setting up this [d2/April 4, 2011 Tentative Conclusion letter to Palmisano, Wilzbach, Lidow, and Prouty] at any point in time in the ten-plus years he has believed that Defendant [IBM] was using his invention...However, Plaintiff alleges that, in response to the letters he *did* [emphasis in original] send Defendant, Defendant always affirmatively denied its use of his invention until he sent his April 4, 2011, letter...Obviously, there are numerous reasons why Defendant justifiably might have ceased responding to Plaintiff's letter-writing campaign, but that speaks to the merits of Plaintiff's claims[.]

118. There is a saying that "A threat to justice anywhere is a threat to justice everywhere." If the Caltech Honor Code is not *voluntary* for any *one,* then it may not be *voluntary* for everyone. This includes Rometty-*Personation* who makes clear it never signs up for Caltech's Honor Code to constrain itself with a federal proceeding or the 14ᵗʰ Amendment.

119.   Relative to Rometty-*Personation*'s claim of immunity from Caltech's Honor

Code, the Court states in its d2/March 5, 2012 finding, after Rometty-*Personation* rejects

attending the d5/March 1, 2012 status-hearing:

> The Court's review of the copy of the Honor Code attached to [Sneed's amended filing]
> indicates that it answer's the [Court's] self-posed question "Is it an Honor System
> offense to not report a suspected violation?" with the answer -- in part -- "Strictly
> speaking, yes." [Caltech Honor Code], Exh. 1, at 25. [Sneed] would seemingly also
> have to address whether there is any provision -- either explicit or understood (and, in
> particular, understood by Wilzbach) -- for whether any such reporting duty is affected
> by whether the "reporter" might understand the alleged violator of the Honor Code to
> have a good faith -- even if incorrect -- belief in his or her assertions. The answer to the
> same question quoted above continues on to advise that "cases where violations are not
> reported will have to be investigated for special circumstances." *Id.*; *see also id.* at 13
> ("A *conscious failure* [emphases added by the Court] to report suspected violations may
> itself be considered an Honor System violation.")

120.   In Sneed's final Opposition filing d5/August 30, 2012, Sneed states:

> Defendant-IBM is given [Caltech's Honor Code] notice because of its membership on
> the Caltech board of trustees. The Court only stipulates [Sneed's] deeds and actions
> conform to Code. This doesn't include Defendant-IBM. Nor does it relieve [Sneed] of
> his past and present obligation not to take unfair advantage of Defendant-IBM. If
> Defendant-IBM saved some $1.86 billion with Specification, [Sneed] has properly
> performed to the standard set by the Court for [this case].

121.   Rometty-*Personation* makes clear it will not disclose its Y2K remediation

methods and it rejects Caltech's Honor Code as applying to it as an IBME™ or to its

employees, beyond Caltech.

122.   It demands the Court rule as follows: (1) the Gerstner-Condition does not

exist; (2) if it does exist, then it is time-barred and therefore still invalid; and (3) Sneed's legal

existence ends, to bring this matter before the Court ever again. The Court mercifully grants

none of these three demands.

123.   The Court rules the statute of limitations does not begin until Sneed has notice of use. The Court rules there is none, so the statue of limitations has yet to start. It leaves option for Sneed to re-file upon receiving notice of use that conforms to *Twombly*.

124.   Honorable President Jean-Lou Chameau, please accept this for review through oversight you exercise with Caltech's Honor Code, but only *voluntarily*. During the decade and a half of the matter, this alum limits participation in Caltech's public events, including Seminar Day or other festivities, by invitation to those enjoying the school's favor and legacy thereof.

125.   Three Caltech alums retain meaningful insight: Wilzbach, Lidow, and Prouty. Rometty-*Personation* disparages them as having little merit beyond hearsay. It is wrong. These are three *gentlemen* of considerable stature in every measure as Caltech alumni. They all promote resolution. This alum is responsive in every amicable way to such outcome.

126.   However, if Rometty-*Personation* persists in its present course, beyond Passover, and disregards good faith proposal for resolution, then this alum shall seek, as Heaven may permit, the divesture of 1/3rd of its capitalization by institutions and *persons* in this society who disagree with the precedent it attempts to set through the 14th Amendment as a disadvantaged *person*. Thereafter, should this matter remain in its present state, and this alum receives probative notice of use, then the second phase of litigation will begin, accepting aid from the press, government, and social media experts, by civil or criminal venues therewith. Set forth during the first month of the Emancipation Proclamation's 150th anniversary.

With prayerful regards and respectfully yours,

Anthony Sneed

Anthony Sneed   Attachment: Court Docket Discovery of Parties' Correspondence
cc: Honorable President Barack H. Obama c/o Honorable Attorney General Eric H. Holder,
     Mr. Pete Wilzbach, Dr. Alex Lidow, Dr. Dale Prouty

a *7ᵗʰ-Gift* page

1

2

3   ANTHONY SNEED, PRO PER
    2058 NORTH MILLS AVENUE
4   CLAREMONT, CA 91711

5

6   Pro Per Plaintiff,
    ANTHONY SNEED
7

8

9

FILED

2012 JUN 19 PM 4:00

CLERK U.S. DISTRICT COURT
CENTRAL DIST. OF CALIF.
LOS ANGELES

10   UNITED STATES DISTRICT COURT

11   CENTRAL DISTRICT OF CALIFORNIA - WESTERN DIVISION

12

13

14   ANTHONY SNEED, An Individual,   )   CASE NO. CV 11-010217GW (PJWx)
)
15   Plaintiff,   )   **REQUEST TO EXTEND DISCOVERY**
)   **AND DEPOSITION CUT-OFF DATE**
16   v.   )   **DUE TO INTERNATIONAL BUSINESS**
)   **MACHINES CORPORATION'S**
17   )   **DISREGARD OF FRCP 34**
)
18   INTERNATIONAL BUSINESS   )
19   MACHINES CORPORATION, A New   )   Status Conf., IBM Discovery: Jun. 20, 2012
     York Corporation, and DOES 1   )   2ⁿᵈ Amend. Complaint: Jun. 27, 2012
20   THROUGH 10, Inclusive.   )   Discovery Cut-Off: Aug. 15, 2012
)   Trial Date: Dec. 4, 2012
21   )
22   Defendants   )   Judge: Hon. Judge George H. Wu
)   Ctrm: 10
23

24

25   1. FRCP 34(a)(1) states a party is obliged to produce all specified, relevant and non-

26   privileged documents or other things which are in his or her "possession, custody, or control" on

27   the date and place specified in the request. See <u>Rockwell Int'l Corp. v. H. Wolfe Iron & Metal</u>

28   <u>Co.</u> (WD PA 1983) 576 F. Supp. 511,512]. If there are no documents responsive to the request,

─────────────────────────────────────
1
REQUEST TO EXTEND DISCOVERY AND DEPOSITION CUT-OFF DATE
DECLARATION OF ANTHONY SNEED REGARDING RULE 11 AND IBM ASSERTIONS

the responding party need only say so in a certified response pursuant to Rule 26(g). See _Buchanan v. Consolidated Stores Corp._ (D MD 2002) 206 FRD 123, 125.

Defendant International Business Machines Corporation (IBM) has exhibited documents, in its DECLARATION OF BRIDGET A. HAULER IN SUPPORT OF IBM CORP.'S MOTION TO DISMISS, March 12, 2012 (Declaration), that IBM never produced in discovery or compliance with 34(a)(1). IBM continues to refuse to produce these documents, their contemporaneous companions, or derivatives thereof (Materials). With IBM's non-response to any discovery request by Plaintiff, IBM objects on a number of grounds to producing Materials, which through Declaration it makes ready-reference to make out-of-context, and false assertions to the Court.

2. Discovery by Plaintiff therefore evidences IBM's possession of contemporaneous documents to those in Declaration's exhibits, and derivative documents thereof. IBM's conduct, early motions and asperities suggest that IBM may no longer retain such documents, or their whereabouts are unknown due to the passage of time. Such statements through motions to this Court are without merit regarding the intersection of Discovery by Plaintiff and Declaration exhibits, or any privilege IBM alleges regarding these documents and their derivatives, especially from 1997.

3. Rule 26(g) states that if there are no documents responsive to the request, the responding party need only say so in a certified response pursuant to Rule 26(g). See _Buchanan v. Consolidated Stores Corp._ (D MD 2002) 206 FRD 123, 125.

2

REQUEST TO EXTEND DISCOVERY AND DEPOSITION CUT-OFF DATE
DECLARATION OF ANTHONY SNEED REGARDING RULE 11 AND IBM ASSERTIONS

1

2

3

4. FRCP 6 states if the responding party cannot conduct a "careful and thorough" search

4

for all responsive documents within the 30-day period, it has an obligation to seek appropriate

5

extensions, either by agreement or, if necessary, by a motion for time under this rule. [*Novelty,*

6

Inc. v. Mountain View Marketing, Inc. (SD IN 2009) 265 FRD 370, 375-376]. Yet such

7

8

documents, their derivatives, and related reports or memoranda (Materials) regarding Plaintiff's

9

specification (Specification) to remediate the Y2K problem at IBM, make their only appearance

10

in Declaration, absent their companion and derivative documents' context, in violent and

11

egregious disregard of FRCP 34. This may injure the Court's impartiality in these proceedings,

12

especially by casually referencing documents out of context in disregard of FRCP 34 and

13

Materials IBM now exposes exist through Declaration.

14

15

16

5. Since IBM has made no FRCP 6 request, Plaintiff's request for time extension is made

17

for IBM to produce Materials exposed through Declaration yet not produced under FRCP

18

34(a)(1). Materials are essential for deposition, both in narrowing the list to specific deponents,

19

and refining interrogatories before deposition.   Declaration's exhibits are consistent with IBM's

20

pattern and practice to retain and index relevant Materials. Those not produced, yet proven IBM

21

retains, may be construed by the Court under 12(b)(6) in the light most favorable to Plaintiff.

22

23

24

6. Any assertion by IBM of vagueness, relevance, time-barred limitations, or destruction-

25

due-to-age, regarding Materials, is demolished by IBM's own Declaration and use of excerpts

26

from Materials. IBM is denying Plaintiff corroboration of what it received from Plaintiff,

27

28

REQUEST TO EXTEND DISCOVERY AND DEPOSITION CUT-OFF DATE
DECLARATION OF ANTHONY SNEED REGARDING RULE 11 AND IBM ASSERTIONS

1

2

3 essential for deposition and preparation thereof, adversely impacting deposition by IBM's

4 disregard of FRCP 34 and production of Materials.

5

6   7. The attached production by Plaintiff is witness that IBM's assertion of vagueness,

7 relevance, time-bared limitation, or destruction-due-to-age of Materials, is made in willful

8 disregard of FRCP 34, while presenting said documents out of context from Materials it

9 possesses, to the detriment of impartiality in these proceedings.   Therefore, Materials IBM

10

11 exposed in Declaration to fabricate IBM's time-barring scheme, with a whispered, yet

12 discernable, admission of use, may under 12(b)(6) be construed favorably to support Plaintiff's

13 pleading, especially if so done out of context with willful disregard of FRCP.

14

15

16   8. Declaration and nonperformance under Rule 34 brings into question IBM's conduct

17 regarding what IBM is now withholding from Plaintiff and this Court, except for out-of-context

18 use of IBM's own choosing.   IBM alleges in DEFENDANT IBM CORP'S REPLY IN

19 SUPPORT OF ITS MOTION TO DISMISS PLAINTIFF'S COMPLAINT (Reply), 6:25-28, 7:1-

20 5:

21

22   That IBM refused payment [to PLAINTIFF in 2001] was certainly sufficient to heighten

23   plaintiff's suspicion, as confirmed by his admission of "repeated" additional inquires [sic] to

24   IBM over the course of many years.   Had plaintiff been *truly ignorant* of any potential

25   breach, he would not have continued to demand   payment from IBM.   Plaintiff made

26   constant and repeated inquiries *precisely* because he was suspicious that IBM had a

27   contractual obligation to compensate him for use he believed had occurred.   Because

28

plaintiff *indisputably* had "sufficient information to put [him] on notice of *the possibility*" of his claims, he cannot invoke the delayed discovery rule.

9.   IBM has consistently denied (Denunciation) using Sneed's specification (Specification) for solving the Y2K problem, and that in fact IBM did not use Specification. For IBM to allege suspicion is warranted that IBM made production-use of Specification in 2001, and therefore only a *truly ignorant* person would not be suspicious of such use as of 2001, is to ascribe to Sneed supernatural powers to know something that IBM in fact never did, by IBM's own, sustained, and repeated Denunciation, including to this Court in 2012. Why would IBM retroactively attempt to plant such suspicion, when IBM has consistently denied use from 1997 to 2001? How does this contradict representations IBM made to Sneed and the Department of Justice (DOJ) in 1997 and afterward, including representations to this Court in 2012? Declaration exposes existence of contemporaneous documents regarding contact initiated by Sneed between IBM and DOJ regarding Specification.

10. For Sneed to assert, prior to and in or about 2001, that IBM made production-use of Specification could have no merit in reality, and be libelous as IBM warned Sneed in 1997, because, as IBM *emphatically* affirms, it never made production-use of Specification. IBM's assertion that Sneed or *any other human being on Earth* could have *reasonable* suspicion about what IBM *knows*, as it affirms, never happened, is libelous against *anyone* IBM makes the assertion.

11. Prior to Mr. Pete Wilzbach, a retired IBM Executive Vice President, giving Sneed *written-probative-notice* of production-use in his d1/June 5, 2011 letter to Sneed, IBM had never provided such notice, nor authorized anyone to do so, as would be its obligation under its d6/June 20, 1997 Letter of Corporate Intent. Non-payment, in this context, is only consistent with non-use. It abuses the English language in the extreme for IBM, under these conditions, to assert non-payment is indicative of production-use, especially after memorializing that IBM would not "expropriate" Sneed's intellectual property with its June 20th letter to Sneed. If IBM did otherwise, it would constitute the dictionary definition of *willful* fraud and concealment.

12. IBM also alleges in Reply 7:6:

Nor was plaintiff prevented from *learning facts* through reasonable diligence.

If this means *learning facts* regarding IBM's non-production-use, such facts were *abundantly* provided by IBM to Sneed through Sneed's *diligent* efforts in 1997, 2001, 2004, 2006, 2007, 2011 and 2012 (during these proceedings), because, as IBM repeatedly asserts, the *fact* is that IBM didn't make production-use of Specification. There were no *other facts* to be learned, based on the reality IBM projects in its Denunciation to this day. There was no way of *learning facts* regarding IBM's production-use, because, as IBM makes abundantly clear, including in sworn statements to this Court, IBM never made production-use of Specification. Diligence can not produce non-existent *facts* about something IBM declares never occurred.

13. But if Reply is an inadvertent, though belated, admission of production-use, in 2012, as it seems to conform itself, IBM owes Sneed $165 million, since it has never disputed

a 7th-*Gift* page

1

2   this figure, even now, by belatedly and perhaps inadvertently admitting production-use. It would

3   be for the Court to determine if IBM is cognizant of what it admits through argument.

4

5

6   14. IBM in asserting that Sneed *could* know, "what IBM did not do," also libels Sneed as

7   receiving such information outside of the terms of IBM's d6/June 20, 1997 Letter of Corporate

8   Intent IBM sent to Sneed. This would constitute a potential unilateral breach by Sneed, though

9   the agreement could still survive a "breach" of this kind. Such a breach would have given Sneed

10   notice, allowing Sneed to file a complaint in 1997, which IBM agrees Sneed was, and is more

11   than willing to do with any such notice. IBM also would have reason to file a complaint against

12   Sneed, for receiving production-use information from other than the designated IBM contact,

13   outside of the terms in the d6/June 20, 1997 IBM Letter of Corporate Intent. IBM would readily

14   file such action if it ever occurred, as would Sneed if any party ever came forward to disclose

15   production-use—which, from Denunciation, is impossible because IBM never made production-

16   use of Sneed's Specification.

17

18

19

20   15. Rather, Sneed strictly adhered to the terms of the d6/June 20, 1997 IBM Letter of

21   Corporate Intent, which limited Sneed to a single-point of contact within IBM as a condition of

22   performance. Sneed performed and made periodic contact, within approximately every 24 to 36

23   months, to meet this condition of performance, in good faith, because IBM, in good faith,

24   declared it would not expropriate Specification. This is not *unrelenting* and *frequent* contact, as

25   IBM defectively alleges in Reply, but rather conduct that is courteous, occasional, measured and

26   conforms to IBM's own condition of performance on Sneed in its d6/June 20, 1997 letter.

27

28

REQUEST TO EXTEND DISCOVERY AND DEPOSITION CUT-OFF DATE
DECLARATION OF ANTHONY SNEED REGARDING RULE 11 AND IBM ASSERTIONS

16. In the period IBM alleges Sneed or others somewhere on Earth could have reasonable suspicion that IBM did what, "IBM declares it did not do," IBM also told DOJ that there was no merit to Sneed's inquiry about IBM using Specification. This correspondence between Sneed, IBM and DOJ is retained by all parties, yet was not produced by IBM under FRCP 34, except out of context from Materials in Declaration, and IBM's assertion about *truly ignorant* parties.

17. To with, for IBM to assert in its Reply that Sneed had reasonable suspicion of IBM's production-use, while declaring in the same time-frame to DOJ that there was no basis for any inquiry about production-use--because IBM alleges it didn't make production-use of Specification--is a pattern the Court may be familiar regarding perjury or intentionally misrepresenting a material fact. As such, it may be example of an egregious violation of Rule 11, and could be subject to sanction. It may also constitute an obstruction of justice under USC 242.

18. To compound this in 2012 with its Reply, in hearings before this Court, IBM argues dismissal under *Twombly/Iqbal*. That is, IBM demands the Court *reject* Sneed's complaint because it is based on *suspicion of production-use* as a flawed cause of action, while arguing that *suspicion of production-use* is a wholesome reason to *accept* IBM's time-barred postulate. In fact, Sneed's complaint is based on *written-probative-notice* from Mr. Pete Wilzbach (Wilzbach), retired IBM Executive Vice President and fellow Caltech alum of Sneed. But without blinking, IBM declares its *recent* (2012) *self-suspicion of production-use* time-bars Sneed's complaint, and that somehow IBM's *recent* (2012) *self-suspicion of production-use*

could be known by *someone* on Earth, in 1997, including Sneed, though IBM denies any such

production-use *ever* occurred. This, too, is a pattern the Court may already be familiar regarding

perjury, violating Rule 11, or misrepresenting a material fact.

19.   The quantum mechanical, "Schrodinger's cat" juxtaposition between IBM's

suspicion-position in 1997 to time-bar Sneed's complaint, as reasonable because IBM asserts *its*

self-suspicion of production-use is probative, and its contradictory *suspicion*-position under

Towmbly/Iqbal in 2012, that *any* suspicion of production-use *(including IBM's alleged, belated*

self-suspicion) is not probative, is repeated in its motions. Its representations to DOJ in 1997 are

dramatically different from its statements to this Court in 2012 regarding production-use, or its

own, belated, probative, self-suspicion of production-use. I, Anthony Sneed, further declare that

I am not aware of any disclosure IBM made about production-use, and was not aware of

production-use on any specific IBM machine or general facility, from 1997, when IBM took

custody of Specification, to 2011--until Wilzbach's *written- probative-notice* within his d1/June

5, 2011 letter to Sneed gave reasonable cause to file.

20. Such *written-probative-notice* from Wilzbach is at least at the threshold to file a

complaint, if IBM now belatedly asserts its own *unwritten-unspoken-notice* of production-use,

revealed through IBM's probative self-suspicion to fabricate its time-barring scheme. Such

unwritten-unspoken-notice of production-use has appeared prominently after IBM hired outside

counsel, subsequent to Sneed's complaint filed d6/October 28, 2011. The *unwritten-unspoken-*

notice contradicts IBM's written denials from 1997 to 2012, including to this Court.

21. In fact, IBM rejects even its own *unwritten-unspoken-notice*, regarding production-use of Sneed's Specification, in its denunciation of Sneed to DOJ, referenced in Declaration, and denunciation of Sneed's complaint to this Court--unless the matter involves IBM's time-barred scheme. Then IBM's *unwritten-unspoken-notice* becomes its probative champion. Finally, IBM admits to potential production-use in Reply 6:25-28. Therein IBM asserts that *any party,* even a *truly-ignorant plaintiff,* who makes repeated demand for payment, in 1997 or afterward, is doing so because the demand has merit, whether the party is *truly-ignorant* [Reply 6:25-28] or *not.* The merit is not in the ignorance or non-ignorance of the inquiring party, but in IBM's admission within this context that IBM made production-use of Specification without payment, to promote its time-barred scheme. And that scheme constitutes an inadvertent probative admission, harmonizing with IBM's own, bleated self-suspicion of production-use, that IBM teaches in its 2012 motions. The above paragraphs may support sanctions for Rule 11 violations from IBM's Motions to Dismiss Sneed's complaint regarding production-use, beginning in 2011, and IBM's appearances before this Court. Yet IBM is still given opportunity for good faith, which the Court allows by briefly entertaining voluntary mediation between the parties during the d5/May 24, 2012 hearing. Sneed concurs.

I declare under penalty of perjury the foregoing is true and correct.

Executed this 3rd day of the week, 19th of June, 2012, in Claremont, California.

ANTHONY SNEED, Plaintiff

REQUEST TO EXTEND DISCOVERY AND DEPOSITION CUT-OFF DATE
DECLARATION OF ANTHONY SNEED REGARDING RULE 11 AND IBM ASSERTIONS

EXHIBIT 1

WILZBACH/SNEED/IBM

COMMUNICATION

1997 May: Sneed correspondence with personnel at IBM subsidiary Integrated Software Services Corp (ISSC); IBM's first custody of copyright-restricted, evaluation-only version of Sneed's Y2K Specification

1. Mr. Pete <u>Wilzbach</u> (Wilzbach), as a friend, encourages Sneed's patenting efforts of his specification for solving the Y2K problem (Specification), called "A Serial Encryption System-Bypass of Year 2000 Date Malfunction."; a copyright restriction is clearly set forth on its cover page.
   1.1 Wilzbach does so after receiving a copyright-restricted, evaluation-only version of Specification.
   1.2 Wilzbach's title at ISSC is IBM VP/Ameritech Alliances; he and Sneed are fellow Caltech alumni.
   1.2 At or about this time, Wilzbach, as IBM VP, closed the largest information technology services contract in IBM's history, out-performing all of his IBM executive management peers.
   1.3 Wilzbach was Sneed's hiring and training manager when Sneed started work at IBM in the 80s.
   1.4 Both worked on, and made notable contributions to NASA planetary missions while undergrad interns.
   1.5 Both worked at Hughes (Boeing) Space and Communications Group after graduating from Caltech.
   1.6 Wilzbach and Sneed have maintained amicable, friendly relations from 1980 to the present.
   1.7 With Wilzbach's encouragement, Sneed performs and continues efforts to patent Specification.
   1.8 At or about this time, ISSC reorganizes as IBM Global Services, an IBM subsidiary.
   1.9 At or about this time, IBM Global Services is 1/3ʳᵈ of IBM's revenues and 2/3ʳᵈˢ of IBM's profits.
   1.10 The presidency of IBM Global Services is a possible career-track to become CEO of IBM Corporate.
   1.11 Mr. Dennie Welsh (Welsh) is president of IBM Global Services during intake of Specification.
   1.12 He is replaced by Mr. Sam Palmisano (Palmisano); Wilzbach is a runner-up for the top spot.

2. Mr. Don Logan (Logan) receives a copyright-restricted version of Specification from Sneed.

3. Logan accepts the copyright-restricted version of Specification for evaluation-only

4. Logan is an IBM Program Director and IBM AS/400 "Y2K Czar" for IBM Global Services (then ISSC).

5. After testing, Logan confirms the merits from his evaluation-only copy of Sneed's Specification:
   5.1 The invention [Specification] is original, unique, and unlike anything IBM was considering.
   5.2 It is superior in all aspects to "windowing" and "expansion," the extant methodologies.
   5.3 It's simplicity, time and cost to implement is a fraction of other methods.

6. Sneed contacts IBM Global Services' legal depart, and is referred to Mr. Dick McDonough (McDonough).
   6.1 McDonough is IBM Global Services Vice President and Assistant General Counsel.
   6.2 At the time, McDonough was just moving into his office at IBM Global Services.
   6.3 McDonough's possessions were still in boxes, he notes to Sneed; they discuss the Y2K problem.
   6.4 McDonough agrees to receive a copyright-restricted, evaluation-only version of Specification.
   6.5 Sneed sends the copyright-restricted, evaluation-only version of Specification to McDonough.
   6.6 The parties are clear the copy is for evaluation-only.
   6.7 Sneed sends it on or about d6/May 29, 1997 with a cover letter.

Ex1 p1

a *7th-Gift* page

ISSC

May 13, 1997

Mr. Anthony Sneed,
ISB
2058 Mills Ave. Ste. 455
Claremont, CA 91711

Tony,

You've obviously done well in your career and have kept your technical edge. I can't even understand your manuscript on Year 2000. Unfortunately, my contract with Ameritech doesn't have me involved in Year 2000 application fixes – they have contracted with BellCore for that.

I wish you the best in your endeavors and hope you get both the patent and the commercial success you are looking for.

Sincerely,

Pete Wilzbach
VP, IBM/Ameritech Alliances

PW:spm

Ex1 p2

ISB
2058 N. Mills Ave. Ste. 455
Claremont. CA 91711

909-626-1777
Fax 909-626-4977

Mr. Don Logan
IBM, Home Office
13760 Belletarre Drive
Alpharetta, GA 30201

d3/May 27th, 1997

Mr. Logan:

I very much enjoyed the business fellowship with you by phone, including:

    o Your early brush with TCAM—and a more nimble solution with VS1, twix and paper tape.

    o Your successful 24 years in the business, and celebrating your Jubilee on the third planet from the Sun.

    o Your proud acknowledgment that you not only know what a plug board is, but you have also wired a few.

Like you, I, too, have had an abiding interest in finite automata. The attached background information and reference letter may affirm that I am just as serious about making systems work the first time, at minimum expense, as you are. Your twix/VTAM solution sounds like it saved a bundle of cash, while making the entire workflow easier to implement.

Your concise analysis of the *Year 2000* anomaly was as crisp as they come, right down to detecting level 0, 1, or 2 DSV's. As you noted, assessment and conversion are important steps in addressing the issue, along with the emphasis for some type of exhaustive testing. I agree that not all systems will have "soft fail" logic when a "00" some how gets ingested. Finally, your perspective on trade-offs between "date expansion" and "windowing," and the new trend toward the latter, is appreciated. Until the "silver bullet" is found, CSI will continue to make due with the tools now available, as you pointed out.

After reviewing the attached agreement, simply sign it and return it by fax (909-626-4977). The invention/patent claim that will promptly follow may far exceed the requirements you shared, including fail safe capability that prevents a "00" condition from occuring—ever.

I look forward to our follow-up contact, and enjoyed the enlightening and educational conversation with you very much. I hope to reciprocate it with the invention to follow.

With Regards,

Anthony Sneed

Anthony Sneed
Life-Time Member, Caltech Alumni Association, ISB Information Services Management Analyst
Attachment Correspondence: Dr. T. Everhart, Caltech; Dr. U.V.d. Embse, Hughes Electronics
CC: Mr. David O'Reilly, Patent Attorney

Ex1 p3

ISB
2058 N. Mills Ave. Ste. 455
Claremont, CA 91711

909-626-1777
Fax 909-626-4977

Mr. Don Logan
IBM, Home Office
13760 Belletarre Drive
Alpharetta, GA 30201

d3/May 27th, 1997

Mr. Logan:

Mr. David O'Reilly, my patent attorney, has informed me that acceptance of the patent claim by the Patent Office and the issuance of a serial number is sufficient protection for the invention.

It has been over 48 hours since our last contact, and the clock is ticking. As this may be one of a multitude of things crossing your desk, I will increase the number and level of contacts in organizations that may benefit from the invention, including yours.

If we do not have further contact this week, I hope that you, your management and staff have a restful weekend.

With Regards,

Anthony Sneed

Anthony Sneed

Ex1 P4

ISB
2058 N. Mills Ave. Ste. 455
Fremont, CA 91711

809-526-1777
Fax 809-628-4977

d5/May 29th, 1997

Mr. Dick McDonough
IBM Global Services
Route 100, Mail Drop 4427, Building 4, Route
Somers, NY 10589

Mr. McDonough:

After returning from vacation, Mr. David O'Reilly, my patent attorney (who also handles intellectual property issues for Caltech), advised me that the protection provided by the U.S. Patent Office is sufficient for disclosing the invention to you. The U.S. Patent Office has given notification by phone to expect a patent serial number, by mail, for the invention.

Based on research from the Gartner Group (http://www.gartner.com) and from a Year 2000 web site (http://www.y2k.com and year2000.com) this problem has the potential to cause very serious dislocations of the federal government, as well as the private sector. The anomaly was brought to my attention when a DOD computer system in Ohio failed prior to mid-February. The failure went undetected for a period of days, and later found to be the Year 2000 Malfunction.

Though I had tracked this problem for several years, I was confident that Caltech, MIT, Xerox, Batelle, IBM, Stanford or some other national research facility would tackle this issue. The failure of a *DOD computer complex* startled me, since if any agency on Earth could buy its way out of this problem, DOD is it. So, by inference, other more marginally funded agencies (HHS, IRS, Social Security, DOI, DOJ, DOA, etc.) are even more exposed.

This precipitated the development of the attached patent claim and its successful submission to the U.S. Patent Office. It has been reviewed by fellow members of the Caltech Alumni Association in their professional capacities and confirmed as sound. The invention has also been tested with senior MIS managements and staffs to confirm its efficiencies and cost savings in "real world" situations. In each case, the economies and time savings have been significant.

I am including reference material that outlines related background to affirm I am very serious about solving high visibility problems like this one, and have consistently done so. This is being faxed to you because there are now less than 140 weeks left to solve this (d5/December 31, 1998 is a key trigger date that will begin a cascading of system failures like the one in Ohio).

If we do not have further contact today, I hope that you, the management and staff of your company have a restful weekend. I look forward to contact with you before d4/June 11th, 1997.

With Regards,

Anthony Sneed

Anthony Sneed
Life-Time Member, Caltech Alumni Association
Reference Enclosures: Dr. T. Everhart, Caltech; Dr. U.V.d Embse, Hughes Electronics

Ex1 p5

1997 June: Sneed correspondence with IBM personnel; the meeting of the minds on price and use

7. Logan requests pricing for three computers at an IBM Global Services evaluation site within IBM.

8. Sneed assures Logan there is no charge for evaluation, and the fee is only $49,764.55 for any
     production-use of Specification at IBM's first site.
     8.1 Logan discloses there are three computers at Logan's facility where Specification may be applied for
           production-use after evaluation.
     8.2 Sneed assures Logan that the fee is only $49,764.55 because the computers are within the same site.
     8.3 Logan creates goodwill with Sneed by making the voluntary disclosure about the three computers.
     8.4 Logan sees that Sneed is straight-forward in supporting Logan's desire to remedy the Y2K problem.
     8.5 Sneed memorializes the no-charge provision for evaluation-only, and the $49,764.55 site-fee for any
           production-use.
     8.6 Sneed forwards the written terms to Logan d3/June 3, 1997.

9. Additionally, in good-faith, Logan voluntarily offers Sneed a royalty of up to 50 cents per-line-of-code.
     9.1 The royalty is a percentage of the up to $2 per-line-of-code price IBM Global Services charges clients.
     9.2 Logan remarks that Specification is so easy and quick to apply that customers might do so without IBM.
     9.3 Sneed responds that, if the whole "city" is on fire, there will be plenty of work for the "fire department."
     9.4 The analogy uses "city" for the U.S., "fire" to be the Y2K emergency, and IBM as the "fire department."
     9.5 Sneed says to Logan, "Giving people the ability to help put out 'fires' around their own property may be
           beneficial to the 'fire department,' especially for diverse and many places the 'fire department' may not
           reach in a timely fashion."
     9.6 Logan responds, "Good analogy!"
     9.7 Sneed contacts Mr. Dan Sullivan (Sullivan), IBM Corporate VP, and Sneed's marketing manager when:
           9.7.1 Sneed graduated with honors from IBM's national training center in New York, completing every
                   requirement of IBM training in Los Angeles, Dallas, and New York, spread over 24 months.
           9.7.2 Sneed achieves 400% of quota and wins a regional sales contest for IBM's entire Western Region.
           9.7.2 Sullivan, by 1997, is located in New York and congratulates Sneed on Specification.
           9.7.3 Sullivan says he knew about Sneed's Specification at IBM's New York office before Sneed called.
           9.7.4 Sullivan advises Sneed to simplify how lines of code are counted in any agreement with IBM.
           9.7.5 This makes it easier to get paid if IBM uses Sneed's Specification, Sullivan notes.
           9.7.6 Sneed performs and sets a price of 14 cents per line, and no charge after a billion lines of code.

10. Written agreement formation proceeds: 14 cents per line of code, and no charge after a billion lines.

11. On d3/June 3, 1997 Sneed affirms, in writing, the $49,764.55 per-site fee to Logan.

Ex 1 p6

ISB
2058 N. Mills Ave, Ste. 455
Claremont, CA 91711

909-626-1777
Fax 909-626-4977

Mr. Don Logan
IBM, Home Office
13760 Belletarre Drive
Alpharetta, GA 30201

d3/June 3rd, 1997

Mr. Logan:

After conversation with you last week, my patent attorney, Mr. David O'Reilly, who also handles intellectual property issues for Caltech, said the serialization of the invention by the U.S. Patent Office is sufficient protection for commercial disclosure of the invention. Currently, a fee of $49,764.55 per computer site is charged to use the invention. This is the presumption, as well, in forwarding a copy of the invention to you.

Three fellow members of the Caltech Alumni Association (Mr. Jim Celoni, Mr. Chris Jensen, Mr. Joe Rayhawk) have confirmed the efficiencies and economies of the invention, one being the director of a major Southern California datacenter. The methodology in the invention addresses the Year 2000 Malfunction in as little as seven weeks, even for some complex installations.

Don, the clock is ticking for all of us on this issue. A "silver bullet" may not be sufficient now. We may require a "gold," "platinum" or "titanium" bullet at this point. Test "fire" this invention under the most adverse conditions, to confirm that it exceeds the "ballistic" requirements needed for the battle you are in.

With Regards,

Anthony Sneed

Anthony Sneed
ISB Information Services Management Analyst
CC: Mr. David O'Reilly, Patent Attorney

Ex1 p7

a *7th-Gift* page

ISB
2058 N. Mills Ave. Ste. 455
Claremont, CA 91711

909-626-1777
Fax 909-626-4977

d3/June 10th, 1997

Mr. Don Logan
IBM, Home Office
13760 Belletarre Drive
Alpharetta, GA 30201

Mr. Logan:

Per our agreement, the attached invoice permits you to use the invention on a single system site to test customer conversions. It does not permit code converted with the invention, or re-engineered to comply with the methodology of the invention, to be transferred to any other system site whatsoever unless a fee is remitted for such a site. This is true whether such a site is inside of IBM or outside of IBM proper, or at a customer location.

This invoice is submitted without favoritism, consistent with precedent at a few other test sites. To do otherwise would leave me open to the allegation of favoritism toward you and Global Services by Primary Counsel and management at the few other test sites. If for any reason this invoice is not paid, or there is reservation to promptly remitting payment, you are to cease and desist the use of the invention in any form and in any way whatsoever and document you have done so.

On a personal note, Mr. Logan, which I freely say without interference or direction from Primary Counsel, it has been my pleasure to meet you, to experience your intellectual integrity, and to respect your competence in practicing the system and marketing arts. I have openly expressed my high regard for you, Mr. Welsh, Mr. McDonough and the IBM Global Services organization. You clearly were acting in the interest of many unseen people who may never know you.

Your decisive action placed IBM days ahead in evaluating the invention, and in the position to seize the moment before anyone else. I can think of no more that you and your organization could have done, even as the lead you established diminishes with each passing day in "deliberations." I have been informed, in detail, of the process that is occurring within your corporate headquarters now. Events and efforts are now being construed by individuals with much different motives than Global Services. In the not too distant future, this will be disclosed in full to you.

Nevertheless, as you and I both know, events may not always achieve our fondest hopes, or strengthen the bonds of business fellowship that have heretofore been delicately nurtured through you and Global Services. Primary Counsel has access to multiple individuals well-placed throughout your organization, who have tracked the progress of the invention since custody was taken from Global Services' hands. Primary Counsel is aware of the directions given to you and your organization. Therefore, at this time, having been given such insight, I take no offense at your unreturned phone calls, or from any unreturned call by an individual or individuals directed to do the same. I respect you for submitting to the authority that now directs all of your actions.

Ex1 p8

In no way do *post-custody* events reflect the character and values I only had opportunity to glimpse in you. Yet, I did feel the edges of the keen marketing and business effectiveness of you and your organization through your *pre-custody* actions. You projected the values and initiative of an organization that I would recommend to fellow members of Caltech's Alumni Association and other peers, even now. There is no damage to my high regard for Global Services through recent events beyond Global Services' control.

Therefore, I will do anything that you may ask me that does not require me to break my Father's Law (Exodus 20:1-17), or a written law of the government of the United States of America, or a written policy, practice or procedure of your company. And I will do it promptly, cheerfully and without reservation. Anything. This also applies to Mr. Welsh, Mr. McDonough and others in Global Services who reflect the same commitment to success, ethical conduct and excellence that you have shared. I write this to you in the Testimony of the Name of my Father's One and Only Son, because what will soon occur may require us all to patiently endure (Matthew 5:38-42).

For the rest of your natural life, beyond the 50 years that I shall hope to one day wear as well as you, we may have no further contact. At this point, we are not permitted to even greet one another (Matthew 5:43-48). It may be to my Father's purpose in the Testimony of the Name of his One and Only Son to permit someone within IBM to use the awesome authority granted IBM under heaven to affect you and I this way, because it remains written,

(Romans 13:1) Everyone must submit himself to the governing authorities, for there is no authority except that which God has established. The authorities that exist have been established by God.

And since such authority can and has been used to do much worse, surely such an individual using the same authority may readily severe a bond of respect and affection for the values and person you have been for 24 years, and continue to severe such a bond for the rest of our natural lives, or as long as it pleases the individual to exercise such authority in this way. I know that it is pleasing to my Father and his Son that I, too, submit to the same authority. As I have done in the past, I shall do the same now and in the future, by the grace of the Name.

So in submission to such authority, I bid you a fond farewell. It may now be inappropriate that we should ever meet, unless we give the individual cause to accuse us of ignoring Romans 13:1 and transgressing Exodus 20:3, which would be a very serious allegation when I stand before my Father and his Son in heaven. So let us be satisfied in our historic contact, completed before anyone could accuse us of rebelling against the authority granted, for a time, on Earth.

2

Ex1 P9

There remains a time in the future, where the authority of heaven may be withdrawn from the individual who prohibits us from freely, openly and cheerfully greeting each other. So it is encouraging that what now exists can not last forever, especially not beyond our natural lives.

I wish you and Global Services well in your endeavors, Don. And I wish you success in your personal endeavors, too.

With Regards,

Anthony Sneed

Anthony Sneed

P.S., Primary Counsel just informed me that a "diversity meeting" was completed in the New York area that controls $8.2 billion in revenue for your company. One of the topics covered by a woman named "Teddy Walberg" [sic] (who had two presenters with her) had to do with the "internal intellectual work force resources IBM can draw on," and its "sensitivity to diverse intellectual resources in the marketplace." There was also discussion about demographic sensitivity, which I hope *includes* you and I, even as we are simultaneously *excluded* from even greeting each other for the foreseeable future and beyond.

Don, there is more research I have to share with you about the invention. Just as by your own admission you and your organization did not know *anything* about the invention before Mr. Welsh's office permitted us to have contact, there is additional research that may prove very useful. This may reflect one of the dangers in letting a private individual restrict the First Amendment in an attempt to achieve a short-term business objective—it retards innovation and discovery. Nevertheless, we do well to respect such authority for as long as it is permitted to be exercised, for reasons covered earlier.

3

Ex1 p10

ISB
2059 N. Mills Ave. Ste. 455
Claremont, CA 91711

909-626-1777
Fax 909-626-4977

Mr. Don Logan
IBM, Home Office
13760 Belletarre Drive
Alpharetta, GA 30201

d3/June 10th, 1997

Invoice 7287

Fee to use "Serial Encryption System-Bypass of Year
2000 Date Malfunctions" at a single (1) system site
within the entire IBM organization for
testing customer or internal IBM conversions or other uses,
per agreement on fee prior to d4/June 4th, 1997.

$49,764.55

Due and payable upon receipt to "ISB" at the above address.
After 30 days, a 1% assessemnt will be charged per week,
compounded, on the outstanding balance, if any.

Exl p1)

12. On d4/June 4, 1997 Sneed forwards written terms to Logan for production-use, if any, of Specification.

    12.1 On d3/June 10, 1997 Sneed invoices Logan for production-use of Specification, if any, on the computers Logan identified at the first IBM site.

    12.2 Sneed includes a cease and desist, and for Logan to document he has done so, if no fee is remitted for production-use, if any, by IBM.

    12.3 Sneed begins drafting agreement (Agreement) for 14 cents as acceptance of Logan's proposal of 50 cents per-line-of-code, and no charge after a billion lines-of-code, with all due respect to Sullivan's advice.

    12.4 On or about d4/June 4, 1997 Sneed prepares a draft of Agreement. Sneed sends Agreement, with the $49,764.55 per site terms and invoice to Logan on d3/June 10, 1997.

ISB
2058 N. Mills Ave. Ste. 465
Fremont, CA 91711

909-626-1777
Fax 909-626-4977

Mr. Don Logan
IBM, Home Office
13760 Belletarre Drive
Alpharetta, GA 30201

d4/June 4th, 1997

Mr. Logan:

As we discussed in passing, yesterday, if the whole "city" is on fire, there will be plenty of work for the "fire department." Giving people the ability to help put out "fires" around their own property may be beneficial to the "fire department," especially for diverse and many places the 'fire department" may not reach in a timely fashion.

Remitting the attached invoice will initiate a retainer period of my services for a minimum of three months, retroactive to yesterday. If warranted and mutually agreed to, the retainer can be extended beyond three months, pending availability at such time.

I recognize that with the written material sent to you yesterday, along with our phone conversation, our brief exchange may permit you to move ahead without retaining my services. I would consider this a compliment, since it reflects the confidence quickly imparted to you, as has been done with fellow members of the Caltech Alumni Association, to successfully achieve the advantages of encryption. This may permit you to do the same for others internal to your organization, after remitting the $49,764.55 single site fee to apply the invention at the IBM datacenter you suggested as a possible future site.

A separate single-site fee would apply for all other individual system sites, whether inside or outside of IBM, where the encryption invention is applied to mitigate the Year 2000 Malfunction. Also, as you proposed, a royalty agreement can be negotiated which may permit a plethora of internal or external system sites to engage the invention at a lower per-system-site cost. This may also permit embedding the invention within extant commercial products, on a per-product basis or through a master product royalty agreement.

Therefore, the retainer will permit me to allocate time resources, by phone, for specific challenges you may soon face. There is useful information that has been garnered from installations on the West Coast that could expedite a few implementation scenarios you have yet to encounter, if you ever encounter them at all. Some of these sites may require more support than I am prepared to provide through a consulting practice near the Claremont colleges.

Therefore, it is entirely appropriate that we proceed, with your concurrence. I look forward to follow-up contact with you before one week from today.

With Regards,

Anthony Sneed
Anthony Sneed

CC: Mr. David O'Reilly, Patent Attorney

EX1 P13

a *7ʰ-Gift* page

1997 June: IBM Corporate's unsolicited call to Sneed on or about d5/June 5, 1997; Sneed's written response

13. Mr. Al Torressen (Torressen) is IBM Corporate's Intellectual Property Licensing Program Manager.
   13.1 Somehow, IBM Corporate management directs Torressen to make an unsolicited call to Sneed, beginning on or about d4/June 4, 1997; Torressen leaves repeated messages for Sneed to call Torressen over a two day period.
   13.2 On or about d5/June 5, 1997, Sneed returns Torressen's unsolicited call.
   13.3 Torressen is unaware of Sneed's contact with McDonough at IBM Global Services' legal department.

14. Sneed did not know Torressen, and had never heard of Torressen before Torressen's unsolicited call.

15. In the call, Torressen informs Sneed that:
   15.1 IBM Corporate "took control" of Sneed's Specification from IBM Global Services (it in fact had not).
   15.2 IBM "wants to work with you directly" to avoid "non-engineers"; Sneed doesn't know to whom Torressen is referring as "non-engineers."
   15.3 Torressen humbly notes he is making direct contact so this "won't cost IBM a whole lot of money."
   15.4 Torressen declares he is now Sneed's single-point of contact for all of IBM regarding Specification.

16. Sneed asks Torressen to identify what part of IBM employees Torressen:
   16.1 Torressen responds that he is with IBM Corporate's legal department.
   16.2 Sneed asks if IBM Global Services has its own legal department.
   16.3 Torressen remarks that Sneed is very clever; at the time Sneed is unclear why Torressen said this.

17. Since Torressen declares he represents IBM's legal department, Sneed responds in good faith:
   17.1 Sneed provides Torressen a copyright-restricted, evaluation-only version of Specification.
   17.2 Sneed provides Torressen a copy of Agreement and terms developed with, and proposed by Logan.
   17.3 Torressen asks if Sneed will work directly with IBM Corporate technical staff as he has with Global's.
   17.4 Sneed responds affirmatively, that he will help IBM Corporate as he has helped IBM Global Services.
   17.5 At the time, awareness of the Y2K problem as a potential national crisis was increasing.
   17.6 Sneed was doing all he could to minimize IBM's internal politics to rapidly advance Specification.
   17.7 Public and private sector computers, throughout the U.S. economy, could soon be affected.
   17.8 IBM had the corporate credibility to present Specification to a plethora of companies and institutions.

Ex1 p14

ISB
2058 N. Mills Ave. Ste. 455
Claremont, CA 91711

909-626-1777
Fax 909-626-4977

d6/June 6th, 1997

Mr. Al Torressen
IBM Corp
500 Columbus Ave
Thornwood, NY 10594

Mr. Torressen:

This is to acknowledge your call yesterday to inform me of IBM Corporation's possession of the invention received from IBM Global Services, and the forwarding of the invention to corporate technical staff to affirm the favorable feed-back already received from personnel in IBM Global Services. The invention is titled a "Serial Encryption System-Bypass of Year 2000 Date Malfunctions," and addresses the *Year 2000 Malfunction*. You also asked if IBM corporate technical personnel could contact me directly, which I agreed to with a retainer to cover time away from my local West Coast responsibilities.

In your conversation you expressed an enthusiasm shared by others for the invention (possibly due to your engineering background from Manhattan, and your early technical career in applied physics during a 20 year career with IBM). Still, you seemed a little anxious about how much the invention *might* cost your company to use to meet internal needs. You noted that when individuals with non-engineering backgrounds, unlike you and this Caltech graduate, get involved in licensing issues, expenses can go up for a company like yours. This may sometimes be true.

Your very phone call, unsolicited, disclosed how the invention had moved through your company, and expressed the simple hope that you and your management prefer to work directly with this Caltech graduate, instead of through a hired intermediary or intermediaries. This has not been discouraged to date, and I see no reason to discontinue the approach, unless later events dictate otherwise. The course you and your company are on, as of today, is appropriate.

Now, as with all matters in my discourse with you and others regarding the matter at hand, whether in the past, the present or the future, I will continue to engage Primary Counsel, and the considerable resources Primary Counsel draws on beyond my view, both within your company and outside of it. This includes reviewing the content of conversation, which requires an ear for administrative nuance, which I may lack, though Primary Counsel is quite adept at responding to, having gone through the process before.

Therefore, I have been moved to submit the attached agreement through you for review and consideration. It is prepared for your company, with no greater intent than to remove your projected uncertainty about cost, in light of perceived merit and function. And this is done to assuage unreasonable delay to begin applying the invention to your own needs, without ambivalence or hesitation about affordability. The greatest danger, at this time, is the very ambivalence, hesitation and doubt that impedes a multitude of others who may deserve to have access to this technology in a timely fashion.

Ex1 p15

It is hoped that the enthusiasm you and your company may share through the invention might proceed forward without trepidation, in the realization that an inexpensive and highly profitable pathway has been set before you. This may help create a strong economic incentive and the business wherewithal to not only begin addressing your own, considerable internal needs with the invention, but to rapidly impart what you know to as many of your fellow business, government and academic institutional entities as possible. It is important that they, too, soon enjoy restored vigor, like you or a person who receives a powerful preventitive vaccine that abates a pernicious virus before it has a chance to do any real harm to the body.

As a president once said (the only president to hold a patent, which is still in the U.S. Patent Office files, and is used in the design of every submarine on Earth, whether research, military or nuclear):

> Next came the Patent laws. These began in England in 1624; and, in this country, with the adoption of our constitution. Before then, any man might instantly use what another had invented; so that the inventor had no special advantage from his own invention. The patent system changed this; secured to the inventor, for a limited time, the exclusive use of his invention; and thereby added the fuel of *interest* to the *fire* of genius, in the discovery and production of new and useful things. Abraham Lincoln, d5/February 11th, 1859

Now, let this *added interest* and reward within the attached agreement *fire* the desire we all have to solve this problem for the peoples of this nation, and those who share this Earth with them, who may deserve not to have their lives dramatically inconvenienced.

The terms of the attached agreement, and the guarantees within, shall remain in effect until sunset d6/June 20th, 1997, at which time they will expire, unless extended in writing prior to that date by this party.

If I do not have further contact with you today, I hope that you, Mr. Rosenthal, Mr. Phelps, Mr. Ricciardi, your management, staff and company have a restful weekend.

With Regards,

Anthony Sneed

Anthony Sneed,
Life-time Member, Caltech Alumni Association
Inventor of "Serial Encryption System-Bypass of Year 2000 Date Malfunctions"

CC: Lee Browne, Emeritus Faculty Member, Caltech
David O'Reilly, Patent Attorney

2

EX1 *p16*

INVENTION LICENSE AND DISCOVERY AGREEMENT   Initial ____

This Invention License and Discovery Agreement (herein "Master Agreement") is made and entered into as of the 5th day of June, 1997, by and between Anthony Sneed, whose business address is 2058 N. Mills Avenue, Claremont, California 91711 (herein "Inventor"), and IBM Corporation (herein "Licensee"), whose principal place of business is Old Orchard Road, Armonk, NY 10504.

In consideration of the mutual covenants and agreements contained herein the parties hereto agree as follows:

1. Inventor makes available to Licensee the invention in accordance with the provisions of Exhibit "A" attached hereto and made a part hereof.

2. Inventor has applied best efforts to identify and deliver forth an invention to address the Year 2000 Malfunction (herein "Invention") that the Licensee deems satisfactory to achieve business objectives unique to Licensee's own company, as well as with Licensee's current or future customers.

3. The compensation as set forth in Exhibit "A" is confidential between Inventor and Licensee and is not to be intentionally released to any other party for any reason whatsoever. In addition to such compensation, Licensee agrees to reimburse Inventor for expenses incurred at the request and with the approval of Licensee.

4. Licensee agrees to pay all invoices from Inventor within thirty (30) days of the date of invoice. A service charge of three percent (3%) per month will be assessed on invoices not paid within sixty (60) days from the invoice date. This charge will be added to and reflected on future invoices.

5. Inventor agrees to use best efforts in developing other inventions or discoveries that the Inventor may deem beneficial to Licensee. This may include additional research and discovery to increase the economy and useful life of Licensee's programs and equipment.

6. The application of Invention shall be under the direct control, and applied at the sole discretion of Licensee to the Licensee's benefit. The determination of fitness of Invention to be used in any particular situation is exclusively Licensee's.

7. Licensee will designate a management level person to: (a) coordinate and supervise the administrative requirements of Invention under this Master Agreement; and (b) interface with Inventor to affect the overall management of the relationship between Inventor and Licensee.

8. Inventor reserves the right to license Invention to another party or parties at the sole discretion of Inventor.

1

Ex.| p17

9. All confidential data relating to Licensee's business or use of Invention submitted by Licensee to Inventor shall be pursuant to this Master Agreement. Inventor will safeguard said data to the same extent that Inventor safeguards confidential data relating to Inventor's own business. If, however, such data is publicly available, or already in Inventor's possession or known to Inventor, or is rightfully obtained by Inventor through third parties, Inventor shall bear no responsibility for Inventor's disclosure or use of such data, inadvertent or otherwise.

10. Inventor shall not be liable to Licensee or to any third party for consequential or incidental damages, lost profits, or any other similar damages under any circumstances, even if Inventor has been advised in advance of the possibility of such damages or losses. Except as set forth above, Licensee hereby agrees to indemnify and hold Inventor, his employees, agents and his company harmless from and against any and all claims, costs, expenses including attorney's fees and cost of suit, which might arise out of or relate to this Master Agreement or the Invention and services to be provided hereunder.

11. This Master Agreement shall, in all respects, be governed by the laws of the State of California.

12. All terms and provisions contained herein shall ensure to the benefit of, and shall be binding upon, the parties hereto and their respective heirs, personal representative, successors and assigns.

13. Any dispute arising out of or in connection with this Master Agreement or any breach of this Master Agreement will be determined and settled by arbitration by submission of such dispute to a disinterested arbitrator agreeable to both parties. In the event the parties fail to agree on the arbitrator within ten days after notice from one party to the other requesting arbitration, such arbitration shall be held in accordance with the rules of the American Arbitration Association. The award rendered thereon by the arbitrator will be final and binding on the parties thereto, and judgment thereon may be entered in any court of competent jurisdiction. In settlement, the losing party shall pay the arbitration expense incurred to reach settlement.

14. This document constitutes the entire understanding and agreement of the parties. Any and all prior agreements, understandings or representations of any kind are hereby superseded by this Master Agreement and are of no further force or effect. No amendment or modification of this document shall be valid unless it is in writing and signed by all parties hereto.

IN WITNESS WHEREOF, the parties hereto have entered into this Invention Licensing and Discovery Agreement as of the day and year first above written.

IBM   Inventor

By:_____

Title:_____   Anthony Sneed

Date:_____   Date:_____

2

Ex1 p18

INVENTION LICENSE AND DISCOVERY AGREEMENT   Initial _____

EXHIBIT "A"

1. This document (herein "Exhibit") becomes part of the master Invention License and Discovery Agreement (herein "Master Agreement") between Anthony Sneed (herein "Inventor") and IBM (herein "Licensee"), which takes effect coincidence with the initiation of the Master Agreement.

2. The terms herein are specific to the invention described as a "Serial Encryption System-Bypass of Year 2000 Date Malfunctions" (herein "Invention") which mitigates the adverse effects of the year 2000 date anomaly for Licensee's widely-used, stored program machines and their associated data storage devices, programmed to use two digits to represent years.

3. The Master Agreement shall be for a period of four years, terminating at sunset, d2/June 18th, 2001, or until all remittances from Licensee are received, whichever is later. Licensee shall provide Inventor at Inventor's business address, by mail, a written report of monthly totals of usage of Invention across any line of code specified under Paragraph 4 of Exhibit. The report shall be due before the end of each fiscal quarter of Licensee, starting from the commencement of the Master Agreement. Each report shall specify the applicable time interval and accumulated royalties for the time period, in accordance with generally accepted accounting practices.

4. The cost to Licensee for license or discovery of the Invention is based on SCHEDULE A, below. SCHEDULE A encompasses the aggregate number of lines of code contained within each and every program Licensee applies the Invention, inclusive of programs developed for internal use by Licensee, its subsidiaries, affiliates or various lines of business to control its own processes and resources; programs Licensee markets as application packages to customers; programs in use on Licensee's customer's machines in government, academic or commercial sectors; and programs Licensee develops through its out-sourcing or contract programming services for Licensee's current or future customers in government, academic or commercial sectors.

SCHEDULE A

1. The first billion lines of code	14 cents per line of code
2. All lines of code beyond the first billion	FREE

5. A non-refundable payment of $755,000 (herein "Good Faith Payment") on the first billion lines of code under SCHEUDLE A is due and payable to "Anthony Sneed" at 2058 N. Mills Avenue, Claremont, California 91711, not later than four weeks after all parties sign the Master Agreement. The balance of remittance for the first billion lines of code under SCHEDULE A is due no later than the following schedule:

25% (less Good Faith Payment)	d5/June 18th, 1998
25%	d6/June 18th, 1999
25%	d2/June 19th, 2000
25%	d2/June 18th, 2001

3

Ex1 p19

a *7th-Gift* page

18. Logan goes silent and does not respond to subsequent correspondence from Sneed.

19. McDonough goes silent and does not respond to subsequent correspondence from Sneed.

20. Torressen goes silent and does not respond to subsequent correspondence from Sneed.

21. Sullivan goes silent and does not respond to subsequent correspondence from Sneed.

22. At or about this time, Sneed contacts Wilzbach to find out the status of Specification evaluation at IBM.
    22.1 Wilzbach tells Sneed that IBM Corporate has seized Specification from IBM Global Services.
    22.2 Wilzbach notes that this was against IBM Global Services' leadership's wishes.
    22.3 Afterward, Sneed reviews the status of IBM's copyright-restricted, evaluation-only efforts.
        22.3.1 There has been only evaluative-use under the no-charge provision Sneed set forth with Logan.
        22.3.2 There has been no disclosure of production-use of Specification by anyone at IBM.
        22.3.3 Specification is installed, for evaluative-use only, at the IBM computer site Logan identified.
        23.3.4 Evaluative-use has been initiated at both IBM Corporate and IBM Global Services.
        22.3.5 Sneed has invoiced Logan $49,764.55 for any non-evaluative use at the site Logan identified.
        22.3.6 Sneed disclosed and memorialized 14 cents per-line-of-code for production-use in Agreement.
        22.3.7 Torressen accepted a copyright-restricted, evaluation-only version of Specification on
            behalf of all the IBM Corporation and its business units, including IBM Global Services.
        22.3.8 Torressen received a copy of Agreement with Specification on or about d6/June 6, 1997.
23. Sneed contacts IBM CEO Mr. Louis V. Gerstner (Gerstner) by letter, before sunset, d5/June 19, 1997.
    23.1 Sneed reviews Specification, Agreement formation, and Logan, McDonough, and Torressen contact.
    23.2 Gerstner responds by designating Torressen to answer Sneed's letter to Gerstner.

ISB
2058 N. Mills Ave. Ste. 455
Jaremont, CA 91711

909-626-1777
Fax 909-626-4977

Mr Louis Gerstner, Jr.
IBM
Armonk, NY

d5/June 19th, 1997

Happy 5th day, Mr. Louis Gerstner, Jr.

A few weeks ago, I received a call from one of your Corporate staff members, who noted he was calling on behalf of your firm and the authority that extends from your name throughout it. He said he wanted to establish a "direct relationship" with this Caltech graduate. There was strong interest expressed by your Corporate staff member in an invention developed by this writer that quickly makes IBM system products impervious to the *Year 2000 Malfunction*.

The invention had been favorably evaluated by your outstanding Global Services group before the IBM Corporate staff member "took control of the invention from Global Services." I never found out if Global Services was a willing participant in this process or if the invention was simply expropriated, notwithstanding Global Services wishes. I continue to have a very high regard for IBM Global Services, which quickly assessed the tremendous efficiencies of the invention and relayed this to your Corporate staff, which your Corporate staff confirmed.

In any event, since the Corporate staff member was acting through the authority of your name over IBM, he was given complete cooperation. All questions your staff member asked were fully responded to, and discussion included a tutorial on the invention that permitted your staff member to become an instant expert, himself, in applying the invention with any IBM platform. Unlike Global Services, however, there was no follow-up by the Corporate staff member who preempted the entire Global Services group. One of his last comments was, "We want to work with you directly so that using your invention won't cost IBM a whole lot of money."

Therefore, I will not be held responsible for any malfunction that may occur to systems the invention is applied to. This is especially true of information withheld from your Corporate staff member outside of the written patent description he secured from Global Services. The attempted expropriation of this intellectual property can not be done without dire consequences to those systems where it is mis-applied. The dubious result that may now occur extends directly from marketing negligence, contractual indifference, and clerical myopia the last three weeks.

The action or inaction by your Corporate staff member has not diminished in any way whatsoever the high regard I have for your name over IBM, as well as for Global Services. Not everyone realizes the responsibility attached to using your name. My Primary Counsel placed other parts of the invention under separate filings than the one now at the U.S. Patent Office.

Happy 5th day, Louis Gerstner. With Regards,

Anthony Sneed
Anthony Sneed
Life-time Member Caltech Alumni Association
Inventor of *"A Serial Encryption System-Bypass of Year 2000 Date Malfunctions"*

Ex 1 p2/

ISB
2058 N. Mills Ave. Ste. 455
Claremont, CA 81711

909-626-1777
Fax 909-626-4977

Mr. Louis Gerstner, Jr.
IBM
Old Orchard Road
Armonk, NY

d4/June 25th, 1997

Executive Summary *(Year 2000 Services Technology)*

Earlier this month, I became aware of a $7 billion-plus business need/opportunity within IBM for a *services technology* that would accelerate century transition planning and implementation acceptance by senior managements, both inside and outside of IBM's installed systems base. I was briefed on the reluctance of many large and small customers to apply IBM's existing methodologies (called "windowing" and "expansion") to address the problem. After reviewing *complexity, testability and expense* (CTE) issues intrinsic to these methods, the "let's-wait-for-something-better" mind-set of many customers reflected a natural tendency to delay making any decision for as long as possible whose outcomes are all unfavorable.

Yet, this very mind-set may be counter-productive to each customer's self-interest. And the longer decision is delayed to act, the higher the risk to IBM or any other organization contracted to "fix" the problem. "Business-as-usual" in these circumstances, including staff reluctance to present an untenable situation to senior management, could lock everyone into a no-win situation, which after a period of further delay could become intractable.

Your personnel's review of the problem with this Caltech graduate revealed psycho-political-business dynamics that were more startling than the underlying technology issues. Any solution presented at this late date would have to far surpass the negative perceptions attached to extant methodologies. It would have to achieve extraordinary CTE efficiencies to dislodge negative feedback throughout all levels of a multitude of boards and managements.

The BEST Tool to Meet Very Stringent Century Transition Requirements

This Caltech graduate developed a tool meeting these unique specifications by d6/February 28th, 1997, and it was forwarded to Mr. David O'Reilly, a senior patent attorney who also provides representation for Caltech. Mr. O'Reilly immediately initiated the process to protect the invention as an encryption technology, a well established U.S. Patent Office category for similar inventions in military and space applications, and now applied to business. U.S. patent protection is now in force for the invention under Article I, Section 8, Paragraph 8 of the U.S. Constitution and patent protection legislation extended from this article.

The invention (herein "BEST" for *Basic Encryption Services Technology*) is a *strategic services technology* for your company and many others. After d2/June 2nd, 1997, BEST was critically analyzed by IBM Global Services personnel, sponsored through the office of Mr. Denny Welsh, president. Mr. Dick McDonough, of counsel to Mr. Welsh, was legal liaison, and Mr. Don Logan, a year 2000 program manager, was the technical management liaison. Both men did an outstanding job of facilitating rapid CTE (complexity, testability and expense) assessment of BEST and its benefit to IBM and its customers.

Ex1 p22

A Solution that Enhances Revenue and Success, Sometimes in as Little as 4 to 7 Weeks

Mr. Logan articulated the unique efficiencies BEST brings to IBM's efforts, as well as its attributes compared to "windowing" and "expansion." Mr. Logan's enthusiasm led to immediate arrangements, including pricing, to commercially apply BEST through an IBM system site containing three processors supporting customer conversion projects. Mr. Logan was honest in pointing out that BEST could permit many customers to address the year 2000 problem alone.

Yet, even in acknowledging this, Mr. Logan realized that BEST would let Global Services serve a larger conversion base, especially firms needing project management expertise in complex environments. Thus, BEST leverages Global Services' strengths (project management) while helping increase the percentage of the installed-base that could rapidly become year 2000 compliant over the next 12 to 24 months. In doing so, Global Services' revenues might be enhanced while its year 2000 project success rate approaches 100%. Global Services' century transition services market could simultaneously expand and perpetuate IBM's decades-long tradition of protecting customer assets, no matter the emergency, natural or man-made.

The BEST Possible Customer Services

As these plans were being put in place, which included invoicing Global Services under agreed arrangements set by d4/June 4th, 1997, IBM Corporate preempted Global Services' efforts. On d6/June 6th, 1997, Mr. Al Torressen, speaking for Corporate he noted, and, therefore, for your office, said that Corporate had taken control of BEST from Global Services on behalf of all of IBM. This expediency, Mr. Torressen noted, was being attempted to use BEST internally at IBM for the least cost. To the extent that Mr. Torressen represented your interests as IBM CEO, he received full cooperation, even while his cost-cutting short-cut raised questions:

> IBM's belief of *respect for the individual* includes Mr. Welsh, Mr. McDonough and Mr. Logan. What efforts were made to maintain this precedence for Global Services in Mr. Torressen's and Corporate's effort to take BEST from Global Services? How did his subsequent actions reflect IBM's other core values, including *"Fair and impartial treatment of suppliers"* and the *"BEST possible customer service."* Global Services, in all of its actions, never wavered from these beliefs, beginning with first contact through Mr. Welsh's office. If Mr. Torressen's actions were economically expedient with his *brother* IBMers, what economic expediencies might be attempted with his *non-brother* IBMers (customers and suppliers)?

These perceptions through brusk statements made early on by Mr. Torressen did not project the fundamental traditions that have helped make IBM successful. Therefore, as a safety device, Primary Counsel advised that any contract negotiations with Mr. Torressen have a well defined terminus, as brief as possible. A contract structured to meet Global Services' and its customers' needs was forwarded to Mr. Torressen by d2/June 9th, 1997, with a guarantee period ending d6/June 20th, 1997.

2

Ex 1 p23

Conclusions

As of d2/June 23rd, 1997, the attached contract for BEST is under your complete control, to reject or accept at your pleasure, without interference from any other party in your company. The measures taken and the situation that now exists fully respect the right of IBM to control its own processes, procedures and methods, consistent with the authority your name, and no other, currently has under heaven throughout all of IBM (Romans 13:1). As of your receipt of this letter, I have fulfilled all of my obligations to you to design BEST, deliver it to you, and to remove any interference to your freewill to do whatever you may now be moved to do. I know that this is very pleasing to my Father, his Son and the Spirit of truth. .

It is my hope that in your own way, in your own time, and to your own benefit, you may agree that BEST is focused on IBM's *three "T's": Software Technology, Services Technology and Hardware Technology.* With your concurrence, which I seek, BEST is a tool for Global Services to expand its operations management market share, which is easiest done with working sites. BEST is consistent with the $19 billion of success Global Services already achieves through its 1/3rd revenue and 2/3rds profit contributions to IBM. Yet, in addition to this, BEST can help protect asset investments within IBM, used every day to achieve aggressive business objectives, including protecting market equity and future demand for IBM products and services.

The remittance of the attached invoices is now required, specifically to avoid the perception of favoritism between your company and other sites similarly invoiced. Without remittance forthwith, you and your company are to cease and desist using BEST in any way whatsoever, and confirm such in writing. Remittance after one week from today, if not before, is satisfactory, and avoids a perception that benefits neither of us.

Finally, the president of the Caltech Alumni Association, Mr. Ed Lambert (and a fellow member of Fleming house when we were undergraduates), is developing my candidacy to be nominated for Caltech's Distinguished Alumnus Award, the institute's highest honor. He and four other members of the Caltech Alumni Association (celebrating its centennial this year) have reached the same conclusion as IBM Global Services. Your favorable comments to Mr. Lambert, if any, by fax (206-232-4734) or phone (206-232-4735) would be appreciated for my candidacy. There would also be a debt of gratitude if you took the time to do so.

Happy 4th day, Louis Gerstner. With Regards,

Anthony Sneed

Anthony Sneed
Life Member Caltech Alumni Association
Inventor of "*A Serial Encryption System-Bypass of Year 2000 Malfunctions*" (herein as "BEST")
Attachments: (1) Outstanding BEST Invoices (2) Contract for your signature
CC: Lee Browne, Caltech Emeritus Faculty Member; David O'Reilly, Patent Attorney,
     Ed Lambert, President Caltech Alumni Association

Ex1 p24

Mr. Louis Gerstner, Jr.   d6/June 27th, 1997
IBM
Armonk, NY   <u>PERSONAL</u>

Happy 6th, Mr. Gerstner:

Primary Counsel has relayed to me that when you were with McKinsey & Company, all or part of one of your fingers was severed off by a lawnmower while doing chores at home. *Yet you completed a scheduled client progress review within a day or so of having your finger rejoined to the rest of your body.* I must pray about this, because it represents a kind of personal sacrifice reflecting, in some degree, the Testimony of my Father's One and Only Son. It clearly underscores dedication and commitment that can never be adequately compensated in this world.

The closest I have come to such testimony, which you have personally endured in some measure, was to have no food or water touch my lips for seven consecutive days. These seven consecutive days were imbedded within 40 days of no food or water touching my lips over a 55 day period. During the 55 days, 40 of which no food or water touched my lips, I, too, reported for work on a regular schedule. I did this in submission to the Testimony of my Father's Son in John 6:27 and John 6:55. I voluntarily endured this, just as you voluntarily went to work. At the time, it created small budget surpluses for my wife and children, which were put to good use.

A couple days ago, Harrison "Jack" Schmitt (Caltech '57 and Apollo 17 crew member) and I were discussing the difficulties of the *7th Apollo mission*, known as Apollo 13 since Apollo 7 with Walley Schirra, in October of 1968, was the *first* Apollo mission. We covered Fred Hayes' kidney infection from reduced water consumption during the *7th Apollo mission*. Jack Schmitt was curious about any side effects I might have experienced during seven consecutive days when no food or water touched my lips. I told him that there were none, even while continuing to work 50 to 60-hour weeks at the office (most office jobs have no heavy lifting, making it easier).

In turn, I asked former senator Schmitt why Frank Borman, Jim Lovell, Jr. and Bill Anders read the entire chapter of Genesis 1, and Genesis 2:1, 2 from lunar orbit d5/December 25th, 1968. He said that it was neither Jim's nor Bill's idea. Since Frank was commander, they simply followed his lead when he was moved to do it. It should be noted that only a minor firing error of the Service Module rocket could have caused Apollo 8 to crash into the moon when entering lunar orbit. Had it not fired when time to leave, they might have become orbiting memorials.

If there are sacrifices that you may need on the part of this Caltech graduate, let me know. When commitment and dedication are manifest, as mentioned above, a way will be found to sustain, perpetuate and enhance them. I have sacrificed before [John 12:26], just let me know. I hope you and your family, and your company have a restful 7th day after sunset.

Happy 6th day, Louis Gerstner. With Regards,

Anthony Sneed
Anthony Sneed

EX/ p25

a *7ʰ-Gift* page

24. On d6/June 20, 1997, Torressen sends Sneed an IBM Letter of Corporate Intent (LCI) which states:
  24.1 "Thank you, <u>again</u>, for your unsolicited, non-confidential letter of May 29, 1997 to Dick McDonough concerning your provisional patent filing, 'A Serial Encryption System By-pass of Year 2000 Date Malfunctions.'"
    24.1.1 Sneed and Torressen never discussed Sneed's letter to McDonough; so Torressen's '<u>again</u>' is disingenuous and unnecessary; Torressen's comments on or about d5/June 5, are noted above.
    24.1.2 Sneed's letter is non-confidential <u>between</u> IBM Corporate and Global Services.
    24.1.3 Torressen states nothing new; Sneed is already working with both Corporate and Global Services, so Torressen's attempted distinction between the two may have no legal merit.
  24.2 "In accordance with your direction that the materials were non-confidential and could be freely distributed <u>within</u> IBM, we conducted a technical review."
    24.2.1 Torressen is correct that evaluative-use, only, may occur anywhere <u>within</u> IBM at no charge.
    24.2.2 There is no confidentiality presumed between IBM Corporate and IBM Global Services since Sneed is working with both of them.
    24.2.3 Sneed stipulated that there is no charge for evaluative-use, only, anywhere <u>within</u> IBM.
    24.2.4 Torressen never declares, discloses or suggests production-use in any way.
    24.2.5 Torressen's evaluative-use at IBM Corporate parallels Logan's evaluative-use at Global Services.
  24.3 "As I told you in our June 5, 1997 phone conversation, your letter was forwarded to me for coordinating IBM's review and response."
    24.3.1 Sneed never discussed, with Torressen, Sneed's letter to McDonough during Torressen's unsolicited d5/June 5 call; his assertion that Sneed had is disingenuous and unnecessary.
    24.3.2 Torressen's unsolicited call was to declare IBM "took control" of Specification for all of IBM.
    24.3.3 If he knew of Sneed's letter to, and contact with McDonough, he may not have called Sneed.
  24.4 "IBM has now completed its review and concluded it has no interest in pursuing a license described in your fax to IBM June 6, 1997, nor does IBM have any interest in pursuing your June 10, 1997 proposal to IBM's Don Logan."
    24.4.1 He admits custody and awareness of written contract formation between Sneed and Logan.
    24.4.2 This meets <u>a</u> requirement of contract formation: identify pricing, terms and the item of value:
      24.4.2.1 No charge for evaluative-use, only, of a copyright-restricted version of Specification.
      24.4.2.2 A $49,764.55 per-site-fee for any production-use of Specification, like Logan's site.
      24.4.2.3 14 cents per-line-of-code, for the aggregate number of lines-of-code in programs Specification is applied, as stipulated in the letters Torressen references and Agreement included thereof.
      24.4.2.4 Torressen only declares evaluative-use of Specification, and, as agreed, there is no charge; he makes no admission of production-use in any way whatsoever.
  24.5 "In the interest of facilitating any <u>further interaction</u>, I request that you deal with me rather than other individuals at IBM."
    24.5.1 This meets another requirement of contract formation: identifying the parties of the contract.
    24.5.2 The only reason for "<u>further interaction</u>" is to disclose production-use, if any, and payment.
    24.5.3 It is consistent with Torressen's declaration, in the d5/June 5, 1997 call, that Torressen is designated as Sneed's single-point of contact for all of IBM regarding Specification.
  24.6 "As the individual responsible for coordinating this matter, your fax of June 19, 1997 to Mr. Gerstner was referred to me for response."
    24.6.1 Gerstner conferred authority on Torressen, making Torressen's actions, statements and deeds binding upon on all of IBM.

Ex1 p26

24.7 "Two points require clarification (1) IBM has no intention of an 'attempted expropriation of this intellectual property[.]'"

    24.7.1 Torressen and IBM can only fulfill this intention by paying for production-use, if any, within contract formation terms Torressen admits custody on behalf of Gerstner and all of IBM.

    24.7.2 Torressen admits to evaluative-use in Sneed's contract formation terms stipulate as no charge, a provision of contract formation that Torressen and IBM *readily accept from Agreement*.

24.8 "[A]nd (2) I never made a statement in any way resembling the quote you attribute to me at the end of the third paragraph."

    24.8.1 Sneed recalls Torressen made the statement, memorialized in multiple contemporaneous letters: "We want to work with you directly so that using your invention won't cost IBM a whole lot of money."

    24.8.2 Sneed stands by his recall, as memorialized in contemporaneous letters.

    24.8.3 Sneed met Torressen's requirement, with 14 cents per line of code instead of up to $2 per line.

24.9 "There are inaccuracies in many of your 10 letters to IBM, but it does not seem productive to address each of them."

    24.9.1 The d6/June 6, 1997 and d3/June 10, 1997 letters memorialize contract formation terms.

    24.9.2 They accurately reflect Sneed's disclosure of pricing and terms for contract formation.

    24.9.3 Since they express Sneed's pricing and terms to Logan and Torressen, there is no inaccuracy.

    24.9.4 Torressen makes specific reference to these letters without disparaging their authenticity or accuracy in his second paragraph, while making general, unsubstantiated assertions about other letters in his third paragraph.

24.10 "For example, your most recent letter of June 18, 1997 addressed to Dan Sullivan (which is only one of 9 identified letters over the last 3 weeks) also contains several inaccuracies and misstates what has occurred."

    24.10.1 The d4/June 18 letter cites correspondence with Dr. David Baltimore, Caltech president, and Sneed's successful efforts designing an on-board operating system for the NASA Galileo mission to Jupiter; Dr. Baltimore's predecessor, Dr. Thomas Everhart, had sent a letter to Sneed complimenting Sneed's effort, and encouraging Sneed to contact Caltech about progress on other endeavors. Sneed performed by including communication about Specification.

    24.10.2 The letter to Sullivan included a copyright-restricted, evaluative-use-only version of Specification.

    24.10.3 The letter affirms the $49,764.55 site-fee for production-use, if any, and an invoice.

    24.10.4 The letter acknowledges Sullivan's favorable comments regarding Specification.

    24.10.5 It requests those comments be forwarded to Sneed's boarding school, Cate, and to Caltech.

        24.10.5.1 Caltech emeritus faculty member Lee Browne was promoting Sneed's candidacy for the Caltech Distinguished Alumnus award, based on Specification.

    24.10.6 To Sneed's recollection, there are no inaccuracies in any of the disclosures in the letter or the request therein to Sullivan.

24.11 "We appreciate the importance of your effort to find a solution to the challenge of Year 2000 which will have worldwide affect [sic] on customers and industry. Although your proposal is not currently of interest, we would encourage you to contact me in the **future** if a patent issues. We wish you well in the **future**."

    24.11.1 Herein is the final meeting of the minds for contract formation between Sneed and IBM.

    24.11.2 In good faith, Sneed disclosed all pricing and terms in the d6/June 6 and d3/June 10 letters.

    24.11.3 Sneed signs those letters, in good faith.

    24.11.4 Torressen declares his signature is authorized by Gerstner to represent IBM's position.

Ex 1 p27

24.11.5 IBM 's only stipulation is the timing of Agreement going into force, and no issue of pricing, terms, and conditions acknowledged as received and vetted by Torressen.

24.11.6 Torressen signs the d6/June 20, 1997 IBM Letter of Corporate Intent in good faith.

24.11.7 IBM, through Torressen, stipulates a patent issuing completes contract formation.

24.11.8 IBM sets a "future" performance date (Date) for Sneed to initiate contact with IBM.

24.11.9 IBM admits only to evaluative-use prior and during final contract formation.

24.11.10 IBM sets the condition to disclose production-use of Specification, if any, after Date.

24.11.11 If IBM doesn't, IBM violates its own self-imposed condition not to "expropriate" Specification.

24.11.12 The "future" Date becomes d3/May 22, 2001, when the US Patent Office issues patent no. 6,236,992 to Sneed for Specification, "A Serial Encryption By-Pass of Year 2000 Date Malfunctions."

24.12 IBM admits to evaluative-use prior to d3/May 22, 2001, within "IBM."

24.12.1 Torressen retro-actively defined "IBM" to mean IBM Corporate and IBM Global Services, with no confidentiality or distinction between them regarding evaluative- or production-use.

24.13 IBM never admits to production-use, in any way, before or after d3/May 22, 2001.

24.13.1 The Specification, when installed, has a useful life of at least 28 years, or at least until 2025.

24.13.2 The minimum 28-year useful life is fully disclosed and memorialized within Specification.

Attached exhibit is admitted from Exhibit 5,
IBM Motion to Dismiss First Amended Complaint
Declaration of Bridget A. Hauler, March 12, 2012

E4/ p28

Office of the Director of Licensing
Intellectual Property & Licensing

500 Columbus Avenue
Thornwood, New York 10594

CERTIFIED MAIL

June 20, 1997

Mr. Anthony Sneed
I.S.B.
2058 N. Mills Avenue, Suite 455
Claremont, CA 91711

Subject: *Your letter dated May 29, 1997 to Mr. Dick McDonough*

Thank you, again, for your unsolicited, non-confidential letter of May 29, 1997 to Dick McDonough concerning your provisional patent filing, "A Serial Encryption System By-Pass of Year 2000 Date Malfunctions." In accordance with your direction that the materials were non-confidential and could be freely distributed within IBM, we conducted a technical review within IBM. As I told you in our June 5, 1997 phone conversation, your letter was forwarded to me for coordinating IBM's review and response.

IBM has now completed its review and concluded it has no interest in pursuing a license described in your fax to IBM on June 6, 1997, nor does IBM have any interest in pursuing your June 10, 1997 proposal to IBM's Don Logan.

In the interest of facilitating any further interaction, I request that you deal with me rather than other individuals in IBM. As the individual responsible for coordinating this matter, your fax of June 19, 1997 to Mr. Gerstner was referred to me for response. Two points require clarification (1) IBM has no intention of an "attempted expropriation of this intellectual property", and (2) I never made a statement in any way resembling the quote you attribute to me at the end of the third paragraph. There are other inaccuracies in many of your 10 letters to IBM, but it does not seem productive to address each of them. For example, your most recent letter of June 18, 1997 addressed to Dan Sullivan (which is only one of 9 identified letters over the last 3 weeks) also contains several inaccuracies and misstates what has occurred.

We appreciate the importance of your effort to find a solution to the challenge of Year 2000 which will have worldwide affect on customers and industry. Although your proposal is not currently of interest, we would encourage you to contact me in the future if a patent issues.

We wish you well in the future.

Sincerely,

Al Torressen
Program Manager, Licensing

AT:gm

cc: Dick McDonough

Ex 1 p29

ISB
2050 N. Mills Ave. Ste. 455
Claremont, CA 91711

909-626-1777
Fax 909-626-4977

Mr. Dan Sullivan
IBM
White Plains, New York

d4/June 18th, 1997

Happy 4th day, Dan.

Attached is correspondence with the new president of Caltech, Dr. David Baltimore. He has been previously apprised of the successful efforts by this Caltech graduate on the Galileo mission, which your organization also received a written copy.

A copy of the invention titled "*A Serial Encryption System-Bypass of Year 2000 Date Malfunctions*" is included. This is forwarded with full respect to the stipulation of the $49,764.55 invoice received by IBM Global Services limiting the use of the invention to a single (1) IBM system site. This invoice is independent of any extant contract under consideration, unless such contract is actually consummated between IBM Corporate and this Caltech graduate.

Whether a commercial contract is with IBM Corporate, another part of IBM, or with another firm in the industry, there is an academic component to what has been achieved. Please forward to Dr. David Baltimore (Office of the President, Caltech, Pasadena, California 91125) a brief summary of the research value of the invention to your company, government, academe and the private sector. This may simply reflect the already favorable representations made by IBM Global Services staff and IBM Corporate staff.

All parts of your organization have been forthright in recognizing the pure research value of the invention, independent of its considerable commercial attributes. Your letter regarding the discovery result of the invention to each individual on the attached list will ensure academic recognition at my boarding school and university, independent of commercial interest in the invention expressed by your company and other firms. Consideration of this personal request is much appreciated, which has been made to me by Mr. Lee Browne, Caltech emeritus faculty member. As an aside, I have received an offer for my life story from an associate of Mr. Francis Ford Coppola, and a book contract offer from a major publisher. These, and other requests, if consummated, will reflect the outcome of events now under consideration.

Happy 4th day, Dan. With Regards,

Anthony Sneed

Anthony Sneed
Life-time Member Caltech Alumni Association

CC: Mr. Lee Browne, Caltech Emeritus Faculty Member
     Dr. Sanderson Smith, National Council of Teachers in Math (NCTM)
     President of Council on Presidential Awardees in Math.

Ex1 p30

Please forward your summary of the invention's possible impact to the following:

Mr. Paul Denison
Chairman, Board of Trustees
The Cate School
1960 Cate Mesa Road
Carpinteria, CA 93014-5005

Mr. Peter Thorp
Headmaster
The Cate School
1960 Cate Mesa Road
Carpinteria, CA 93014-5005

Mr. Fred Clark
Former Headmaster
c/o Mr. Peter Clark
Oregon State University
Geosciences Dept/Wilkinson Room 114
Corvalis, OR 97331

Dr. Sanderson Smith
NCTM Board Member
The Cate School
1960 Cate Mesa Road
Carpinteria, CA 93014-5005

Mr. Lee Browne
Emeritus Faculty Member
Caltech
Pasadena, CA 91125

Dr. Carver Mead
Office of the President
Caltech
Pasadena, CA 91125

Ms. Betty Woodworth
Emerita Archivist
The Cate School
1960 Cate Mesa Road
Carpinteria, CA 93014-5005

Ms. Judy Goodstein
Caltech Archives
Mail Code 015A-74
Pasadena, CA 91125

a *7ʰ-Gift* page

1997 July: Sneed contacts the Department of Justice (DOJ) to open an inquiry about IBM's conduct

25. Torressen made unsupported statements in the IBM Letter of Corporate Intent dated d6/June 20, 1997.
    25.1 Torressen falsely claimed that Sneed called Torressen, first, with the d5/June 5, 1997 call.
    25.2 Sneed did not know who Torressen was when Torressen called Sneed on d5/June 5, 1997.
    25.3 Torrssen had never been mentioned or introduced to Sneed prior to Torressen's unsolicited, d5/June 5, 1997 call to Sneed.
26. As Sneed reflected on previous information from Wilzbach, the Torressen call appeared to be a ruse to bypass the 1956 Consent Decree "firewall" between IBM Corporate and IBM Global Services.
    26.1 Torressen, supported by Gerstner, attempted to retro-actively construe the call as being initiated by Sneed; it wasn't.
    26.2 Sneed was already working with McDonough and Logan, before Torressen's unsolicited call.
27. Sneed contacted DOJ to review IBM's and Torressen's conduct.
    27.1 Mr. Ken Brown (Brown), Senior DOJ Attorney in the Antitrust Division, opened the inquiry.
    27.2 Brown was knowledgeable about the 1956 Consent Decree between DOJ and IBM.
    27.3 Brown vetted the matter with Sneed and took statements from Sneed, which included reviewing the nature of the Specification and contact with various parties at IBM.
    27.4 Brown noted that if called upon, Sneed *may* have to testify as a witness. IBM later protested that it seemed DOJ was scheduling a hearing on matters about which IBM wasn't informed.
    27.5 Brown provided Sneed informal, anecdotal advice about contract negotiation with IBM, as well as free market leverage to gain acceptance of Specification if IBM proved uncooperative.
    27.6 Brown noted, and Sneed agreed, that the advice was officially meaningless, yet Sneed found it exceedingly helpful, especially Brown's imperative to get IBM to put their position in writing by any means necessary. Brown and DOJ were concerned about a potential Y2K litigation surge.
    27.7 Brown also made clear that DOJ would reflect impartiality in all communication with the parties.
28. Brown vetted the matter with Sneed and determined that Sneed had no liability whatsoever under DOJ's 1956 Consent Decree and conduct between IBM Corporate and IBM Global Services.
29. Brown also determined that there may, in fact, be no presumed confidentiality between IBM Corporate and IBM Global Services in some instances, thus making Torressen's ruse perhaps unnecessary and harmless to Sneed.
    29.1 Torressen's ruse call was harmless to Sneed, since Sneed was working in good faith with parties at IBM Corporate and IBM Global Services, inclusive of Logan, McDonough, Sullivan, Gerstner, and Torressen.
    29.2 Torressen's call did memorialize that IBM CEO Gerstner was aware of Specification and Gerstner authorized Torressen to act on all of IBM's behalf, inclusive of IBM Corporate and IBM Global Services. This demurred any issue of confidentiality between the two regarding Specification.
    29.3 Brown noted that IBM Corporate has authority to close the IBM Global Services subsidiary, and therefore has all lesser authority a corporation may exercise over a subsidiary.
    29.4 Brown also remarked to Sneed that if IBM were a car company, DOJ could not compel IBM to make fuel-efficient cars to the detriment of gas-guzzlers' higher-profits, even if IBM has the technology and know-how to do so. Logan had previously warned Sneed that alternatives to Specification, though less efficient, time-consuming, and labor-intensive to apply, do increase demand for more of IBM's resources, whether personnel, hardware, upgraded software, and so forth. Sneed surmised that if Y2K represents a once-in-a-life-time billing opportunity, the other methods would maximize IBM's profits. Y2K remediation contracts were a "must" for virtually every large IBM installation, giving IBM some leverage for this type of business with its most vulnerable customers.

Ex1 p32

30. Brown and DOJ compelled the first written response to Sneed from a legal officer of IBM, McDonough, IBM VP and Assistant General Counsel, who previously had gone silent and stopped responding to correspondence from Sneed after d6/May 29, 1997.

  30.1 Prior to DOJ's involvement, only Torressen had put IBM's position in writing, on behalf of Gerstner.

  30.2 McDonough's d6/July 11, 1997 letter to Sneed included the following admissions:

    30.2.1 Evaluative-use of Specification, within IBM, had occurred under provisions of Agreement that permits IBM's evaluative-use, within IBM, at no-charge, for the copyright-restricted, evaluation-only version of Specification IBM admits having custody from Sneed.

    30.2.2 McDonough makes no declaration of production-use in any way whatsoever.

    30.2.3 McDonough affirms that Torressen's signature legally represents IBM.

    30.2.4 McDonough is critical of contact with Logan, Sullivan, Gerstner, DOJ personnel and himself.

    30.2.5 McDonough asserts that Sneed's communication with others and McDonough contains assertions, innuendo, or misstatements, though McDonough does not substantiate or provide particulars of communication he may or may not be referring, and parties he may or may not be identifying, other than Torressen and contract formation memorialized in IBM's d6/June 20, 1997 Letter of Corporate Intent to Sneed.

    30.2.6 McDonough alleges he is acting in good faith, yet there was no response whatsoever to Sneed's last written letter to McDonough, d6/May 29, 1997; McDonough was silent for almost a month and a half, until DOJ contacted IBM.

    30.2.7 McDonough affirms contract formation in IBM's d6/June 20, 1997 Letter of Corporate Intent.

    30.2.8 Both McDonough's letter and the Letter of Corporate Intent reference the "future," defined in the Letter of Corporate Intent as after a patent issues.

    30.2.9 However, McDonough's d6/June 11, 1997 letter is self-defeating and possibly duplicitous.

    30.2.10 McDonough, in good faith, affirms the terms of IBM's d6/June 20, 1997 Letter of Corporate Intent that declares Torressen as the single-point of contact between Sneed and IBM regarding Specification, and no other; Torressen is specifically designated as such by IBM CEO Gerstner, whom Torressen explicitly references in IBM's Letter of Corporate Intent.

    30.2.11 McDonough's letter, without reference to Torressen or Gerstner, may be unauthorized.

    30.2.12 Since neither Torressen nor McDonough nor Gerstner nor Logan nor Sullivan nor anyone at IBM discloses production-use, before or after McDonough's d6/June 11 letter, IBM's no-charge evaluative-use of Specification is within Agreement and terms set forth by Sneed.

    30.2.13 There could only be "legal matters," to use McDonough's term in his d6/June 11 letter, such as IBM contradicting its own self-imposed condition not to expropriate Specification, if IBM is engaging in non-evaluative use.

    30.2.14 Non-evaluative-use, if any, would be disclosed by Torressen under Agreement.

    30.2.15 Furthermore, Torressen stipulated a "future" Date of d3/May 22, 2001 as a condition of contract formation and completion of contract formation.

    30.2.16 McDonough's recommendation for Sneed to contact McDonough regarding Specification and inquiry about use, outside of written stipulation within the d6/June 20, 1997 IBM Letter of Corporate Intent, is exactly the scenario McDonough construes as bad faith conduct, and perhaps a breach of contract formation. Setting forth that scenario seems the whole purpose of McDonough's d6/June 11 letter to Sneed.

    30.2.17 Since there has been no production-use disclosure by Torressen, McDonough, or anyone else at IBM, and contract formation has a specific "future" trigger-Date, there are no "legal matters" of which Sneed is aware between the parties.

Ex1 p33

U. S. Department of Justice

Antitrust Division

Bicentennial Building
600 E Street, NW
Washington, DC 20530

Rcvd 2:30 PM about PST

July 8, 1997

Dick McDonough, Esq.
IBM Global Services
Route 100, Mail Drop 4427, Building 4
Somers, NY 10589

Re: Complaint of Anthony Sneed to U.S. Department of Justice

Dear Mr. McDonough:

Yesterday, I received a copy of a seven-page letter from
Anthony Sneed of ISB to you. In his letter, Mr. Sneed attributes
to me or the Department of Justice a number of statements and
conclusions that were not made by me or anyone else at the
Department.

As you know, Mr. Sneed complained to the Antitrust Division
about the way IBM and IBM Global Services have conducted
negotiations for IBM or IBM Global Services' potential license
and use of an invention for addressing Year 2000 malfunctions.
We advised Mr. Sneed that Section VIII of the 1956 Consent
Decree, which specified the conditions under which IBM could
operate its service bureau business, was no longer in effect and
had no application to his complaint. We also advised Mr. Sneed
that neither the Consent Decree nor the antitrust laws impose any
obligation on IBM to license and employ Mr. Sneed's invention and
that IBM is therefore free under the Consent Decree and the
antitrust laws unilaterally to refuse to license or use the
invention.

Contrary to Mr. Sneed's assertions, neither I nor anyone
else at the Department reviewed or reached any conclusion as to
the value of his invention. Nor did we express any opinion as to
whether IBM Global Services has entered or is able to enter into
any agreement that is binding on IBM.

We advised Mr. Sneed that we would be willing to consider
additional information concerning alleged antitrust violations
and that, if he provides credible evidence of such a violation,
he likely would be a witness in any enforcement action. However,
we did not tell Mr. Sneed that he would need to respond to a
deposition request, appear before a grand jury or appear before a

1) Consistent though issue concerned misappropriation not licensing

2) Agreed invention with attributes described should be made commercially available through competitive m forces

Availability accepted to all of above, though only asked to consider specific
3) availability as a witness by Justice, if needed and appropriate.

Ex/ p34

congressional hearing. To the contrary, we told Mr. Sneed that
the information he provided, without more, not did not disclose
any violation by IBM of the Consent Decree or antitrust laws.

Finally, we did not tell Mr. Sneed that the Consent Decree
has not been actively enforced for over a decade. Rather, we
said that Section VIII of the Consent Decree applied to few of
the computer related services provided by ISSC.

I hope this letter clarifies any misunderstanding about the
Department's discussions with Mr. Sneed, and I regret any
confusion or inconvenience resulting from Mr. Sneed's
misstatements.

First section VIII
time mentioned is
here. Never
specifically
mentioned,
that can be
recalled; in
previous conversation,
otherwise it would
have been documented," the
with the "about 10 years
consent decree was
not actively enforced."

Sincerely,

Kent Brown
Attorney

CC: Anthony Sneed
     Donald J. Rosenberg, Esq.

⑤ All other material
in the complaint
not mentioned here
or annotated in the
margin of this letter
is unaffected and
stands as recalled
and written.
Anthony Sneed
ds/July 8th, 1997

2

Ex.1 p35

ISB
~~5058~~ N. Mills Ave. Ste. 455
Claremont, CA 91711

909-626-1777
Fax 909-826-4977

Mr. Kent Brown   d2/July 7th, 1997
U.S. Justice Department
Washington D.C.   CONFIDENTIAL

Mr. Brown:

1. Every position you presented that is based in law or substantiated by material fact or facts will be maintained and not conceded in any way whatsoever, which I expect of you and Justice as well. In addition, your recommendation to show market development diligence in light of the alleged expropriation (my allegation, and not yours) of technology has been beneficial.

Let it be acknowledged that the specific firms you mentioned were only for illustration. When you asked, for example, if this Caltech graduate had contacted "XYZ" (XYZ being a very prominent national computer services company) this Caltech graduate answered truthfully that it had not been done. However, within 24 hours of your call, their chief technologist was contacted. He was delighted to receive a call, expressed strong interest in the invention, and immediately requested a non-disclosure agreement to receive a complete written description of the invention. This call, alone, may result in over $49,000 in revenue, all because you asked what to you was a very simple question. With the confidence generated from this call, several other firms were contacted within 48 business hours of your call. This is the preliminary follow-up report we agreed to exchange (not included are three other hardware manufactures):

  a. A Federally Funded Aerospace Firm in El Segundo, CA. They are struggling with year 2000 and have concerns about addressing the problem on all of their systems throughout the U.S. within 2 years using existing methodologies.

  b. A Nationally Prominent University System in Oakland, CA. IBM could not guarantee year 2000 compliance on key software they need fixed. According to the contact, the IBM marketing rep said that even though the software IBM provides is year 2000 compliant, the IBM lawyers would not let him say it is due to liability issues. In light of the testability concerns shared with me by IBM Global Services, the "it is but it isn't" posture is a classical way to make a sale without accepting liability if the product fails.

  c. The CEO of a $30 million Software Manufacturer in San Francisco, CA. He signed the non-disclosure agreement and had his technical team analyze the invention. He felt IBM might not cooperate in letting them apply the invention on their IBM AS/400. The CEO felt the invention was superior to either method IBM had offered, which IBM does encourage them to use.

  d. The Information Systems Division Head of a Major Medical Center in Duarte, CA. After signing the non-disclosure agreement and analyzing the invention, this division head agreed that the invention is the type of cross-platform solution they were looking for. The division head immediately secured a second non-disclosure agreement for their outsourcing company to sign to begin planning to apply the invention. There was no question about the $49,000 fee because of the cost savings compared to other methodologies.

Ex I p36

e. <u>An Internationally Known University in Los Angeles, CA.</u> The year 2000 program manager is in the early assessment phase. After signing the non-disclosure agreement, the invention description was forwarded. Their needs involve rapid modification of their administrative systems that track academic and other accounting functions.

f. <u>A Nationally known Accounting Firm in Chicago, IL.</u> Discussions are preliminary. The partner involved in year 2000 requested the non-disclosure agreement after initial assessment of the invention's potential.

g. <u>An Internationally Known Telecommunications Firm, in Murray Hill, NJ.</u> Discussions are preliminary. An intellectual property analyst has been assigned, with follow-up expected after 24 hours.

2. All recommendations you made of a general nature, that are beneficial, will be followed, including letting free-market forces engage the invention through diligent marketing efforts. Each and every such effort shall be under the non-disclosure agreement you have reviewed, containing the same initial pre-conditions established with IBM Global Services.

3. If tangible information exists of collusion to exploit wide-spread corporate vulnerability and gouging of CEO's through the year 2000 crisis by certain industry groups, it will be forwarded to you forthwith. From our discussion today, benign economic exploitation is not sufficient. In addition to a virtual statistical certainty of collusion, due process evidentiary requirements must be met to bring a valid claim. Even speculative statements between companies may not be sufficient in the absence of tangible documentation that reflects willful intent, and purposeful execution, of collusion to the detriment of free-market forces and societal norms.

4. Unless otherwise and specifically noted, lifting the 1956 consent decree does not necessarily remove constructive barriers between a subsidiary and its parent company. In this particular case of IBM Global Services and IBM, based on conversation with you this afternoon, contracts entered into by a subsidiary do not necessarily pierce the parent company's corporate veil. Specific case law involving contracts may shed more light on the subject, which, as you noted, may be outside of Justice's general jurisdiction regarding this and similar claims.

5. Just because topics are discussed with Justice, in general, does not imply that such topics and discussions are based in case law, or constitute anything more than opinions that may or may not have legal merit. Unless a specific reference to case law is made, or unless one or more specific items in a discussion relate to a known ruling or documented outcome (like the tangible fact that the IBM 1956 consent decree was lifted within the past two years, and has not been enforced for about ten years) all other information is to be treated as merely conversational, unless otherwise noted.

6. The Justice department attempts to maintain neutrality to all parties until facts from each party regarding a claim or claims can be presented or discovered as useful to decide if due process remedies are warranted, and if Justice's involvement is even appropriate.

Ex1 P37.

a *7th-Gift* page

ISB

2056 N. Mills Ave. Ste. 455
Claremont, CA 91711

909-626-1777
Fax 909-626-4977

7. Finally, I shall in no way waiver in respecting Justice's authority, within federal statutes, to require parties with useful information to its purposes, and maintaining control of its processes, procedures and methods, to come forward and be available at the time and place Justice or any other federal agency or department of government may deem mutually appropriate. Your minor suggestion of availability for Justice's needs, *if necessary*, invoked the humility I would have toward all such government agencies. Since I am in contact with other parts of the government, the comments you noted were inclusive of them as well. Your comments during our conversation only related to requirements Justice may stipulate, and no other agency of government.

Your call helped me appreciate the tremendous oversight you can bring to a matter. Yet, I am somewhat comforted that if the roles were reversed, your restraint in exercising oversight would be the same. The careful image of neutrality Justice attempts to maintain in all matters was apparent in our conversation this afternoon. I now realize that it was your very sense of neutrality that permitted me to take an unbiased look at IBM's conduct and consider parallel, but unnamed, legal venues to seek remedy. The good faith interaction with you compromised nothing, and showed no favoritism for or against IBM's conduct and a claim or claims concerning it.

The sum total of all of the conversation with you was that if I (and not you) feel that an individual or individuals within IBM used a ruse to interfere with free-market forces, or to expropriate an invention against its owner's will, then don't be discouraged. The market place is huge, with many willing partners who can see the same merit in an invention that IBM may (not that you agreed it did) expropriate for its own needs and not make commercially available to others. If this constitutes a breach of contract, there are civil venues available.

Finally, in reviewing the written correspondence to IBM Global Services, except concerning the above supplemental material received from you today, I would not change a word, and the document stands as my best recollection of the substance and implications of the conversation we had. As to your advice to "hire a lawyer". I am already engaged. My Primary Counsel (he must remain unnamed, for now) reviews all contact with you and others.

With Regards,

Anthony Sneed
Anthony Sneed

CC: Lee Browne, Caltech Emeritus Faculty Member
Patrick Quinlan, Congressional Staff Member
David O'Reilly, Patent Attorney
J. Nicholas Gross, Patent Attorney
Joseph Boyce, Wall Street Journal
John Greaney, U.S. Justice Department

3

Ex1 p38

ISB
2058 N. Mills Ave. Ste. 455
Claremont, CA 91711

909-826-1777
Fax 909-826-4977

Mr. Kent Brown
U.S. Department of Justice
Washington, D.C.

d4/July 9th, 1997

Mr. Brown:

This is to formally Answer your fax received d3/July 8th, 1997. The fax was annotated and returned until time permitted a full Answer. Following up on other suggestions you made, which include EDS, KPMG, Anderson, American Airlines, NASA, etc., delayed me until now.

1. (Second Paragraph of letter): The issue of the complaint has more to do with <u>expropriation</u> of an invention rather than the refusal to license an invention under antitrust laws.
The expropriation involved a ruse through the exchange of insider information between IBM Global Services and IBM. As such, the original question poised to the Department through Mr. Brown concerned the legality of insider trading of information and the 1956 Consent Decree. In the informal conversation with Mr. Brown on d3/July 1st, 1997, he noted that the Consent Decree was lifted about "a year and a half ago," and had not been enforced for about "10 years." This witness specifically asked if there were vestiges of the Consent Decree that might prohibit insider trading of information, as in this complaint. Mr. Brown assured this witness there were no such constraints from the Department since the Department's position is not to interfere in relations between a parent company and its subsidiary. His argument was persuasive when he pointed out that if a parent company can completely close a subsidiary, then other attributes of the relationship were subordinated as well.

Mr. Brown did suggest, however, that if IBM Global Services reached an agreement then such an agreement could be binding upon IBM as well. This witness concluded from this that any ruse originating from IBM to demur such a contract with IBM Global Services was in fact an attempt to disengage from a contract that already bound it. Before including this in the written complaint, this witness specifically asked Mr. Brown how long had he been with the Department. If he were a first or second year attorney, less confidence would have been placed in Mr. Brown's view concerning this aspect of the complaint. Yet, Mr. Brown acknowledged he had 20 years with the Department <u>and</u> familiarity with the Consent Decree. Though the information he shared was an extraordinary revelation to this witness, nevertheless this witness respected Mr. Brown's 20 years of experience and familiarity with the recently lifted Consent Decree and its consequences. Only then was this included in the complaint.

On d2/July 7th, Mr. Brown reversed himself, and stated that a contract entered into by IBM Global Services was not necessarily binding upon IBM. This was the original view of this witness before the d3/July 1st conversation with Mr. Brown, especially since the ruse call by IBM at no time suggested that IBM was bound by the agreement between this witness and IBM Global Services. Mr. Brown suggested that the contract terms that established IBM Global Services as a subsidiary may have bearing on the matter, but that this came under contract law and, therefore, outside the scope of our conversation. He also noted that IBM has a "corporate veil" which an IBM Global Services contract may not be able to pierce.

Ex1 p39

On the same day, d2/July 7th, this witness documented to Mr. Brown and Mr. Grannie at the Department the receipt and new material Mr. Brown shared, especially since it represented the original view held by this witness prior to d3/July 1st. Since the Department's fax machine was powered off, the material was not sent until the following morning. This witness could only infer Mr. Brown's intent to be helpful with his original comments on d3/July 1st. As to the effect of the reversal on the complaint, it did not diminish IBM's use of a ruse, independent of the Consent Decree's position about agreement reached between this witness and IBM Global Services. Therefore, this complaint will proceed without being prejudiced by the contradictory actions IBM projected in its ruse to confuse its relationship with IBM Global Services. In fact, Mr. Brown's reversal is an acknowledgment of that very confusion and reflects the dubious position IBM has created for all parties in molesting this witness.

2. (Third Paragraph) Concerning Mr. Brown's comment herein; this witness is extremely reluctant to add more than a few benign statements. At the time this witness contacted Mr. Brown, it appeared that IBM could act with impunity. This witness was aware from other sources within IBM Global Services, unrelated to this complaint, that the Department had lifted the Consent Decree. And this witness felt that IBM's actions regarding this witness' agreement with IBM Global Services reflected what the Consent Decree had formerly restrained. Without conceding a clear wrong by IBM, Mr. Brown did suggest that IBM could not act with complete indifference to contracts IBM Global Services enters into, and that IBM's attempted molestation may have caused a breach (Mr. Brown's term) of agreement between IBM Global Services and this witness. Mr. Brown noted that even this possibility was conditioned upon actual agreement and terms reached between IBM Global Services and this witness. After reviewing all documentary evidence of the pre-conditions and written terms established before letting IBM Global Services apply the invention, the position based on such supporting documentation was included in the complaint.

Regarding the commercial value of the invention, Mr. Brown did agree that an invention with the attributes reviewed by him would directly stimulate free-market forces if IBM was unresponsive. Like a Caltech professor reviewing a lab project, Mr. Brown questioned this Caltech graduate about his diligence in engaging other available channels. Mr. Brown quickly saw that this witness had focused on hardware companies instead of computer services companies. Mr. Brown specifically asked, in illustration, if a prominent services firm had been contacted, and he was answered truthfully. From Mr. Brown's insight and penetrating questions of this witness, the perception existed that Mr. Brown had internalized the value of the invention. Like a CEO or a marketing executive, Mr. Brown spontaneously created an excellent marketing plan, which this witness continues to execute to this very day. It is now realized that such marketing and technical sophistication on the part of Mr. Brown reflects only his professional objectivity based on the facts presented to him, and no more.

3. (4th Paragraph) Mr. Brown's suggestion that this witness may have to respond to the Department's summons to provide useful testimony reminded this witness of his first job at NASA/JPL in the summer of his sophomore year at Caltech. This, of course, involved working for the United States of America and its government, which lead to extraordinary efforts resulting in a later assignment to trouble-shoot part of the Galileo mission to Jupiter.

2

GX1 P40

In a similar way, Mr. Brown's suggestion meant there may be an appreciation of the societal impact of withholding useful technology that could help the nation and the government avoid disruption to its processes and institutions. The material Mr. Brown references in this paragraph acknowledges the mental willingness to respond to whatever venue the Department's suggestion might lead, within federal statutes, without reservation.

4. (5th Paragraph, last page). The first time this witness became aware of a "Section VIII" of the Consent Decree was when it was read in Mr. Brown's d3/July 8th letter. Had it been known the prior day, for example, it would have been included with all other specific information Mr. Brown shared, such as "year and a half", "10 years", "corporate veil", etc. Furthermore, as a new detail, it would have been included in the complaint and the summary provided Mr. Brown on d2/July 7th. It was with some disappoint that it was first discovered on d3/July 8th. Furthermore, it appears that there are vestiges, though few, that relate to services provided by ISSC. This also is new and useful information, but this witness does not recall it before reading it in Mr. Brown's letter, nor is it in this witness' notes. Nevertheless, there may be a purpose and subtle benefit to all parties for its eventual introduction here. Again, it is very difficult to infer any intent to Mr. Brown, except his purpose to resolve the complaint constructively. If there is a negative intent, then this witness shall recognize it too late, because of Mr. Brown's clear sense of fairness. It is IBM's molestation of an agreement that has warped a routine acquisition of a strategic technology, not Mr. Brown's efforts.

Outside of this Answer, specific to Mr. Brown's letter d3/July 8th, 1997, nothing shall be construed to change the complaint concerning IBM's motives, methods and objectives in executing its ruse, and its pretext to withhold a useful invention from government, academic, commercial and nonprofit institutions. All other written material not specifically referenced herein remains unaffected and represents the best recollection of the substance and implication of events from this witness' perspective. Furthermore, nothing in Mr. Brown's actions to date diminishes the regard for him and the Department in their attempt to resolve this complaint.

Nor does this witness consider Mr. Brown's reversal of a previous position unusual in light of new information he discovered through his own research. This is not to be construed as Mr. Brown's misstatement, but a reflection of diligence in a timely fashion to ensure this witness has accurate information. It is acknowledged that Mr. Brown updated his position, in writing, at the earliest opportunity, minimizing possible injury to this witness and his complaint.

With Regards,

Anthony Sneed
Life Member, Caltech Alumni Association
CC: Lee Browne, Caltech Emeritus Faculty Member
     David O'Reilly and Nick Gross, Patent Attorneys, Joseph Boyce, Wall Street Journal
     Patrick Quinlan, Congressional Staff Member
     John Greaney, Department of Justice

3

Ex 1 p41

Route: 701
Somers, NY 10589

July 11, 1997

Mr. Anthony Sneed
ISB
2058 N. Mills Avenue. Suite 455
Claremont, CA 91711

Dear Mr. Sneed:

I am writing in response to your letter dated July 7, 1997 to me.

At your initiative and request, IBM agreed to review your provisional patent filing for a possible solution to the computer challenge posed by Year 2000. As you and I discussed, based on advice of your patent counsel, we did not consider the material you provided as confidential because of the pending patent application.

IBM has attempted to deal with you in good faith. We have asked for a technical evaluation of the materials you furnished to me and based on that assessment, concluded IBM is not interested in pursuing any form of license with you. This was communicated to you in a letter dated June 20, 1997 from Mr. Al Torressen, of the Intellectual Property and Licensing function, who was assigned by IBM to address this matter.

Your dealings with IBM since our initial contact have been less than forthright. You have contacted numerous individuals, both within and outside IBM, and based on the correspondence, it is absolutely clear that you are misrepresenting or distorting facts to suite your own purposes. You have chosen not to inform me about other IBM employees you are actively communicating with.

Your letter to me of July 7, 1997 - headed Executive Summary - contains numerous misstatements. I will not attempt to refute each of them, but as has been clearly stated, IBM has not agreed to license the "invention" you furnished to me via FAX in May, 1997. At this point, it does not serve any purpose to respond point by point to each of the assertions, the innuendo or misstatements you have made in the numerous letters you sent to various individuals in IBM.

Ex1 p42

July 11, 1997
Page 2

In the future, you should deal directly with Al Torressen for all matters relating in any manner to licensing.

If you seriously believe there are legal matters you need to address with IBM, please have your lawyer contact me.

Sincerely,

Richard A. McDonough IIII
Vice President and Assistant
General Counsel

cc: Mr. A. Torressen, Esq.

a *7ᵗʰ-Gift* page

Route 100
Somers, NY 10589

July 11, 1997

Mr. Kent Brown, Esq.
United States Department of Justice
Antitrust Division
555 4th Street, N.W.
Washington, DC 20001

Dear Mr. Brown:

Thank you for your letter of July 8, 1997 clarifying and correcting the record relative to several of the statements Mr. Sneed made in a letter dated July 7, 1997, which was addressed to me wherein you were one of several cc's.

Unfortunately, our experience in dealings with Mr. Sneed has been very similar to that expressed in your letter to me. His letter of July 7th to me is replete with inaccuracies, misstatements of fact, and exaggerated distortion of what has occurred.

I do not believe any of our dealings with Mr. Sneed pose an issue under the Consent Decree, but I am certainly willing to discuss and review it with you should you deem it appropriate.

Sincerely,

Richard A. McDonough III
Vice President and Assistant
General Counsel

cc: Mr. Donald J. Rosenberg, Esq.
    Mr. Anthony Sneed

Ex1 p44

30.3 Sneed responds to McDonough the same day, d6/July 11, 1997, now having additional, written Evidence and declaration of only evaluative-use by IBM counsel, after DOJ contact.

    30.3.1 Sneed's response is highly speculative, and projects possible meanings of McDonough's letter if, in the "future," IBM admits to anything more than evaluative-use in 1997, as memorialized in McDonough's letter.

    30.3.2 The highly-speculative implications are explored, in the event McDonough has not reflected upon his and IBM's actions in the historical context of IBM's past, both recently and from its founding in 1913, and its profit-taking from 1933 to 1944 through its German subsidiary.

30.4 McDonough goes silent and does not respond. Sneed accepts that McDonough's d6/July 11, 1997 letter, and admissions therein, may be sincere regarding only evaluative-use within IBM.

    30.4.1. Otherwise, McDonough is memorializing an obstruction, which even IBM counsel would never knowingly do, to Sneed's knowledge.

Attached exhibits are admitted from Exhibits 6 and 1, IBM Motion to Dismiss First Amended Complaint Declaration of Bridget A. Hauler, March 12, 2012

Ex/ 45

ISB

2058 N. Mills Ave. Ste. 455
Claremont, CA 91711

909-826-1777
Fax 909-826-4977

Mr. Dick McDonough   d6/July 11th, 1997
IBM, Somers, NY

Mr. McDonough:

In the appropriate place, at the appropriate time, when everything said is recorded and included
in an accurate forum, I will respond to the hostile, scurrilous, distorted, malicious view of events
you attempted to project through your letter dated today.

As I have stated before, you are to cease and desist using in any way whatsoever the invention
you have attempted to expropriate through ruse and disregard of written conditions agreed to
prior to your attempted expropriation. *And provide written verification you have done so.* You
are delinquent in providing such verification in "good faith." *Why write me today without it?*

Any interaction with you and your organization regarding expropriation or settlement thereof is
to be in writing, by fax or otherwise—your word in a phone conversation has no merit
whatsoever. It seems that you and your company prefer to create unwritten phone messages or
telephone conversations of the methods you apply in expropriating another's intellectual property.
It disgusts me to have to first endure your and other's phone conversations and then read the
transcribed notes later. If you put your own misdeeds in writing, I only have to receive such
reprehensible conduct once, instead of twice. This includes the ruse attempted by "Al
Torressen," a *stranger* who initiated an unsolicited call with the sole purpose of expropriating
the technology in question. Using the term "good faith" with what you know is a vulgar farce.

You suggest that "dealings have been less than forthright." If you are some how referring to the
expropriation of this inventor's intellectual property, then I will concur. The value of this
technology may exceed $7 billion to your company, technology you were unable to create in the
last decade, and, by your staff's own admission, was nothing you had ever seen or thought of
prior, even though you spend $5 billion a year on research. Possibly this is misinformation, too.

As to unsolicited contact, it was an IBM Global Services employee who *first* strongly urged
immediate contact with the office of your president when I mentioned the invention's capability.
And it was another IBM Global Services employee who strongly recommended bypassing the
administrative arm of your company (which includes your department) for unspecified reasons.
It serves no purpose for me to review what has already been documented, not one word of which
will change based on written correspondence, written voice transcripts, recorded messages and
dates, and other material from personnel within your company. Though you express irritation
in not having your own employees identified, so did another organizational structure, like yours,
complain from 1933 to 1944 while robbing an entire people, taking even their fillings.

But you are attempting to take even more, through an *intellectual* genocide under cover of
"corporate necessity." It was this same "corporate necessity" that permitted your company to
have some of the most perverted employment demographics in the history of the world from
1913 to 1963. Such perversions were actually rewarded and internationally recognized by the
most ruthless regime ever to disgrace the face of the earth in the 1930's—until 1944.

Ex 1   46

So, I shall watch, record and document your present day continuity of such perversions, but the vulgar stench of what you are doing is no less foul and repulsive today as it was in the past, even your attempt to suppress First Amendment rights of employees in your company who continue to act in good faith, though I have cautioned them of the dangers from your regime. Who knows, you might yet pull it off if you are willing to hold to the BIG BLIGHT strategy that you are working to perfect. And what's one more inventor, the son of a race you whipped clean from your corporate history from 1913 to 1963. Primary Counsel has advised me of the ends you may be willing to go to perpetuate the intellectual genocide that you became expert in for 50 years. You have been given opportunity to stop, but you seem to be unwilling.

Only this time, it will not be in a back room, or concealed from the public conscience. Like Goliath, you will have to appear on an open field with multiple witnesses. The invention you have attempted to expropriate is the stone that is now embedded within your forehead—you just haven't realized it yet. By the grace of the *Name* of my Father's One and Only Son, you will. You may think, "No one will know, so let's just take it." My Father's seven-fold Spirit, the Spirit of truth, knows. Possibly you and others will be rewarded handsomely by your company for your attempted theft. And you may even for a season get to enjoy what may be seen with the eye, tasted with the tongue, worn on the flesh, or made by the hand. Enjoy it well, because it is only for a season, just as it was from 1933 until 1945—a period that contained uninterrupted pleasure and the acquisition of what belonged to others. The interest on all that was taken continues to grow to the point that the whole world can't meet the "balloon payment" soon due.

As for me, my wife and four children, we shall enjoy the simple pleasures we have as a family. I shall forget you until the appointed time in Court. But at that time, I will bring to bear every source of testimony as to who you are and what you have attempted to do in secret, under the cloak of an American company that claims "*Respect for the individual*" and "*Fair and impartial treatment of suppliers*."

My wife and four children shall not know, now, of these things, nor of the expropriation of technology you now suppress for your own reasons. They shall continue to know me as the inventor who helped the Galileo mission achieve success. And if computers you make are used by the public and work correctly for the services and products delivered to my wife and children, then in this small way they shall enjoy the fruits of their husband's and father's efforts. When they are old enough, they will know your name. And as long as there is a Sneed in my line on Earth, the testimony of the service rendered shall be known. There is nothing you or anyone in your company can do to stop it. You can tarnish it, you can distort it, you can wound it. But can't destroy the testimony, in the contest of a thousand years—and it won't take that long.

With Deep Regrets Evening Typing Your Name on the Opening of This Letter,

Anthony Sneed
Anthony Sneed
CC: Lee Browne, Caltech Emeritus Faculty Member
　　Joseph Boyce, Wall Street Journal
　　Kent Brown, U.S. Department of Justice

31. In light of Logan's disclosure that other methods may be more revenue positive to IBM than Sneed's Specification, and McDonough's discontent and desire to avoid any further inquiry from DOJ, Sneed accepts that it is in IBM's profit-making interest not to make production-use of Sneed's Specification.

    31.1 Sneed realizes the very efficiency and affordability of his Specification may make it less desirable.

        31.1.1 It would be as if IBM were a doctor, and had a low-cost remedy for a fatal medical affliction that could quickly be made available to the public; for itself, IBM prescribes the low-cost remedy that is curative the morning after it is taken; but for its patients IBM prescribes a complicated regimen spread over weeks, that allows its personnel to bill patients while they stay at patients' homes for weeks applying the treatment, which may also require patients to buy new medical equipment from IBM.

        31.1.2 IBM's PR muscle would avoid such public perception and future adverse effect on its stock.

    31.2 Other Y2K methods may not involve any potential DOJ oversight or future inquiry of IBM's conduct.

    31.3 However, Sneed is still under a condition of performance to contact IBM in the "future" per the d6/June 20, 1997 Letter of Corporate Intent signed by Torressen on Gerstner's and IBM's behalf.

Ex/ 48

2001 June: Sneed contacts Gerstner and Torressen per IBM's d6/June 20, 1997 Letter of Corporate Intent's condition of performance on Sneed.

    31.4   The U.S. Patent and Trademark Office issues U.S. patent 6,236,992 ("992") to Sneed on d3/May 22, 2001 for Sneed's Specification.

    31.5   Sneed notifies Gerstner on d3/June 19, 1997, by letter, of the condition of performance placed on Sneed to contact IBM, as stipulated by Torressen in IBM's d6/June 20, 1997 Letter of Corporate Intent.

    31.6   Sneed notifies Torressen the same day, by letter, identifying Mr. J. Nicholas Gross as patent attorney of record, regarding any non-evaluative use of Specification and disclosure obligation of production-use, if any, of Sneed's Specification.

    31.7   Sneed does so *pro forma*, remembering the preferences IBM expressed for alternatives, the higher profit-making with alternatives, and IBM's disinterest to have anything to do with DOJ.

    31.8   As is pattern and practice for IBM regarding Sneed's claim for anything beyond evaluative-use, there is no response; there is no obligation to disclose anything if IBM made no production-use of Specification.

        31.8.1   Sneed recalled notice of libel or defamation, to which McDonough alluded, for claiming anything more than evaluative-use by IBM.

a *7ʰ-Gift* page

FAX   FAX   FAX   FAX   FAX   FAX   FAX

I.S.B.
2058 North Mills Avenue
Claremont, CA 91711
909-523-5736   FAX 909-626-4977

FOR: Mr. Louis Victor Gerstner, Chairman, IBM
FIRM: IBM
FAX: 914-499-7030
PAGES: ___2___ (including this page)
FROM: Anthony Sneed   (include "*1771" in numeric message)
DATE: d3/June 19, 2001

<u>Via Facsimile</u>

Happy 3rd day, Mr. Gerstner.

Long before this Caltech graduate was awarded the corporate encryption patent referenced in the attached, your firm had already extended consideration and open communication regarding its potential benefit and merit thereof.

IBM legal personnel sent correspondence requesting notification of the patent's issuance, which is completed effective today via this facsimile.

With regards,

Anthony Sneed

Anthony Sneed
Life-member Caltech Alumni Association
U.S. Patent Holder, #6,236,992

Attachment: Response to request by Mr. Al Torressen, IBM Intellectual Property Licensing

1

Office of the Director of Licensing
Intellectual Property & Licensing

500 Columbus Avenue
Thornwood, New York 10594

CERTIFIED MAIL

June 20, 1997

Mr. Anthony Sneed
I.S.B.
2058 N. Mills Avenue, Suite 455
Claremont, CA 91711

Subject: *Your letter dated May 29, 1997 to Mr. Dick McDonough*

Thank you, again, for your unsolicited, non-confidential letter of May 29, 1997 to Dick McDonough concerning your provisional patent filing, "A Serial Encryption System By-Pass of Year 2000 Date Malfunctions." In accordance with your direction that the materials were non-confidential and could be freely distributed within IBM, we conducted a technical review within IBM. As I told you in our June 5, 1997 phone conversation, your letter was forwarded to me for coordinating IBM's review and response.

IBM has now completed its review and concluded it has no interest in pursuing a license described in your fax to IBM on June 6, 1997, nor does IBM have any interest in pursuing your June 10, 1997 proposal to IBM's Don Logan.

In the interest of facilitating any further interaction, I request that you deal with me rather than other individuals in IBM. As the individual responsible for coordinating this matter, your fax of June 19, 1997 to Mr. Gerstner was referred to me for response. Two points require clarification (1) IBM has no intention of an "attempted expropriation of this intellectual property", and (2) I never made a statement in any way resembling the quote you attribute to me at the end of the third paragraph. There are other inaccuracies in many of your 10 letters to IBM, but it does not seem productive to address each of them. For example, your most recent letter of June 18, 1997 addressed to Dan Sullivan (which is only one of 9 identified letters over the last 3 weeks) also contains several inaccuracies and misstates what has occurred.

We appreciate the importance of your effort to find a solution to the challenge of Year 2000 which will have worldwide affect on customers and industry. Although your proposal is not currently of interest, we would encourage you to contact me in the future if a patent issues.

We wish you well in the future.

Sincerely,

Al Torressen
Program Manager, Licensing

AT:gm

cc: Dick McDonough

Ex1 p51

FAX   FAX   FAX   FAX   FAX   FAX   FAX

I.S.B.
2058 North Mills Avenue
Claremont, CA 91711
909-523-5736   FAX 909-626-4977

FOR: Mr. Al Torressen
FIRM: IBM
FAX: 914-765-4390
PAGES: 1   (including this page)
FROM: Anthony Sneed (include "*1771" in numeric message)
DATE: d3/June 19, 2001

<u>Via Facsimile</u>

Happy 3rd day, Mr. Torressen.

Per your correspondence request, the serial number of the corporate encryption patent is
6,236,992.   The registered patent attorney is Mr. J. Nicholas Gross, who is listed on the
patent, as well. He has power of attorney for licensing or settlement and can be reached at 415-
551-8298.

With regards,

Anthony Sneed

Anthony Sneed
Life-Member Caltech Alumni Association
U.S Patent Holder, #6,236,992

cc: Mr. Louis Victor Gerstner, Chairman, IBM

1

Ev1 52

2004 June: Sneed has attorney Robert J. Spitz (Spitz) contact IBM per IBM's d6/June 20, 1997 Letter of Corporate Intent; Sneed does further follow-up, himself, in 2006

32. IBM's d6/June 20, 1997 Letter of Corporate Intent stipulates when Sneed may <u>begin</u> calling IBM, but sets no limit on how long to keep calling thereafter in the event IBM belatedly discovers unauthorized production-use, somewhere within IBM.

     32.1 The useful life of Sneed's Specification, when installed, is a minimum of 28 years.

     32.2 This obligates Sneed to call IBM until at least 2025, or 28 years after 1997.

     32.3 With an organization as large as IBM, and expressed good faith to complete contract formation through the IBM d6/June 20, 1997 Letter of Corporate Intent, IBM could discover some remote part of the company that may have used Specification, and disclose that to Sneed even in 2004 or later.

33. Spitz, on his own and unilaterally, makes a purely speculative assertion of use, and makes a demand based on Spitz's own speculative assertion in a letter Spitz unilaterally composes and sends to Torressen d2/June 14, 2004.

**Attached exhibit is admitted from Exhibit 2,
IBM Motion to Dismiss First Amended Complaint
Declaration of Bridget A. Hauler, March 12, 2012**

Ex1 P53

06/14/04   36-23-04 P12.37

Al Torressen
Program Manager, Licensing
IBM
500 Columbus Ave.
Thornwood, NY 10594

RE: Anthony Sneed
       BEST 2000

Dear Mr. Torressen:

Please be advised that I represent the above named with regard to his claims against IBM with regard to its use of the technology entitled Serial Encryption System-Bypass of Year 2000 Date Malfunctions and better known as BEST 2000. My client has recently obtained information that IBM implemented the software remediation techniques described in the claims of his patent.

There were negotiations with my client over the consideration for use of this technology, prior to its implementation by IBM. My client submitted a licensing agreement whereby IBM would pay $ 0.14 per line of code. However, the negotiations over this licensing agreement were suspended when IBM Corporate concluded that my client would be unable to obtain a patent on his invention. IBM Corporate informed my client that they would enter into an agreement with him for their use of this technology after he received his patent. As you know, Mr. Sneed was granted the '992 patent for this methodology.

Nevertheless, IBM Corporate proceeded to utilize the solution to the Y2K problem that is described in the claims of my client's patent and in the text that describes BEST 2000 that he provided to IBM.   This unique solution to the Y2K problem was used by IBM for reprogramming a large number of its computers with great efficiency.

Computers containing millions of lines of code have been changed by both IBM Corporate and IBM GS utilizing my client's patented solution to the Y2K problem. Furthermore, it is believed that IBM was compensated by its clients and customers for use of this economical and efficient Y2K solution. IBM widely utilized my client's BEST 2000 technology on its own computers. There is no question that IBM has made use of my client's invention without the payment of any consideration.

1

Ex 1 p54

EXHIBIT 2
PAGE 4

Therefore, I demand that IBM immediately do the following:

    1.   Provide me with copies of documentation associated with the implementation of my client's BEST 2000 methodology on IBM computers.

    2.   Provide me with documentation showing the income received by IBM for the implementation of BEST 2000.

    In the alternative, I IBM may choose to enter into good faith negotiations for the full and fair compensation of its use of BEST 2000. Please be advised that my client considers the amount of $140 million to be the estimated compensation that would be due him under the licensing agreement that was discussed by the parties. While my client might be willing to accept a lesser amount in settlement of this matter, there is no question that a lawsuit could bring a much higher recovery.

Your prompt response to this request is imperative. In the event you fail to contact my office within ten days to discuss a resolution of this matter, you will leave my client with few alternatives. My client will be forced to seek appropriate remedies for the IBM use of this technology.

In the unlikely event that IBM takes the position that there has been no use of my client's patent, please provide me with copies of computer code showing each solution used by IBM for solving the Y2K problem. If there was no use of BEST 2000, there should be no problem in providing copies of the code or a complete description of the methodology used in order to verify your claim.

Your prompt efforts to resolve this matter is in the best interests of our clients. I am looking forward to your response and further discussions of this matter.

Sincerely,

Robert J. Spitz

2

EXI p55

EXHIBIT 2
PAGE 5

a *7ᵗʰ-Gift* page

34. Mr. James E. Schreiber (Schreiber), IBM Intellectual Property & Licensing Program Director, responds to Spitz with a letter dated d5/June 24, 2004, acknowledging Spitz's d2/June 14, 2004 letter and its speculative demand.

     34.1 Schreiber discloses that Torressen and IBM have separated and that Schreiber is now the designated contact stipulated in the d6/June 20, 1997 IBM Letter of Corporate Intent to Sneed.

     34.2 Schreiber directs other IBM personnel to formulate a response for Schreiber to answer Spitz.

35. Three months later, in a d3/September 28, 2004 letter, Schreiber responds, consistent with McDonough and Torressen disclosures of only evaluative-use within IBM and no use meriting payment to Sneed.

     35.1 Schreiber specifically references the d6/June 20, 1997 IBM Letter of Corporate Intent, and that Schreiber's response is in accord with Agreement formed through the IBM Letter of Corporate Intent.

36. Sneed makes another *pro forma* inquiry in 2006, IBM also declaring only evaluative-use within IBM, and nothing beyond evaluative-use.

Ex1 56

OCT 04 2004 12:23 FR IP LAW   914 765 4390 TO 919093959535   P.01

Office of the Director of Licensing
Intellectual Property and Licensing

North Castle Drive
Armonk, NY 10504

September 28, 2004

<u>**VIA MAIL & FACSIMILE: (909) 395-9535**</u>

Mr. Robert J. Spitz, Esq.
204 North San Antonio Avenue
Ontario, CA 91672

<u>Reference: U.S. Patent Number 6,236,992</u>

Dear Mr. Spitz:

This letter is to advise you that we have completed our review of U.S. Patent Number 6,236,992 which you identified to IBM in your letter of June 14, 2004.

After giving careful consideration to your information, we have determined that there is no present or past need for IBM to pursue rights to the referenced patent. Please understand that this decision does not reflect adversely on the merits of the invention, but represents our current perception of IBM's needs and requirements.

With respect to your comments regarding past licensing discussions between IBM and your client, the only reference we have located is contained in a June 20, 1997 letter from Al Torressen of IBM to your client, stating: "IBM has now completed its review and concluded it has no interest in pursuing a license described in your fax to IBM on June 6, 1997, nor does IBM have any interest in pursuing your June 10, 1997 proposal to IBM's Don Logan."

Thank you for your interest in IBM and for giving us the opportunity to reconsider this patent. Please continue to use me as your focal point for communications in this regard.

Sincerely,

James E. Schreiber
Program Director,
Intellectual Property & Licensing

JES:lcb

EX / P57

d3/October 24, 2006

Mr. Al Torressen
Program Manager, Licensing or party acting on his behalf
IBM 500 Columbus Avenue
Thornwood, NY 10594

RE: Anthony Sneed, BEST 2000, Filing of LA Superior Case No. BC296142

Mr. Torressen:

Your June 1997 correspondence to this Caltech alum committed your company to the following:

       A. ("POINT-A") Confidential distribution of BEST 2000 within your company
       B. ("POINT-B") Economic remuneration for BEST 2000 as an original invention

1. You already performed POINT-A, in writing, with broad distribution of BEST 2000 within your company. Your declaration thereof acknowledges the implicit merit BEST 2000 delivered. In admitting wide distribution of BEST 2000 within IBM corporate ("CORPORATE"), as initiated through your role as Corporate Licensing Program Manager, and by no other, such voluntary admission, in writing, reflects a good faith intention to compensate this Caltech alum for what provided benefit to your firm. The only terms approved by me were and are $49,764.55 per site for any use of BEST 2000.

2. You made no effort to conceal the wide distribution of BEST 2000 within your company, which is corroborated by parties within your firm whom I've known for decades, as well as parties within your firm who confirmed the merit of BEST 2000 as being a valuable intellectual asset to your firm.

3. You admitted custody of BEST 2000 which was first received by your IBM Global Services subsidiary ("GLOBAL") in May 1997. Your establishing direct contact with this Caltech alum attempted to avoid exorbitant fees your IBM Global Services subsidiary was entitled to charge you under the vestiges of the 1956 Consent Decree and legacy thereof which GLOBAL could leverage and compel you to pay it. GLOBAL'S plans were to charge CORPORATE between $1 and $2 per line of code for CORPORATE'S application of BEST 2000, anywhere within CORPORATE, or for GLOBAL'S services to remediate CORPORATE'S Y2K needs with BEST 2000. To this end, GLOBAL desired to pay a royalty to this Caltech alum of up to 50 cents per line of code from fees paid GLOBAL from CORPORATE.

4. As was clear from your admission and contemporaneous disclosures thereof, you had no other reason to contact me. You furthermore disclosed apprehension about GLOBAL'S potential fees which CORPORATE would have to pay. Your only contact with me was for the purpose of not being billed by GLOBAL for IBM's internal corporate use of BEST 2000, which you admitted broadly distributing for IBM's corporate-wide benefit within confidential stipulations allowed in our communication.

5. I vetted this matter with the Department of Justice (JUSTICE), since your unsolicited attempt to communicate with me could appear to violate the general intent of the 1956 IBM Consent Decree, specifically excluding GLOBAL from showing undue favoritism to CORPORATE, to the detriment or injury of GLOBAL'S non-CORPORATE clients who would not be shown such favoritism. JUSTICE assured this Caltech alum that he could not be prosecuted regarding collusion regarding the 1956 IBM Consent Decree if any such misconduct were later determined through due process.

6. The terms GLOBAL asserted for a royalty agreement were not agreed to, and did not displace the memorialized $49,764.55 per site stipulated for use of BEST 2000. These are the extant terms for any use by CORPORATE of BEST 2000. Though not required for this Caltech alum for his remuneration, POINT-B stipulated notifying you that BEST 2000 was original know-how, a burden met with U.S. patent #6,236,992. Since this stipulation is perfunctory, it has no impact upon the good faith intent regarding compensation to this Caltech alum, as reflected in your admissions regarding BEST 2000 through CORPORATE. Neither party concealed good faith intent, whether this Caltech alum's *a priori*,

memorializing $49,764.55 per site before any use of BEST 2000 by CORPORATE or GLOBAL, or CORPORATE'S subsequent admissions within such stipulated terms.

7. Therefore, as an offer-in-compromise, regarding CORPORATE'S good faith intent and good faith admissions thereof, and FOR SETTLEMENT PURPOSES ONLY, the following shall constitute liquidated damages and release of all claims of any kind regarding BEST 2000 and/or proprietary know-how, if any, thereof: That payment equivalent to 15 sites at $49,764.55 per site, or $746,468.25, shall constitute payment in full for all and any use of BEST 2000, by CORPORATE or by GLOBAL. Such shall in no way be construed as anything other than resolving good faith intent between the parties and shall not be construed as an actual admission of use of BEST 2000 or that CORPORATE in fact remedied Y2K by any means of any kind. The $746,468.25 is payable to "ANTHONY SNEED" at 2058 North Mills Avenue, Claremont, CA 91711-2812. These terms shall be valid until sunset, d2/October 31, 2006.

8. In the event you elect to molest the good faith intent extant between CORPORATE and this Caltech alum, attached is a copy of Los Angeles County Superior Court case no. BC296142 ("BC2") and California Second Appellate District, Division Five, case no. B176278 ("B17"). This Caltech alum and former IBM finance industry team member (IBM marketing school honor graduate, 1982) successfully completed B17 and BC2 with evidence substantially *weaker* than admissions made orally and in writing by parties within your firm.

9.   BC2 corroborates the $49,764.55 amount memorialized with GLOBAL and CORPORATE. Furthermore, BC2 establishes BEST 2000 as an original work, exceeding POINT-B's burden which you arbitrarily set forth, entertained because of your prior good faith admissions. BC2 also informs the instant action for breach of good faith intent, or *fax* entitlement to breach good faith intent, that can be avoided if you allow the extant good faith intent between CORPORATE and this Caltech alum to remain unmolested.

10. In the event you elect to molest the good faith intent extant between CORPORATE and this Caltech alum, enclosed is a copy of *From Jupiter to Genesis*, ("FJTG") written by this Caltech alum and former IBM finance industry team member. Your firm's name is noted on the book's cover and appears multiple times within the book. BC2 is also addressed in its appendix. FJTG is a first-hit title on Google. It was constructively reviewed before publication by Nobel laureate and former Caltech president Dr. David Baltimore and by Caltech emeritus faculty member Lee F. Browne. It has been confirmed as authentic and original by Dr. Kevin VanHoozer, PhD and professor of Systematic Theology at Trinity International University, and separately by Rabbi Leonard Levy, PhD and professor of Torah studies.   As permitted, FJTG will be updated for a 2007 edition. If you leave the good faith intent between CORPORATE and this Caltech alum unmolested, then the 2007 edition will reflect the same benign reference to IBM as the 2006 edition.   However, if you choose to molest the good faith intent extant between CORPORATE and this writer, it will compel this writer to add an appendix regarding corporate conduct substantially similar to yours, as well as all written evidence precipitating said instant action.

I do not contemplate such action lightly, as I maintain regular contact with multiple IBMers I've know for decades, which includes my tenure as elected president of the West Coast, 6,000 member, IBM Club. Others have been instrumental regarding validating written and oral assertions made by you and your firm, or contradictory evidence thereof. Since any subsequent molestation of the extant good faith intent between this writer and CORPORATE could have unanticipated ramifications, this writer can't anticipate unexpected, adverse affects upon his own mutual fund IBM stock holdings and IBM stock holdings of a large number of his friends and relatives now employed by IBM or outside of IBM.

One of the reasons BEST 2000 was made available to CORPORATE, originally, was to PREVENT such adverse outcomes of a Y2K calamity upon family and friends, as well as upon IBM proper and any business entity that relies on IBM's products and services. This writer has been a student of IBM practically his whole life. In elementary school, IBM-labeled clocks were in all of his classrooms, which originated from the old International Time company headed by G. Fairchild, who later rose to chairman of CTR in 1913, which became IBM in 1924. These clocks welcomed the beginning and end of school days for seven years of this writers life. At 13, the writer learned to program an IBM 360 in a high level language. At 14, the

Ex! p59

writer was accepted by a professional computer institute and learned IBM assembler, JCL and operation of a 360/30. Only 5 years later, the IBM machine language skills this writer mastered were applied at NASA/JPL on the VLBI (Very-Long Base Interferometry) project. Quasar noise wave-fronts were digitized at NASA sites in Canberra Australia, Madrid Spain, and Goldstone California to compute base-line distances between these disparate antennas, to within 5 meters over the earth's surface. The digital correlator programmed by this writer was successful, which later led to other NASA projects, one covered in FJTG. During VLBI, at 19, he aced all three terms of a Caltech computer architecture course, which was based on IBM's machine-language design. Within 5 years of that, came tenure at IBM itself.

The extant good faith intent between this writer and CORPORATE is informed by a long history. Multiple employees in your firm, who own copies of FJTG, are interested in the final outcome of BEST 2000. *They have asked the matter move forward* as events may warrant based on IBM's beliefs, including:

      O Respect for the Individual
      O Responsibility as a corporate citizen
      O Fair and impartial treatment of suppliers

These may seem to some as quaint notions from a bygone era. Yet, as I know from day-to-day contact, they still resonate with many of your current employees, including multiple VP's whom this writer holds in high regard. To a large degree, they resonate in this writer, even after reading *The Lengthening Shadow* by Belden , *IBM Colossus in Transition* by Sobel, or *The IBM Way* by Rogers, and many other works.

Still, as I've learned in dealing with another corporate entity of only a billion dollars a year, referenced in BC2, ignoring good faith intent, concealing evidence, and recanting testimony between witnesses are just ways of doing "business." Therefore, such is not intended as "personal" by a corporation.

If you elect to molest the extant good faith intent extant between this writer and CORPORATE, you will molest more than just me, my wife and four children. If you are a husband of only one wife, as I am, you will know that such molestation will not go unanswered. As BC2 and B17 reflect, whether it takes one year or eight years, such molestation will provoke a response. And, as you are over 70 times larger than the defendant firm in BC2, so have seven times the resources been set aside. That litigation, overall, required only seven attorneys who represented this writer and helped him prevail. I imagine it may require at least that many with you. As circumstances warrant, jurisdiction will be in Los Angeles county, where your firm has a legal presence, and in the California's Second Appellate District, Division Five, for appeal.

In closing, I wish to acknowledge your candor, if any, whether intended or not. Had you not left unsolicited voice-mail messages and followed-up with your subsequent oral and written admissions, it would be a different matter altogether. However, as corroborated with contemporaneous disclosures and extemporaneous evidence, your candor is substantive. Good faith intent is the only presumption inferred, unless subsequent events impeach perpetuating any such inference.

This material is cc'ed to John R. Opel, whom this writer personally considers one of the greatest executives in the history of your company. There is nothing you or anyone else may do to add to or detract from the high regard in which John Opel is held. To this day, any direct request made by John Opel, that doesn't transgress a Commandment (Exodus 20:1-17), or a written U.S. law, or a written IBM policy, practice or belief will be obeyed without hesitation and without complaint, as has been done in the past.

With regards,

Anthony Sneed
Life-member Caltech Alumni Association, U.S. Patent Holder #6,236,992, Author
Enclosures: LA County Court Case BC296142 and Appellate Summary; Copy of *From Jupiter to Genesis*
-cc: John R. Opel, 590 Madison Avenue, New York, New York 10022

Ex1 p60

2007: Wilzbach's settlement initiative with Palmisano

37. Between 2005 and 2007, Dr. Alex Lidow (Lidow), Caltech board member, fellow Caltech alum of Sneed's, and CEO of International Rectifier Corporation (IRC), encourages Sneed not to let the matter drop with IBM.

   37.1 Lidow used Sneed's Specification to remedy the Y2K problem on his IBM AS/400 computers in the U.S. and six other countries, including France, Germany, England, Singapore, China and Japan.

   37.2 Lidow and his firm did so at considerable savings over alternatives offered by Anderson Consulting (now Accenture), IBM Global Services, or others; the savings were as much as 70% on a two-year, up to $10 million "best-efforts" (no guarantee) proposal by Anderson Consulting; one site with 1.2 million lines of code was repaired in six weeks with Specification after agreement was entered into with Sneed for $49,764.55 per site. There were no post-production issues whatsoever. Sneed also negotiated a $400,000 contract for Ms. Florence Lee, a Taiwanese-born relational database expert, to re-architect Lidow's worldwide information technology infrastructure. It allowed Lidow to reconcile inventory and revenues on a daily basis, across his multinational operations. This reengineering effort, too, performed as Lidow expected--he was very pleased.

   37.3 In or about April 1998, Sneed facilitated contact between Lidow's technical management and Logan at IBM; Logan confirmed that Sneed's Specification worked on IBM computers, to Lidow's personnel's satisfaction. Lidow's firm then proceeded with worldwide deployment of Specification, having been assured by IBM, the manufacturer of Lidow's company's computers.

   37.4 Lidow, respecting the Caltech Honor Code, was adamant Sneed receive compensation promised, which Lidow's CFO attempted to block; after litigation, which Lidow assented for Sneed to file, advising Sneed "Don't sell your self short," Sneed received $236,000 from IRC. Lidow made clear Sneed's efforts were a significant contribution to his company, founded by Lidow's grandfather in 1947. IRC is now a $1 billion NYSE-listed firm. The CFO was later removed. Afterward, Lidow invited Sneed to his parent's estate for a Caltech function.

   37.5 Lidow felt that IBM owed Sneed, too, based on Lidow's speculations but no substantive proof.

   37.6 Dr. Dale Prouty (Prouty), Caltech PhD Applied Physics and international venture capitalist, also prevailed upon Sneed, from at least 2003 to 2007, not to let the matter drop with IBM.

      37.6.1 Prouty and Sneed worked together on the NASA Galileo mission to Jupiter, and have known each other for over 30 years. In 2004, Prouty provided sworn testimony of production-use of Sneed's Specification on purchased IBM computers within and outside the United States, at IRC locations in France, Germany, England, China, Singapore, Japan, and El Segundo California.

   37.7 From 2005 to 2007, Lidow and Prouty remained persistent that Sneed not drop the matter with IBM, yet neither could provide anything more than speculative reasons for such a course.

38. In or about the 4th quarter of 2007, Sneed contacted Wilzbach at home through the Caltech alumni office.

   38.1 Sneed and Wilzbach had not been in contact since in or about June 1997.

   38.2 In good faith, Sneed had limited his contact to IBM personnel designated within the d6/June 20, 1997 IBM Letter of Corporate of Intent, regarding any inquiry about Sneed's Specification.

   38.3 The call was personal, since Wilzbach was an IBM employee. Sneed had been careful to limit communication with Wilzbach as Wilzbach was a career IBM executive.

   38.4 During the call, Wilzbach disclosed to Sneed that Wilzbach was no longer employed by IBM.

Ex1 p61

a *7th-Gift* page

38.5 Wilzbach disclosed that before he left IBM, he was an "E-5," which is a small group of executives that are finalists to become IBM CEO candidates; as Sneed recalls, Wilzbach described himself as an Executive Vice President, or as an Executive and Vice President when he left IBM.

38.6 He noted Mr. Sam Palmisano (Palmisano) was a peer E-5 with Wilzbach, and later became IBM CEO.

38.7 Prior to that, Palmisano and Wilzbach were in senior management at IBM Global Services when Logan and McDonough took custody of Sneed's Specification from Sneed in or about May of 1997, on behalf of IBM Global Services. At or about that   time, Wilzbach closed the largest outsourcing deal in the history of IBM, outperforming all of his executive management peers--an extraordinary accomplishment.

38.8 At or about this time, Palmisano was promoted to president of IBM Global Services, a major step to becoming IBM CEO.

38.9 After Palmisano was promoted to IBM Global Services president, Wilzbach later left IBM.

38.10 Subsequent to Wilzbach leaving IBM, Palmisano was later promoted to IBM CEO.

39. In the call, Wilzbach brought Sneed up to speed on Wilzbach's wife, Charlotte, his children, and his business endeavors, among which was building a facility for his classic car collection, including a 1950's Corvette with serial number 3; there was a problem with letting the cement foundation dry in the cooling weather, but otherwise Wilzbach was progressing and prospering.

39.1 Sneed then reviewed the IBM matter with Wilzbach, and Sneed's contact with IBM since 2001, per the d6/June 20, 1997 IBM Letter of Corporate Intent that Sneed do so. Wilzbach rapidly internalized the particulars Sneed disclosed.

39.2 Wilzbach then instantly proposed that Sneed pursue settlement directly with Palmisano.

39.2.1 Wilzbach made it clear that Sneed deal directly with Palmisano, and no one else.

39.2.2 Wilzbach directed Sneed how to compose a settlement letter, limited to one page.

39.2.3 Wilzbach recommended the settlement amount be reasonable enough for Palmisano to easily make the matter "go away."

39.2.4 Wilzbach explained to Sneed that settlement is preferable, because Wilzbach has been on the IBM-side of litigation and recommended Sneed avoid it if at all possible.

39.2.5 Wilzbach made clear he was providing only informal advice as a friend, and no more.

39.3 Sneed carefully considered the course Wilzbach proposed. It could expose Sneed to suit, for making a false settlement claim beyond evaluative-use, with no facts, and speculation at best because Wilzbach shared no information whatsoever about evaluative-use or production-use of Sneed's Specification by IBM.

39.4 However, Wilzbach's forthrightness in the call, regarding everything discussed, allayed Sneed's concerns; their business fellowship and friendship spanned 30 years.

39.5 Sneed performed; he composed a settlement letter following the form Wilzbach proposed.

40. The settlement letter form proposed by Wilzbach and composed by Sneed was sent to Palmisano on d6/October 19, 2007.

40.1 It was cc'ed to Wilzbach, Lidow and Prouty, since all three were in agreement to pursue the matter with IBM.

40.2   Sneed put all three on notice by specifically referencing Caltech's Honor Code, and that communication between Sneed and them is so informed because of the seriousness of what they were proposing Sneed perform.

     40.2.1   This was as close as Sneed could get to an oath regarding the course the parties were proposing to Sneed, or actions the parties were recommending to Sneed.

     40.2.2   Their course of action could lead to litigation which, once begun, could have unexpected outcomes, and require considerable commitment of blood, time, and treasure to complete.

     40.2.3   Sneed wanted any advice from them to be consistent with the Caltech Honor Code, for any criticism of any actions by Sneed to be unfettered, and that any subsequent course of action reflect favorably on Caltech, or no action be taken at all.

     40.2.4   Basically, Sneed wanted any of them to say, "Stop! You are wrong! Don't go forward with this!"

     40.2.5   Not one of them expressed any reservation or doubt regarding the letter Sneed composed on which they were all cc'ed.

     40.2.6   Caltech emeritus faculty member, Lee Browne, who promoted Sneed as a possible Caltech Distinguished Alumnus candidate and followed the matter with IBM, was very pleased that Sneed was proceeding with the course of action recommended by Wilzbach, Prouty and Lidow.

     40.2.7   Following Wilzbach's proposed form, the letter set a modest settlement sum of $1,194,349.20.

     40.2.8   The letter reviewed Lidow's successful use of Specification on IBM computers worldwide.

     40.2.9   The letter referenced only in passing Wilzbach's counsel, some 17 years prior, regarding the IBM Consent Decree.

     40.2.10   The letter expressed Sneed's deference to Wilzbach and Lidow due to their seniority under the Caltech Honor Code as fellow Caltech alums who graduated before Sneed.

     40.2.11   The letter referenced Torressen and IBM's d6/June 20, 1997 Letter of Corporate Intent.

     40.2.12   The letter proposed settlement, liquidated damages, and release of all claims, regarding use, if any, that warrants payment to Sneed under IBM's Letter of Corporate Intent and Agreement with Sneed.

Ex 1 p63

Mr. Sam Palmisano, IBM   d6/October 19, 2007
1 Orchard Road, Armonk, NY 10504

Mr. Palmisano:

1. I am Anthony Sneed, a Caltech alumnus, author and patent holder ("INVENTOR"). IBM hired INVENTOR in 1980 under Mr. Pete Wilzbach ("WILZBACH"). INVENTOR is an IBM marketing school honor graduate and past president of IBM's 6,000-member west coast IBM Club. WILZBACH is copied below out of deference to Caltech's Honor Code, as he is a fellow Caltech alumnus.

2. INVENTOR's know-how, BEST2000™, is U.S. Patent #6,236,992 ("992"). It is a cross-platform encryption system that insulates computers from the Y2K anomaly. Prior to and near 2000, Dr. Alex Lidow, Caltech trustee and retired International Rectifier ("IR") CEO, deployed BEST2000™ on IR's AS/400's in France, Germany, England, Italy, China, Japan, Singapore and the U.S. BEST2000 cut costs some 70% and rolled-out 50% faster than Anderson Consulting's scheme, ending their efforts. Dr. Lidow is a fellow alumnus. Under Caltech's Honor Code this is sent *after* his corroboration of relevant facts.

3. Prior to Y2K, IBM received BEST2000™ from INVENTOR and corroborated its commercial merit and economies. Later, IBM ("herein IBM-G") did the same with Mr. Niel Armstrong, IR's VP of IT, while IR debated Anderson Consulting's proposal. Mr. Armstrong's sworn statement corroborates these facts. BEST2000's™ per-site fee is $49,764.55 ("FEE"), memorialized in writing between INVENTOR and IBM ("TERMS"), prior to any evaluation by IBM. FEE to IBM is identical to FEE IR paid INVENTOR.

4. IBM memorialized orally and in writing custody, evaluation, merit and dissemination of BEST2000™ within IBM under TERMS. IBM never asserted use. IR disclosed use and paid FEE only by Dr. Lidow's vow to Caltech's Honor Code. IBM (herein "IBM-G") disclosed its $1 to $2 per-line-of-code pricing ("PRICING") to clients for Y2K-remediation. IBM-G inquired if INVENTOR would accept a royalty on PRICING rather than a per-site FEE under TERMS. FEE and TERMS remained extant and unchanged. IBM-G said it needed PRICING to bill other parties for BEST2000™ services, including IBM.

5. A different part of IBM (herein "IBM-C") made an unsolicited call to INVENTOR. IBM-C expressed desire to avoid IBM-G's PRICING by working with INVENTOR directly. Relationship between IBM-C and IBM-G was subject to a 1956 Consent Decree ("DECREE"). INVENTOR's extant FEE and TERMS with IBM-G and IBM-C's subsequent unsolicited call created a controversy relative to WILZBACH's COUNSEL 17 years prior regarding DECREE's ruinous effects on IBM. Based on that counsel INVENTOR *instantly* disclosed BEST2000™ to IBM-C, *promptly* memorialized the same by fax, and disclosed IBM-G's PRICING scheme, and BEST2000's™ *pending* patent status, which IBM-G had not.

6. Within Caltech's Honor Code IBM-C, represented by Mr. Al Torressen as IBM director of licensing, memorialized these actions in correspondence ("TORRESSEN") to INVENTOR, on or about d5/June 19th, 1997: IBM-C's custody of BEST2000™, its merit, and its dissemination for evaluation within IBM, consistent with TERMS. IBM-C rejected IBM-G's PRICING scheme leaving TERMS and FEE intact. Finally, IBM-C directed INVENTOR to contact IBM-C after BEST2000™ became a patent, at such time when no other party could make any claim against IBM, *if* IBM elected to use BEST2000™.

7. BEST2000™ became U.S. Patent 992 on d3/May 22, 2001. As of today, no party may make any claim against IBM regarding BEST2000™, meeting the burden IBM-C set in TORRESSEN *after* extant FEE and TERMS were in effect. (For settlement purposes only, in November 2007:) INVENTOR agrees to settlement for liquidated damages and release of all claims--whether involving a patent, or an oral or written agreement--between INVENTOR and IBM for use of BEST2000™, if any by IBM, in any way whatsoever. Settlement is for $746,468.25, the equivalent FEE for 15 sites before IBM-G's PRICING scheme to bill others including IBM-C, plus 10% per year since 2001, for a cumulative of $1,194,349.20. Mr. Torressen can't be reached to corroborate TORRESSEN. According to your firm, he's not available. His whereabouts are undisclosed. With all due respect to you, Mr. Palmisano, INVENTOR prays all executive forbearance might better inform Caltech's Honor Code *and* IBM's Basic Beliefs rather than *positive law* alternatives which, yet formidable, are but temporal shadows of *code and beliefs* reflected in extant *good-faith* by all.

With regards, gratitude, and respectfully yours,

Anthony Sneed
Anthony Sneed   VIA MAIL and FEDEX

Ex1 p24

Mr. Anthony Sneed
2058 North Mills Road
Claremont, CA 91711

cc: Mr. Pete Wilzbach, IBM Executive Vice President (retired)
    Dr. Alex Lidow, Caltech Board of Trustees, International Rectifier CEO (retired)
    Dr. Dale Prouty, Caltech PhD, Venture Capitalist (retired)
    Mr. Lee Browne, Caltech Emeritus Faculty Member (retired)
    Mrs. Florine Harris, Director, Willowbrook Middle School National School Honor Society (retired)

EX1 p65

41. IBM's response, 10 days later, d2/October 29, 1997, made no reference to Palmisano, but simply stated:

> On November 20, 2006, I sent you a letter to advise you that IBM continues to see no need for a license to U.S. Patent Number 6,236,992. As we stated in our letters to you of September 28, 2004, June 20, 1997, November 20, 2006, there are no past, present or future needs for IBM under this patent. Since we have repeatedly told you of our position, we see no reason for you to continue to send us communications regarding this matter. <u>IBM considers this matter closed and we will no longer comment on this matter.</u>
>
> > Winfield J. Brown, Associate General Counsel, Intellectual Property Law
> > Letter to Sneed from IBM, d2/October 29, 1997

41.1 Though not a cease and desist, Winfield J. Brown (J.Brown) as Associate General Counsel to Palmisano did declare there would be no further cooperation regarding disclosure of use, if any, as stipulated within IBM's Letter of Corporate Intent.

41.2 However, J.Brown is not Palmisano, nor did J.Brown reference Sneed's d6/October 19, 1997 to Palmisano. This did not conform to Wilzbach's expectation in his proposal to Sneed.

Ex1 p66

IBM

North Castle Drive
Armonk, NY 10504

October 29, 2007

Mr. Anthony Sneed
2058 North Mills Avenue
Claremont, CA 91711-2812

Dear Mr. Sneed,

On November 20, 2006, I sent you a letter to advise you that IBM continues to see no need for a license to U.S. Patent Number 6,236,992. As we stated in our letters to you of September 28, 2004, June 20, 1997 and November 20, 2006, there are no past, present or future needs for IBM under this patent. Since we have repeatedly told you of our position, we see no reason for you to continue to send us communications regarding this matter. IBM considers this matter closed and we will no longer comment on this matter.

Regards

Winfield J. Brown
Associate General Counsel
Intellectual Property Law

WJB:kk

Ex1 p67

a *7ᵗʰ-Gift* page

42. So Sneed sent a second letter to Palmisano, d2/November 5, 2007, noting there had been no response from Palmisano, and that J.Brown's letter makes no reference to Palmisano.

    42.1 Sneed reviewed the sum and substance of the matter.

    42.2 Sneed noted that J.Brown's unilateral letter could constitute inquiry notice if it breaches IBM's Letter of Corporate Intent that requires Sneed to perform and contact IBM regarding use, if the letter indeed represents Palmisano's position.

Ex1 p68

Mr. Sam Palmisano, IBM CEO   d2/November 5, 2007
1 Orchard Road,   **CONFIDENTIAL**
Armonk, NY 10504

Happy 2ⁿᵈ-day, Mr. Sam Palmisano.

1. A note dated d2/October 29, 2007, from IBM's North Castle Drive office, memorialized Mr. Winfield Brown's own admissions ("BROWN"). BROWN may or may not be in response to d6/October 19, 2007, correspondence with you for settlement consideration under <u>Caltech's Honor Code</u> and <u>IBM's Basic Beliefs</u> ("PALMISANO"), as BROWN doesn't reference PALMISANO in any way.

2. While BROWN disregards PALMISANO, BROWN makes specific reference to Mr. Al. Torresson's d6/June 20, 1997, correspondence and memorialized conditions thereof ("TORRESSON"). Mr. Torresson may or may not be IBM's director of licensing at this time. BROWN evidences no objection whatsoever to PALSMISANO. Alternately, BROWN may be uninformed by PALMISANO. In conjunction with memorialized terms and conditions ("TERMS") prior to and with TORRESSON that set a per-site use fee of $49,764.55 ("FEE"), TORRESSON acknowledges IBM's custody, evaluation and dissemination based on merit of this Caltech/IBM alum's ("INVENTOR's") know-how ("BEST2000™") specific to preventing Y2K failures on IBM and other computers.

3. Neither TORRESSON nor BROWN, nor any communication from IBM to INVENTOR between the dates of TORRESSON and BROWN, asserts IBM's production use of BEST2000™. Under TERMS, FEE is for use, NOT evaluation. BROWN doesn't dispute admissions in TORRESSON, nor has any prior communication from IBM to INVENTOR disputed admissions in TORRESSON.

4. TORRESSON does stipulate INVENTOR contact IBM after a patent is awarded, at such time when no other party could make a claim against IBM regarding BEST2000™ and any FEE. TORRESSON does NOT stipulate IBM contact INVENTOR, but rather INVENTOR *must perform* by contacting IBM. TORRESSON sets no limitation upon such *condition of performance*. BEST2000™ encrypts a computer's programs and datafiles, making them immune to Y2K failures for a period of some 20 years, and typically 28 years in actual application and practice. Furthermore, BEST2000™ allows re-encryption for additional 28-year periods. Therefore, TORRESSON's *condition of performance* for INVENTOR extends to 2025, *at minimum*, based on TORRESSON's date and BEST2000™'s system re-engineering design, well known by IBM legal and technical personnel informing TORRESSON. This fact, alone, obligates INVENTOR contact your firm, in conformance to TORRESSON, as TORRESSON doesn't compel, require, or demand anyone at your firm voluntarily disclose use to INVENTOR, but rather sets when INVENTOR may *begin* contacting IBM.

5. BROWN memorializes pattern and practice of INVENTOR's and IBM's good-faith efforts with on-going annual or bi-annual contact, consistent with TORRESSON's admissions of custody, merit and wide-spread dissemination of BEST2000™ within IBM. Notwithstanding such good-faith, BROWN suggests it might now be better for IBM, *and all parties thereof*, to henceforth make no further disclosures of matter in PALMISANO. BROWN also is completely silent about *positive law* precedent, especially Los Angeles County Superior Court case no. BC296142 ("BC2") and California appellate case no. B176278 ("B17"). Case law cites and findings thereof conclude BEST2000™ in use, today, on IBM AS/400's in at least seven countries, sites PALMISANO memorializes and which BROWN appears oblivious. BC2 memorializes economies, merit and technical signature ("SIGNATURE") of computers operating with BEST2000™ as compared to all other Y2K methods considered – computers and databases technically similar or substantially the same as those within IBM. Furthermore, IBM verified merit and economies of BEST2000™'s *for production use* to senior management at International Rectifier Corporation, a major IBM account with substantially similar potential Y2K defects as IBM, before applying BEST2000™. This, too, is memorialized in PALMISANO. BROWN's postulation ignores these facts.

6. BROWN misstates a patent issue. PALMISANO addresses FEE and TERMS, only, and not a patent license. BROWN may be a first attempt to demur possibility of BEST2000™'s SIGNATRUE on Y2K-remediated IBM computers and those in TORRESSON's admissions regarding BEST2000™'s custody, merit and wide-spread dissemination within IBM for evaluation, to the present day.

7. Rather, foundational to PALMISANO for SETTLEMENT PURPOSES ONLY is the memorialized FEE for use of BEST2000™ under TERMS met. BROWN does not dispute 15 sites as reasonable settlement for FEE and TERMS. Nor under TORRESSON does BROWN dispute 10% per year since 2001, the total settlement amount being $1,194,349.20 for liquidated damages and release of all claims, INCLUDING BROWN's inchoate patent reference. Therefore, NOTHING in BROWN molests extant good-faith settlement for $1,194,349.20, payable to "Anthony Sneed" at the address below, postmarked by sunset d6/November 30, 2007. Please do so now. I'm husbanded to one wife and four children, feeding them from the work of my hands and training at Caltech, NASA and IBM.

With regards, gratitude and respect,

Anthony Sneed

Anthony Sneed   VIA FAX 914-765-4290 AND MAIL   Attachment: Note from North Castle

Ex1 p69

Mr. Anthony Sneed
2058 North Mills Avenue
Claremont, CA 91711

cc: Mr. Pete Wilzbach, IBM Executive Vice President (retired)
    Dr. Alex Lidow, Caltech Board of Trustees, International Rectifier CEO (retired)
    Dr. Dale Prouty, Caltech PhD, Venture Capitalist (retired)
    Mr. Lee Browne, Caltech Emeritus Faculty Member (retired)
    Mrs. Florine Harris, Director, Willowbrook Middle School National School Honor Society (retired)

P.S. Two parables inform contact with you:

PARABLE A

"Therefore, the kingdom of heaven is like a king who wanted to settle accounts with his servants. As he began the settlement, a man who owed him a few million dollars was brought to him. Since he was not able to pay, the master ordered that he and his wife and his children and all that he had be sold to repay the debt. The servant fell on his knees before him. 'Be patient with me,' he begged, 'and I will pay back everything.' The servant's master took pity on him, cancelled the debt and let him go.

"But when that servant went out, he found one of his fellow servants who owed him a few hundred dollars. He grabbed him and began to choke him. 'Pay back what you owe me!' he demanded. His fellow servant fell to his knees and begged him, 'Be patient with me, and I will pay you back.' But he refused. Instead, he went off and had the man thrown into prison until he could pay the debt. When the other servants saw what had happened, they were greatly distressed and went and told their master everything that had happened.

"Then the master called the servant in. 'You wicked servant,' he said, 'I canceled all that debt of yours because you begged me to. Shouldn't you have had mercy on your fellow servant just as I had on you?' In anger his master turned him over to the jailers to be tortured, until he should pay back all he owed.

"This is how my heavenly Father will treat each of you unless you forgive your brother from your heart."   Matthew 18:23-35

PARABLE B

"...In a certain town there was a judge who neither feared God nor cared about men. And there was a widow in that town who kept coming to him with the plea, 'Grant me justice against my adversary.' For some time he refused. But finally he said to himself, "Even though I don't fear God or care about men, yet because this widow keeps bothering me, I will see that she gets justice, so that she won't eventually wear me out with her coming!"

And [I-SHALL-BE-SALVATION] said, "Listen to what the unjust judge says."   Luke 18:2-6

Pete Wilzbach, who remains one of the most respected fellow Caltech graduates I know, has given testimony you are not an unjust judge. In what shall soon occur, determined before there was an IBM, may you and your company enjoy prosperity and goodwill from customers, employees, suppliers and shareholders while receiving the best counsel available under heaven in your deliberations.

Ex1 p70

43. The follow-up from IBM came only three days later, again from J.Brown, d5/November 8, 2007.

    43.1 J.Brown explicitly declares and memorializes he represents IBM and Palmisano's position.

    43.2 He acknowledges the d2/November 5, 2007 letter as being received by Palmisano.

    43.3 Again, he adheres to the admission of only evaluative-use which is at no-charge under Agreement with Sneed, and that nothing is owed to Sneed under IBM's Letter of Corporate Intent signed by Torresson [sic], d6/June 20, 1997.

    43.4 Finally, he obviates the statement that "we will no longer comment on this matter," within his d2/October 29, 1997 letter, recanted by the very existence of his d5/November 8, 2007.

North Castle Drive
Armonk, NY 10504

November 8, 2007

Mr. Anthony Sneed
2058 North Mills Avenue
Claremont, CA 91711-2812

Dear Mr. Sneed,

I represent IBM in this matter. I wrote to you on October 29, 2007 in response to your
letter of October 19, 2007 to Sam Palmisano. I write you now in response to your letter
dated November 5, 2007 to Sam Palmisano. I have considered the allegations in your
recent letters to Mr Palmisano, your other letters to IBM and the letters sent by IBM to
you, including the letter you referenced that Mr Torresson sent you in 1997. IBM does
not owe you anything under any legal theory. We see no reason for you to continue to
send us communications regarding this matter. IBM considers this matter closed.

Regards

Winfield J. Brown
Associate General Counsel
Intellectual Property Law

WJB:kk

Ex1 p72

Mr. Sam Palmisano, IBM CEO
1 Orchard Road,
Armonk, NY 10504

d4/November 21, 2007
CONFIDENTIAL

Happy 4th-day, Mr. Sam Palmisano.

1. A note dated d2/October 29, 2007, ("BROWN") from IBM's North Castle Drive office, memorialized what appeared to be Mr. Winfield Brown's personal disclosures, declarations and admissions ("BROWN"). Mr. Brown subsequently memorialized a sworn declaration on your behalf ("OATH"), that construes BROWN and all findings therein, as binding upon you as IBM chairman and CEO, as well as upon all historical continuity vested in you at this time ("BOARD").

2. OATH also affirms that correspondence to you, dated d6/October 19, 2007 ("PALMISANO"), informs BROWN, specific to this Caltech/IBM alum's ("INVENTOR's") know-how ("BEST2000™") that encrypts IBM and other computers to prevent Y2K failures. BEST2000™ makes IBM and other computers immune to Y2K failures for 20 or more years, and typically 28 years in actual practice.

3. BROWN <u>does not dispute</u> paragraph 2 of PALMISANO ("PALMISANO.2") and the cost-effective, productive use of BEST2000™ on IBM computers in at least seven countries. BROWN <u>does not dispute</u> PALMISANO.3 and IBM's admission to a third-party of BEST2000™'s suitability as a production solution on IBM computers or that third-party's sworn statement to same. BROWN <u>does not dispute</u> PALMISANO.3's extant per-site fee of $49,764.55 ("FEE") is for production for use of BEST2000™ and not for evaluation ("TERMS"), established prior to IBM commencing evaluation. BROWN <u>does not dispute</u> PALMISANO.4 that IBM memorialized custody, evaluation, merit and dissemination of BEST2000™ within IBM based on technical merit. BROWN <u>does not dispute</u> PALMISANO.5 and Mr. Al Torresson's unsolicited call regarding BEST2000™, after TERMS and FEE were established to begin evaluation. BROWN <u>does not dispute</u> IBM has at no time asserted or disclosed discovery of BEST2000™'s production use.

4. OATH, on your behalf, also acknowledges correspondence from Mr. Al Torresson, d6/June 20, 1997 ("TORRESSON"). OATH construes TORRESSON, and all findings therein, as binding upon BOARD and IBM. BROWN <u>does not dispute</u> TORRESSON's admissions of custody, merit, evaluation and dissemination of BEST2000™ within IBM. BROWN <u>does not dispute</u>, to this day, INVENTOR's on-going conformance to TORRESSON's memorialized agreement regarding good-faith, cooperative efforts to discover use of BEST2000™, if any, and good-faith disclosure by IBM during any contact with INVENTOR if so discovered. Such good-faith, annual or bi-annual, cooperative contact began, without limitation, after d3/May 22, 2001, the date set by TORRESSON, which BROWN <u>does not dispute</u>. BROWN <u>does not dispute</u> INVENTOR's obligation to perform in conformance to TORRESSON, six years as of today. Furthermore BROWN <u>does not dispute</u> TORRESSON's good faith discovery provisions as binding on BOARD and IBM, within BEST2000™'s 28-year life-cycle which, according to TORRESSON's date, ends no sooner than 2025 ("MATTER").

5. BROWN under OATH on your behalf <u>does not dispute</u> PALMISANO.6, specific to memorialized agreement of extant FEE and TERMS prior to, and informing TORRESSON. OATH honors BOARD's and IBM's intent, spirit and good faith set forth in TORRESSON. TORRESSON sets reasonable expectation regarding comment or disclosure on MATTER, without limitation, and based on good faith, cooperative efforts. Yet regarding MATTER, BROWN contains a declaration ["BROWN INQUIRY NOTICE"]:

> "[IBM] will no longer comment on this matter."
> Under OATH on your behalf, by Mr. Winfield Joseph Brown, d2/October 29, 2007

6. Whether this notice is accidental, willful, or based on any *potential recent discovery* of BEST2000™ use within IBM as of d2/October 29, 2007, TORRESSON's admissions of custody, merit, evaluation and dissemination of BEST2000™, within a company as large as IBM, doesn't preclude discovery as of d2/October 29, 2007, or afterward. And TORRESSON is without limitation for such discovery and contact with INVENTOR. Therefore, under California Civil Procedures, BROWN now invokes INQUIRY NOTICE.

7. In good faith, and FOR SETTLEMENT PURPOSES ONLY, PALMISANO's settlement terms remain intact up until sunset d6/November 30, 2007. Settlement for $1,194,349.20, as memorialized, is for liquidated damages and release of all claims, whether for any oral agreement, written agreement or patent. Settlement amount of $1,194,349.20 is equivalent to FEE for 15 sites , if any, inclusive of 10% per year since 2001, payable to "Anthony Sneed" at the address below, without admission of liability by any party. Please do so now. The technical signature of BEST2000™ on IBM computers is established by an MIT engineer, expert witness, and nationally known Y2K expert. His nine-point findings contrast BEST2000™ from code of two alternatives already disclosed by IBM.

With regards and best wishes for restful celebrations in the coming days and weeks,

Anthony Sneed
Anthony Sneed   VIA FAX 914-765-4290 AND MAIL   cc: Primary Counsel

Ex1 p73

a *7^{h}-Gift* page

Mr. Anthony Sneed, 2058 North Mills Avenue, Claremont, CA 91711
cc: Mr. Pete Wilzbach, IBM Executive Vice President (retired)
    Dr. Alex Lidow, Caltech Board of Trustees, International Rectifier CEO (retired)
    Dr. Dale Prouty, Caltech PhD, Venture Capitalist (retired)
    Mr. Lee Browne, Caltech Emeritus Faculty Member (retired)
    Mrs. Florine Harris, Director, Willowbrook Middle School National School Honor Society (retired)

P.S. Mr. Palmisano, this is to acknowledge BROWN now invoking INQUIRY NOTICE, which Mr. Brown makes under OATH on your behalf. BOARD AND IBM are, therefore, on inquiry notice prescribed by Acts 5:1-11, if there's no settlement by sunset d6/November 30, 2007, in response to Mr. Brown's OATH on your behalf regarding *any legal theory*, which therefore includes precedent not bound by *positive law* or Earthly tribunals. Mr. Brown's OATH on your behalf construes TORRESSON as binding on BOARD and IBM, by foundational authority in Romans 13:1-14, referenced in response to Mr. Brown's inclusion of *any legal theory*.

In closing, 25 years ago on this day of Thanksgiving week, a three-standard-deviation power-surge from a local municipal power-grid wiped-out a major, west coast bank's datacenter's four IBM 3033 processors. A floor's worth of DASD, some dozen 3890 check processors, a statewide ATM and workstation network were knocked out. The bank, as a whole, handled some $32 billion of federal reserve float every four days. Bags of unprocessed checks rapidly began filling hallways. The IBM branch was without any senior or account marketing personnel, due to the soon-to-begin holiday. INVENTOR was a *trainee* just back from VSM (Virtual System Marketing class, in Dallas), the third of five IBM classes leading to honor graduation from Marketing School in Poughkeepsie, NY.

The VSM class manager said, before end of class, that branch managers ("BM's") would give returning trainees hypothetical scenarios, just to see how trainees might respond. So when INVENTOR's BM described a "hypothetical" three-standard-deviation power surge that wiped out the largest account in the branch and asked, "What are YOU going to do about it?", INVENTOR requested a 3081 be pulled-off the line in Poughkeepsie for stand-by overnight shipment. "You've got it!" the BM responded. INVENTOR then requested double shifts of Field Engineers for 72 hours. "You've got it!" INVENTOR asked to be relieved of marketing training responsibilities for one week. "You've got it!" Then INVENTOR said there was nothing else to do except live at the account until it was back up. "GO DO IT!" the BM ordered. "Yes sir," INVENTOR said and left the BM's office, having nailed the "hypothetical" VSM BM/trainee test scenario. Just to complete the "simulation," INVENTOR called his good friend and large account System Engineer at the bank, and described how well the "simulation" had gone. "That wasn't a simulation," the SE said. "The bank REALLY is knocked out and there's NO marketing personnel here." INVENTOR then realized he'd accepted responsibility to be the principal IBM marketing contact for one of the worst finance industry disasters in IBM Western Region history.

Needless to say, INVENTOR promptly arrived at the account and lived there, in Brooks Brothers suit, wing-tip shoes, VERY conservative tie (exceeding INVENTOR's boarding school standards), and starched white shirts, for all of Thanksgiving and the weekend. INVENTOR provided regular briefings to multiple different customer executives on recovery progress. IBM FE personnel coordinated cannibalizing two of the four 3033's, to get the other two back online. An old 370/168 was re-gen'ed and brought online to help run the check processors. The emergency 3081 arrived as requested. INVENTOR opened its panels and inspected the box inside and out. Two of its 16 TCM slots were blanks. FE installed the 3081 to bring it online. All the while, customer execs were constantly given information briefings, which INVENTOR provided, holding separate update meetings with FE personnel and SE's.

During the four days, there was 100% marketing coverage. The bank previously paid some $800,000 to a backlog bookie out of Denver for an earlier ship-date and might sell its backlog position to someone else. During a tour with five bank execs, INVENTOR noticed water leaking from the bottom of the 3081 TCM complex onto its power supply below it, into a Styrofoam cup. One of the execs asked, "What's that?" INVENTOR queried the FE, who said, "One of the TCM's has a leak. A new one should be here after a day or so." The system, when online, would help process $32 billion in federal reserve float every four days. The customer execs looked on. "Bring the processor down," INVENTOR said. "What??" the FE replied. "Bring it down," INVENTOR said. The FE complied. INVENTOR opened the processor panels exposing the TCM complex and the 3081's internals. "Why are there two blanks?" INVENTOR asked. "This dyadic-processor configuration only requires 12 of 16 possible TCM's," the FE said. "Then take the two hoses from a blank and replace the leaking coupling on the bottom one." Within 15 minutes there was no leak. By the first business day of the week, INVENTOR had a final meeting with customer execs who asked how long could the bank keep the 3081. INVENTOR had signed the annual "Business Ethics and Guidelines" supplied by IBM Legal to all marketing personnel. Recalling its provision on emergency processors, INVENTOR answered "We have it for 90 days." The entire disaster and INVENTOR's conduct were later reviewed by their predecessor, Mr. John Opel. Mr. Opel backed all of INVENTOR's decisions and actions during the disaster, overruling INVENTOR's marketing manager (sent to Japan), branch manager (early retirement), regional manager (reassigned) and division manager (later left the company). For a full year, the bank tried to recruit INVENTOR, for INVENTOR's well being. INVENTOR went on to graduate with honors from IBM Marketing School, was assigned a territory, and won the Western Region fast-start contest with a 400% of quota performance. INVENTOR later left IBM and started a consulting practice. One six-year contract was with an IBM third-party software start-up. INVENTOR helped it grow from near bankruptcy at $4 million, to $26 million/year whereupon INVENTOR was offered the VP of Sales during the firm's successful IPO. Another project is BEST2000™.

Exl p74

44. On or about d4/December 2, 2007, Sneed called Wilzbach to say IBM declined to enter into settlement.

    44.1   Wilzbach instantly proposed that Sneed sue Palmisano and IBM.

    44.2   Sneed was momentarily stunned.

        44.2.1   As Sneed's new training manager when Sneed joined IBM in 1980, Wilzbach gave Sneed assignments that were daunting even for seasoned, experienced IBM marketing and system engineering professionals. Yet they all turned out well, while Sneed matriculated through IBM's corporate training and graduated with honors. Wilzbach even promoted Sneed's candidacy to become president of IBM's 6,000 member Western Region IBM employees club, which Sneed was subsequently elected to by a majority vote.

        44.2.2   Now Wilzbach was proposing Sneed sue an $80 billion+ corporation with tremendous legal wherewithal and years and years of litigation experience against the U.S. government, multination corporations, and whole nations.

    44.3   Sneed asked if Wilzbach might act as liaison between Palmisano and Sneed to resolve the matter.

        44.3.1   Wilzbach said he couldn't, and that his role was only to provide informal, friendly advice.

        44.3.2   Wilzbach proceeded to disclose guidelines for Sneed to follow during litigation.

        44.3.3   Wilzbach provided no information whatsoever regarding how, or if IBM used Specification.

        44.3.4   However, his advice was in complete harmony with Lidow's and Prouty's speculations.

    44.4   After these discussions, Sneed learned that Wilzbach was enduring multiple events in Wilzbach's personal life that almost brought he and Sneed to tears. Sneed was confident that Wilzbach could patiently endure, and in compassion listened, respecting a man any father would be privileged to claim as his own son.

    44.6   Wilzbach had already served, and endured, IBM's internal politics for decades. He was now a free man, to serve his family's needs and his own aspirations, unencumbered by a corporate culture of conflicting purposes. No longer having to move his family from state to state, and city to city, and climb the corporate ladder, his family enjoyed stability in having a hometown with roots, rarely experienced to the same degree during Wilzbach's tenure at IBM.

    44.7   If anyone deserves as much peace and quiet as possible, it is Pete and Charlotte Wilzbach.

    44.8   There has been a pattern to Sneed's professional pursuits during the past few decades to now: Sneed desires to always respond to executive orders and perform; whether it's a management directive at Hughes (Boeing) to design the operating system to pull data from below the cloud-tops of Jupiter, requiring Sneed successfully develop 4,000 lines of code in four weeks; or exceeding the president of Caltech's written expectation to serve in new ways, reflected through Specification; or IBM management's order to be the principal marketing liaison and restore the operations of a bank processing $32 billion in federal reserve float every four days after its datacenter was destroyed, which Sneed completed in four days, as an IBM trainee, and was offered a VP position by the bank; or signing on to a nearly bankrupt software start-up to help it grow from $4 million to $48 million annually in four years, whereupon Sneed was offered a vice presidency--there has been only one objective: perform.

    44.9   Wilzbach was now, through his proposal, issuing an executive order for Sneed to sue IBM and Palmisano, but as no more than informal, friendly advice about how Sneed might accomplish this.

        44.9.1   At the time, Sneed was unaware that one of Palmisano's hand-picked successors to be a candidate for IBM CEO, Mr. Bob Moffet, was indicted and later convicted of insider-information trading; at or about that time he left IBM to serve a federal prison sentence.

    44.10   Only four years prior, Sneed proceeded with litigation against IRC, but only after another fellow Caltech alum, Lidow, assented to Sneed doing so with the sobriquet "Don't sell your self short." Subsequent to that litigation, IRC's CFO, who aided in blocking payment to Sneed, was let go.

Exl p75

Mr. Pete Wilzbach, IBM Executive VP (retired)   d5/December 3, 2008
62 Hospitality Street, Mount Pleasant, SC 29464   **CONFIDENTIAL**

Happy 5ᵗʰ-day, Pete.

1. This memorializes guidelines you shared, in contact yesterday, for effort after November 2007 to determine use, if any, by IBM of my BEST2000™ invention, deployed in at least seven countries, to prevent Y2K failures on IBM computers ("MATTER");
    G1. Respect legal counsel's advice and direction based on MATTER'S merit and IBM's right to a presumption of innocence.
    G2. Adhere to foundational, memorialized agreement ("AGREEMENT") between parties regarding MATTER.
    G3. Secure testimony of BEST2000™ custody/use by IBM from one or more credible witnesses.
    G4. Translate actual "use" into quantifiable dollar amounts owed, based on memorialized agreement.

2. Relative to G2, MATTER was only discussed with you after learning, in October 2007, of your retirement from IBM. AGREEMENT set forth written, good faith expectation that your fellow Caltech alum ("FELLOW ALUM") receive information regarding IBM's BEST2000™ use, if any, from a single, authorized IBM corporate source who declared control of BEST2000™ use, if any by IBM, worldwide ("GOOD FAITH"). Your retirement allowed FELLOW ALUM to speak freely and at length, as we now have.
3. Relative to G3, there's a second, senior IBM executive, whose retirement status affects his in-home deposition availability during the next 12 months, respecting amicable perspective he, too, has of MATTER. Prior to our recent contact, he counseled FELLOW ALUM on translating actual "use", if any, into quantifiable amounts due, respecting the same sensitivity expressed in G4.
4. Relative to G1 and IBM, Caltech's Honor Code informs contact with you; Dr. Dale Prouty, retired venture capitalist; and Dr. Alex Lidow, retired International Rectifier Corporation ("IRC") CEO and Caltech board member. In successful resolution of a similar matter with IRC, litigation only proceeded with Dr. Lidow's concurrence that extant memorialized agreement warranted payment. All litigation ceased upon establishing merit that Dr. Lidow, personally, acted in good faith, and that agreed terms warranted payment based on testimony by IRC's VP of IT, Mr. Niel Armstrong; and IRC's Y2K project manager, Mr. Chris Case. Anderson Consulting personnel ("ANDERSON"), responding to examination during trial, were less than creditable. ANDERSON personnel unsuccessfully attempted to conceal their confusing role in delaying IRC's acceptance and cost-saving use of BEST2000™, or risk from alternatives.
5. Furthermore, the court entertained testimony that ANDERSON may have been technically negligent, itself, not to recommend BEST2000™ for IRC's IBM AS/400 computers in at least seven countries. With BEST2000™, IRC's computers work to this day.
6. FELLOW ALUM similarly helped IBM, itself, without favoritism. IBM, too, was exposed, by circumstances possibly beyond its control, to a similar business disruption scenario as IRC's, also with only 24 months to act before Y2K failures might affect IBM corporate computers. After AGREEMENT was established, Mr. Dennie Welch's unit within IBM exhibited the same élan and respect for Caltech's Honor Code as Dr. Alex Lidow did. Both Mr. Welch's and Dr. Lidow's *technical* personnel acted with candor and decisiveness. Mr. Welch and Dr. Lidow were better served having *all* alternatives available to them, including BEST2000™.
7. I agree with guidelines set forth in G1 to G4 as reasonable and proper. As with Dr. Lidow, FELLOW ALUM deferred to you before proceeding, as you both have seniority under Caltech's Honor Code. IBM's allegation ("ALLEGATION") against FELLOW ALUM is that he is breaking the TEN COMMANDMENTS and specifically violates EXODUS 20:16 ("E2016") as follows:
    A1. Violates E2016 by asserting IBM received or acknowledges first receiving BEST2000™ know-how from FELLOW ALUM.
    A2. Violates E2016 by asserting IBM accepted AGREEMENT'S no charge provision for IBM to evaluate BEST2000™.
    A3. Violates E2016 by asserting $49,764.55 or any other amount was ever set forth as the per site use fee for BEST2000™.
    A4. Violates E2016 by asserting d3/May 22, 2001, as the earliest date IBM *memorialized and set* to pay FELLOW ALUM.
    A5. Violates E2016 by asserting FELLOW ALUM met conditions agreed to in writing by all parties, including AGREEMENT.
    A6. Violates E2016 by asserting FELLOW ALUM indemnified IBM, forever, from suit by third-parties for use of BEST2000™.
    A7. Violates E2016 by asserting BEST2000™'s use and benefits thereof nothing IBM couldn't discover for itself, 1960 to 2025.

The seriousness of ALLEGATION transcends any minor monetary amount due. IBM has memorialized it is prepared to make ALLEGATION to the world, and to heaven itself if it knew how. IBM'S ALLEGATION now asserts its business conduct can not be bound by E2016 or any similar "non-business" Commandment, as its DEHOMAG subsidiary asserted under color of authority, 1933 to 1945. It is by the grace of the Name of my Father's One and Only Son, Yashua Messiah, through the Spirit of Truth, that I respond to ALLEGATION with IBM's own written admissions of d6/June 20, 1997, AGREEMENT, and supporting testimony thereof.

With regards and respectfully yours,

Anthony Sneed
Anthony Sneed

Mr. Anthony Sneed, 2058 North Mills Avenue, Claremont, CA 91711
cc: Mr. Sam Palmissno, IBM Chairman and CEO
   Dr. Alex Lidow, Caltech Board of Trustees, International Rectifier CEO (retired)
   Dr. Dale Prouty, Caltech PhD, Venture Capitalist (retired)
   Mr. Lee Browne, Caltech Emeritus Faculty Member (retired)
   Mrs. Florine Harris, Director, Willowbrook Middle School National School Honor Society (retired)

[P.S. Thanks for sharing the information on the successful Mariner mission and the kilobyte storage system you helped innovate to meet weight, low-power and loss-of power specifications. If done in your junior year as a summer intern, this would allow its use in 1969 or shortly thereafter. If completed in or about the summer of 1968, as a junior, your design was available, possibly for Mariners 6 and 7, and definitely for Mariner 9's extended orbital operations. Floyd Humphrey, the magnetism/EE prof and your Catalina skipper, met with me as a pre-frosh. He made my visit to Throop memorable, that building gone when I arrived in the fall of '73.

One day, Floyd visited an EE lab I was taking. While we chatted, the subject of HP25 calculators came up. I told him about a game called MOO played on the campus (unnamed) timesharing system, similar to MASTERMIND. At Caltech, programming MOO into the HP25's 49-step memory was deemed impossible. Noticing he had his HP25, I asked to see it, so he handed it to me. I punched in all 49 steps by recall, in about 20 seconds, of a MOO program I'd written. After vetting that it worked, he was curious how MOO was possible in 49 steps – so was Arthur Rubin who finished first on the *Putnam* national math exam, and Dr. Howard Rumsey, then a leading JPL information theorist – they KNEW it was impossible. MOO involves guessing a 4-digit (none repeated) number based on "bulls" (correct digit in correct position) and "cows" (correct digit in wrong position – and a "bull" is not a "cow") in each guess. MOO gives results like 2.1 for "2 bulls and 1 cow", or 4.0 meaning your guess was the actual hidden number. After they pleaded, I told them the trick: the essential 3 instruction-steps needed were "if x=y", "*", and "1". "*" is a decimal point.

That code sequence compares the positions (in the x and y registers) for the same single-digit found in both the hidden number and a guess: it adds 0.1 to a result register if the positions are different and 1.0 if the positions are the same. Who would've thought to put a "hanging" decimal point ("*"), by itself, as a conditional "branch" option? Then again, it was economies like this that knocked about 20% off the discrete Fourier transform microcode in the Galileo orbiter's telemetry processor, to keep frequency- and phase- lock on the Galileo probe's 1.4 gigahertz signal, as the probe plunged below Jupiter's cloud-tops. I later took my HP25 apart, modified it to have 1000 steps of paging memory (why not? you built a house didn't you!) and picked up about 18 units of EE credit for doing so.

But before Galileo, like you, I was a summer intern on another NASA project called Pioneer/Venus ("P/V"). When I started on the project, I was shown a nice desk, told where to pick-up my paycheck and then left at my desk. So, I found my supervisor and told him I was on the P/V project. After verifying I knew where my desk was and where to pickup my paycheck every other week, he seemed satisfied and went back to his work, reminding me he was pretty busy. At first I thought this was like a "movie set" where, you know, a LOT of people are "extras." So I told him I needed something to *look* at while "working" at my desk. "Like what?" he wondered. I said, "How about the entire command sequence stored in the Venus orbiter's computer, from launch through cruise, and after it orbits Venus." When he asked why, I said it's probably a lot of material and "looks" good to have on the desk, still working through my "movie set" idea of how aerospace worked. After about a day, he showed up at my desk with multiple three-inch binders, then left.

Since nobody asked me to *do* anything, I had all the free time in the universe, seven days a week. I poured over all the manuals multiple times, committing parts to memory. The spacecraft had two computers for fault tolerance, instruction-times of 125 milliseconds and so on and so forth. In an EE class at Caltech, taught by Carver Mead, I learned "true" fault-tolerance used three computers, the odd one voted out by the two in agreement. P/V's design only ran two computers in parallel. If one stopped, the other would keep issuing commands, consistent with NASA's mission spec that no single-point failure cause loss of mission. Weight constraints limited the spacecraft to two computers, which I bought into as I read further. The only scenario I couldn't find tested, any where, was if the computers got out of sync by only an instruction or so, rather than one computer just flat-out failing. So, I simulated the WHOLE command sequence to see what happened if they got out of sync by an instruction. Sure enough, there was a sequence specific to a safety mechanism and the radial thrusters. The sequence could start a firing and not be able to stop it, even after being shut-down by the "other" computer if they went out of sync but at least 125 milliseconds. And that could cause loss of mission, let alone loss of all on-board fuel. So I asked my supervisor, who was terribly busy, for a schematic of the command processors and their electrical connections to the rest of the spacecraft. After the usual "Why?" exchange, I got the manuals. I was amazed. The schematics showed each processor had its own *independent* "start-switch" which popped open after separation from the launch vehicle. And separation was controlled by *explosive* bolts. From the spec, a 125 millisecond delay between the *independent* start-switch openings was possible when the bolts exploded, separating spacecraft from launch vehicle. In other words, it was a single-point catastrophic mission failure mode. To make a long story short, this led to my inclusion at a design review meeting before NASA took custody of the spacecraft. At the well-attended meeting, the senior NASA official in charge asked who found the single-point failure and what redesign was necessary. Management pointed to me. I answered the question, the redesign was agreed to by all and later implemented. P/V was successful. After graduation, I was promoted to full member of the technical staff and asked to look at another spacecraft called Galileo, that effort written about in *From Jupiter to Genesis*, a world-wide first-hit title on Google (smile).]

Ex 1 p77

45. Between 2007 and 2011, Lidow and Prouty continued to urge Sneed not to drop the matter with IBM.
    45.1 IBM had for a decade admitted to nothing more than evaluative-use, and memorialized that fact.
    45.2 The IBM Letter of Corporate Intent designated a single-point of contact between Sneed and IBM regarding any inquiry about Sneed's Specification; contact with another party could violate the letter and spirit of Agreement between Sneed and IBM, and contract formation made in good faith.
    45.3 Wilzbach had proposed Sneed sue IBM and Palmisano as no more than informal, friendly advice.
    45.4 From the IRC case, Sneed knew not to bring a complaint without reasonable cause, and even with reasonable cause a well-funded corporation can be formidable if it is bent on perfidy.
    45.5 Yet Lidow and Prouty would not let up on Sneed. Whenever they spoke with Sneed, they might inject the coda, "Have you done anything about IBM yet?" providing no more than speculative reasons as to why anything could be "done about IBM" under Agreement between Sneed and IBM.

46. By 2011, it was time for Sneed to contact IBM and make a *pro forma* inquiry about use, made by Sneed every 24 to 36 months under IBM's condition of performance in its d6/June 20, 1997 letter.
    46.1 Sneed included the particular, non-probative conclusions of Lidow's and Prouty's, about IBM using Sneed's Specifications, along with Wilzbach's non-probative advice for Sneed to sue IBM after IBM rebuffed Wilzbach's settlement proposal Sneed executed. This allowed their voices to be heard and memorialized, as all three set non-probative expectation that Sneed sue IBM.
    46.2 The d2/April 4, 2011 letter to Palmisano contained this declaration in its 5ᵗʰ paragraph:

    > By October 2007, Sneed learned WILZBACH retired from IBM. WILZBACH, as a retired IBM EVP, corroborated LIDOW'S opinion, as an IBM customer executive, that [Specification's] use on computers *inside* of IBM merited payment to SNEED. Both WILZBACH and LIDOW are Caltech alums, LIDOW a member of the Caltech board of trustees... J.Brown triggered inquiry notice ("NOTICE") on or about October 29, 2007, declaring ("OPINION") that IBM would no longer respond to any discovery request under AGREEMENT and CODICIL as had Logan, Torressen, and Schreiber leading up to 2007. [Specification's] useful life extends to 2025, which includes computers inside of IBM reengineered with [Specification].
    > Sneed's letter to Palmisano, d2/April 4, 2011

    46.3 The letter included an invoice for $1,492,936.50, or 10% annual accrual since the original settlement effort Wilzbach proposed in 2007.
    46.4 The letter was cc'ed to Wilzbach, Lidow, Prouty, and Caltech President Dr. Jean-Lou Chameau, respecting their witness through the Caltech Honor Code of Sneed's *pro forma* inquiry.
    46.5 Palmisano and IBM were non-responsive. Sneed later contacted IBM. Esther, Sneed's gracious contact for some four years, confirmed that J.Brown was still employed by IBM. Esther is J.Brown's administrative assistant. Sneed and Esther concluded the call amicably.
    46.6 Lidow and Prouty assented to the sum and substance of the letter.
    46.7 For Sneed to perform and file litigation as Wilzbach proposed, there would have to be a written, signed document exhibiting reasonable cause to file a complaint.
        46.7.1 However, if Wilzbach prepared such a letter, affirming even a hint or a declaration of use, if any, by IBM, he could run afoul of IBM's secrecy provisions for former executives.
        46.7.2 If such a letter opens the way for litigation, Wilzbach, having negotiated tens of millions of dollars in contracts, knows that it would see the light of day in discovery, if not earlier.
        46.7.2 Sneed waited for a written response from any party. Only one responded: Wilzbach.

Ex1 p78

FAX TO: Mr. Sam Palmisano c/o James Brown, 914-945-3281
FROM: Mr. Anthony Sneed ISB::Institute Fellow, *Emeritus* ("SNEED")
RE: With gratitude and respect for all of your past consideration.

PAGES: 2 (Including this one) **CONFIDENTIAL**
DATE: d2/April 4, 2011 **FINAL DRAFT**

Happy 2nd -day, Sam Palmisano.

1. Chief Justice Earl Warren, in 1954, a year prior to SNEED'S birth, issued a Supreme Court decision ("WARREN"), informed by John Adams ("ADAMS"), Thomas Jefferson ("JEFFERSON") and Abraham Lincoln ("LINCOLN"). Just 18 years later, Caltech president Dr. Harold Brown ("BROWN") applied WARREN to admit SNEED to Caltech. Afterwards, fellow Caltech alum and future IBM EVP Pete Wilzbach ("WILZBACH") employed SNEED at IBM. After IBM, fellow Caltech alum and International Rectifier CEO Dr. Alex Lidow ("LIDOW") deployed BEST2000™, patented by SNEED, on IBM computers in seven countries, achieving 70% savings on a $10 million-alternative Accenture (Anderson) proposed.

2. After SNEED entered second grade, President Kennedy ("KENNEDY"), through the Department of Justice, directed Mr. Nicholas de Katzenbach ("KATZENBACH"), later counsel for your firm, to also apply WARREN in dismantling federal laws based on a belief system, herein called *predestined "anti-meritism"* ("ANTI-MERITISM"), that had impaired U.S. free-market expansions from 1788, to 1865, and beyond. Historically, nations practicing *predestined* ANTI-MERITISM under color of authority, have endured mixed efficacies: The Dreyfus matter in France; and Germany's genocidal *predestined* ANTI-MERITISM in the 40s, to expunge the life-merit of individuals such as Albert Einstein, Lisa Meitner, Franklin Roosevelt, Thurgood Marshall, Jesse Owens, Joe Louis, soldiers, sailors, Marines and anyone who resisted ANTI-MERITISM.

3. In the U.S., today, anyone may proclaim ANTI-MERITISM as a *"divine" right* or as a *personal belief.* However, WARREN determined *predestined* ANTI-MERITISM to be at odds with government's purpose and any institution conducting business on behalf of government. WARREN leaves unmolested the freedom of private parties to profess ANTI-MERITISM as a *"divine" right* or as a *personal belief.*

4. ADAMS-JEFFERSON-LINCOLN-WARREN-KENNEDY-KATZENBACH-BROWN-WILZBACH-LIDOW define an extraordinary precedent ("PRECEDENT") that bonds us. PRECEDENT informed contact with IBM personnel regarding BEST2000™, from June 1997 to October 2007. IBM's designated personnel ("DESIGNEE") repeatedly changed. IBM Global Service's ("GS") Don Logan and Dick McDonough performed intake and vetting, acknowledging SNEED'S written agreement ("AGREEMENT") included with BEST2000™ which set forth the fair market price ("PRICE") regarding use by IBM. Further acknowledging AGREEMENT'S priority, IBM legal DESIGNEES Al Torressen, James Schreiber and James Brown ("J.BROWN") took custody of BEST2000™ from GS, and performed under AGREEMENT and codicil ("CODICIL") memorialized by Al Torressen, setting May 22, 2001, as the date ("DATE") for SNEED to initiate contact with IBM for payment. Between DATE and 2007, each DESIGNEE declared, in writing, that IBM had not discovered use of BEST2000™ in any way that merited payment to SNEED.

5. By October 2007, SNEED learned WILZBACH retired from IBM. WILZBACH, as a retired IBM EVP, corroborated LIDOW'S opinion, as an IBM customer executive, that BEST2000™ use on computers *inside* of IBM merited payment to SNEED. Both WILZBACH and LIDOW are Caltech alums. LIDOW, a member of the Caltech board of trustees. In addition, Dr. Dale Prouty ("PROUTY"), a third Caltech alum, vetted BEST2000™ with sworn testimony of its use and economy on IBM computers. On or about October 19, 2007, these opinions ("OPINIONS") resulted in a new discovery request to J.BROWN. J.BROWN triggered inquiry notice ("NOTICE") on or about October 29, 2007, declaring ("OPINION") that IBM would no longer respond to any discovery request under AGREEMENT and CODICIL as had Logan, Torressen, and Schreiber leading up to 2007. BEST2000™'s useful life extends to 2025, which includes computers inside of IBM reengineered with BEST2000™.

6. J.BROWN'S OPINION in this matter is at odds with three Caltech alums, including a retired IBM EVP, a retired IBM customer executive, and a venture capitalist. J.BROWN'S motive for OPINION is unknown, though PRECEDENT with WILZBACH, LIDOW and PROUTY endures.

7. No one on earth, Mr. Palmisano, is better positioned to be a decider of facts in this matter than you. A third-party could bring J.BROWN and four Caltech alums together, to publicly chronicle what we know. However, this will not improve upon what you already know as former head of GS, before becoming chairman and CEO of IBM, with PRECEDENT informing WILZBACH'S, LIDOW'S and PROUTY'S OPINIONS, today. And today is April 4, 2011, an anniversary of the violent murder of Dr. Martin Luther King, Jr. by a coward--a murder representing the moral degeneracy of *predestined* ANTI-MERITISM and those who embrace it. Soon, you will retire from IBM, and rightfully so; and, perhaps after a few decades, if not before, your name, as will mine, shall pass on into eternity, our testimony sealed with these Proverbs for all time:

The righteous care about justice for the poor, but the wicked have no such concern. [29:7]

The memory of the righteous will be a blessing, but the name of the wicked will rot. [10:7]

If you say, "But we knew nothing about this," does not he who weighs the heart perceive it? Does not he who guards your life know it? And will he not repay each person according to what he has done? [24:12]

Your imprimatur through the attached invoice concludes the matter, as we each follow our separate paths. As an aside, I had not been to a dentist in 34 years, until a local dentist graciously gave me a free examine and X-rays last quarter. The dentist was astounded I've been able to function, after determining the condition of my bottom, #1 molar, the only major defect. Removing it, at this point, may affect the nerve along my jaw. Your invoice may permit me to attend to related needs of my wife, two girls and two boys, first, and then myself. Considering all the afflictions possible to man, along with the depravity remembered this day, and the celebration of life and liberty two weeks from today, I trust all shall yet turn out well.

With prayerful regards and respectfully yours,

Anthony Sneed

Anthony Sneed
Attachment: Invoice 7277
cc: Mr. Pete Wilzbach; Dr. Alex Lidow; Dr. Dale Prouty; Dr. Jean-Lou Chameau, President, Caltech; Mr. Ken Roberts, Esq., Orrick, Herrington & Sutcliffe

Ex1 p79

a *7ᵗʰ-Gift* page

INVOICE 7247

d2/April 4, 2011

Mr. Sam Palmisano c/o Mr. Winfield James Brown
1 Orchard Road, Armonk, NY, 10504

$1,492,936.50   Due upon receipt

Payable to "ANTHONY SNEED" at 2058 North Mills Avenue, Claremont, CA, 91711, based on fee and terms acknowledged by IBM, in IBM correspondence sent to, and received by ANTHONY SNEED. Payment within 30 days, and receipt by ANTHONY SNEED, shall constitute liquidated damages and release of all claims between ANTHONY SNEED and IBM, and shall settle all matters between ANTHONY SNEED and IBM, including use of BEST2000™ described in U.S. patent #62366992 issued to ANTHONY SNEED May 22, 2001, and any derivative work thereof, applied in any way by IBM, without limit on the number of computers IBM has applied it before today, or applies it after today.

Ex1 p80

47. Wilzbach did something unexpected by Sneed: he affirmed he would stand with the other two Caltech alums for the use declaration, with the title of IBM VP in the d2/April 4 letter--and he put it in writing with a letter to Sneed d1/June 5, 2011. In it:

   47.1 Wilzbach continues to support the proposal that Sneed sue IBM and Palmisano if Sneed chooses.

   47.2 Wilzbach reminds Sneed that Wilzbach's advice is only informal and as a friend.

   47.3 Wilzbach declares his non-involvement regarding any dispute between Sneed and IBM.

   47.4 Wilzbach directs Sneed not to use Wilzbach's name on further correspondence to IBM, to avoid triggering an IBM disclosure provision, which the four previous letters that referenced Wilzbach, since 2007, had not.

   47.5 Wilzbach will allow his name to be cc'ed on the letter to Palmisano, making Wilzbach one of three Caltech alums, the Caltech president a silent witness, who vetted the letter for accuracy and are standing together regarding its contents; Wilzbach affirms he will not give IBM reason to believe otherwise, confirmed when IBM engaged in uncorroborated speculation about Wilzbach in IBM's Reply to Dismiss Amended Complaint, page 6, lines 12-17.

      47.5.1 Wilzbach is true to his word: he is not cooperating with IBM's desire to use him to sway the Court, as it is written, "Those who forsake the law praise the wicked, but those who keep the law resist them." Proverbs 28:4. No one can blame Wilzbach's desire not to be part of such conduct; in this matter he stays clear of IBM. He quit the company, walked out the door and never looked back, effective December 31, 1999--on the eve of the Y2K trigger-date, first disclosed by Wilzbach to Sneed in or about 4Q 2007.

      47.5.2 Wilzbach has not cooperated with IBM or its attorneys, true to his word in his d1/June 5, 2011 letter to Sneed. Sneed continues to trust Wilzbach and his intentions, along with Lidow's and Prouty's , as reflecting favorably on the Caltech Honor Code--a code IBM and its counsel trample, dismiss, belittle and denigrate for their own purposes in these proceedings, as if the title "Honor" defiles anything it touches in their pleadings. They teach away from the Honor Code's sentiment in their hostile vituperations to the Court. Yet it is the term "Honor" that also ennobles the title "Your Honor" and the purposeful privilege all parties are given to be in such presence. IBM and its attorneys unilaterally, by mere whim, elected to consider such privilege with casual disregard, last March.

      47.5.3 Even as this is written, Sneed prays for the Court's forbearance to receive this Motion, prepared by a lay heart and lay hands, that tremble in fear, not of lost enrichment another may now covet, but rather of missing any opportunity to lighten the considerable burden the Court endures every day. The greater fear is of not being responsive and obedient in every proper and appropriate way to the purpose of perpetuating the Court and its impartiality, as much as wherewithal and circumstances may permit. The majesty and reverence the Court inspires for the law, by blending power, strength, wisdom, and grace, already extended to the parties on each side too many times to count--including d2/March 1, 2012 when only Attorney Spitz performed, as the Court tended to IBM's interest with compassion and a tender mercy that touches Sneed's heart to this day--requires no less.

48. Based on Wilzbach's letter and its direct reference to the d2/April 4, 2011declaration of use to Palmisano, and Palmisano's and IBM's non-responsiveness, thus breaching IBM's own Letter of Corporate Intent, Sneed filed complaint d6/October 28, 2011, or within 7 months of the breach.

Tony Sneed   June 5, 2011
2058 N. Mills Ave.
Claremont, CA 91711   *Rcvd Jun 11, 2011*

Dear Tony,

Please do not use my name on any correspondence to IBM regarding your claims. I think I told you
back then that I did not want to be an advocate for you on this matter, and I have no desire to get
involved. I was not involved in any way in this matter when I was working for IBM, and the only
advice I gave you as a friend was that you should get a lawyer involved if you think IBM wronged you
in this situation. I won't copy Mr. Palmisano's office on this note, but if my name appears on any
further correspondence, I will be forced to tell them.

I wish you the best in life, and appreciate your passion for God's Word, but I do not want to get
involved in business matters that don't pertain to me.

Sincerely,

Pete Wilzbach

Ex1 p82

48.1 Wilzbach memorialized a probative act, that he stands with two fellow Caltech alums for a declaration of production-use. That probative act specifically excludes IBM and Palmisano, blunting the sword of IBM's confidentiality agreement. That same probative act includes Caltech's president, Lidow, and Prouty in affirming a declaration that IBM made production-use of Specification.

48.2 In a single stroke, Wilzbach both affirms what he knows while minimizing his own legal exposure to IBM's confidentiality agreement threat.

48.3 Sneed reciprocates under the Caltech Honor Code declaring that Wilzbach made no disclosure about any specific machine on which IBM uses Specification. Sneed memorializes this in a d2/June 13, 2011 letter to Wilzbach.

48.4 The d2/June 13, 2011 letter also memorializes Sneed's adherence to IBM's d6/June 20 Letter of Corporate Intent. However, Wilzbach's probative actions, proposing settlement, then suit, then memorializing a probative act to stand with two fellow Caltech alums regarding a declaration of production-use, establishes a particular pattern that is factually consistent with giving Sneed written probative notice.

48.5 If Sneed does not act on such written notice, IBM could later assert Sneed had reasonable grounds for suspicion that IBM made production-use from Wilzbach's d1/June 5, 2011 letter due to the particularity of referenced communication with Caltech and IBM personnel. IBM sits on Caltech's board, along with Lidow. Lidow, Prouty, and Wilzbach's positions are consistent with IBM's denials of production-use as being misleading.

48.6 It would be the height of injustice for Wilzbach to suffer harm from IBM's confidentiality agreement for acting in good faith; he, with Prouty and Lidow, have not supported IBM's denials.

48.7 Thus, Sneed's d2/June 13, 2011 letter is a witness that Wilzbach has not supported a cover-up through false denials, nor has he breached any confidentiality agreement terms IBM may later assert against him and his family.

Ex1 p83

Mr. Pete Wilzbach   d2/June 13, 2011
51 Ocean Course, Johns Island, SC 29445

Happy 2nd-day, Pete. Grace and peace to you and yours.

     1. A Very Welcome Letter: I was elated to receive your letter on d7/June 11, and, of course, a proverb came to mind: [27:9]
         Perfume and incense bring joy to the heart, and the pleasantness of one's friend springs from his earnest counsel. [27:9]
     2. A Father's Heart: If you had shortened your letter even more (conciseness is your hallmark), the fact you took time to *mail it* is kind; I've *never* logged on to use the Internet for *any* reason. When last we spoke, you were bearing the weight of the world, including being a servant-husband and mending Charlotte, along with Heaven's compassion for all with a father's heart, who patiently endure its pain and joy, abiding in forgiving grace; of securing financing and leading a construction team to lay a cement foundation during cooling weather to shelter a Corvette (serial number 3) and other irreplaceable vehicles; of prospering with real estate investments, being able to better pace your life year to year; and many other memories. Without your concurrence, no one, including me, no one at IBM, and no one else has Heaven's permission to disturb the freedom you enjoy, and all blessings thereof. *No one.*
     3. One Word and No Other; *HONOR*: There are five sentences in your letter. A proverb harmonizes with the first:
                Do not accuse a man for no reason when he has done you no harm. [3:30]
I admit to *honoring* your name in correspondence with Caltech and IBM, along with fellow Caltech alums Dr. Alex Lidow and Dr. Dale Prouty. To have excluded your name would be an example or *predestined* ANTI-MERITISM. Anyone who attempts to construe *honoring* your name as other than that, does so for unknown motives or purposes, which may have nothing to do with you *or* me. The evidence supports such honor; the Ameritech success, alone, reflects inspired leadership in the constellation of American initiative and achievement. You have helped perpetuate the highest ideals of faithful merit that guides any nation. My debt of gratitude is no less.
     4. Dr. Alex Lidow, as CEO of International Rectifier ("IRC"), shepherded BEST2000, resulting in BEST20000's use on IBM computers in seven countries. His friendship is a blessing. He always acknowledged the merit of BEST20000, even while his CFO used litigation to thwart Alex's freedom as a CEO. Litigation was resolved and reflected Alex's integrity, which included exposing opposing counsel's attempt to suborn perjury. After the win, Alex invited me to his parent's estate; I was just relieved we could speak freely again. The CFO was later removed. Six months after trial, the law firm that exploited the matter with up to a thousand billable hours *was removed from the face of the earth.* Alex unilaterally acted with faithful merit throughout, when others didn't.
     5. Dr. Dale Prouty's adherence to doing what is right, and just, and fair goes back to our time together on the Galileo mission to Jupiter, some 25 years before the BEST2000 effort began. Dale discovered use of BEST2000 on IBM systems outside the U.S. His sworn testimony and credibility proved pivotal at trial, even though none of this pertained to him.
     6. Always a Friend; Hold the *Avocados*: I agree with your second sentence, except for the word "THINK"; I'm *certain* you said you could not provide advocacy. And there's good reason. First, you convinced me you have no specific knowledge of *any* machine BEST2000 is or was used on at IBM. And, even if you did, you may not have the same freedom as, say, the MIT-degreed expert witnesses. His nine-point forensic analysis supported proof of BEST2000 use on IBM systems, in seven countries. You only posited marginal, friendly advice that counsel may be appropriate for the IBM matter, harmonizing with advice from Dale and Alex. They would only encourage such for a matter with *significant* merit. As a friend, I believe you *might* feel the same way.
     7. Freedom of Opinions *and* Perceptions: My *opinion* is no more than that—an *opinion*. It's informed by my *perception* of your *opinion*, along with Dale's and Alex's. And that is: pursuing the matter, with counsel, may be a reasonable course of action. The *perception* derived from this is: "Use of BEST2000 on computers inside of IBM" is a reasonable *premise* to pursue, the outcome of such pursuit being *unknown*. My correspondence with IBM reflects that *premise*, without the certainty of a finding of fact. Therefore, my *premise, perception* and *opinion* are not yours—they are mine. Your *opinion*, affirmed in sentence three, represents marginal, friendly advice, with the freedom to disregard it based on what I "THINK" may or may not be true. And your *opinion* has been extant and unchanged for over three years, as has been my *perception* of your *opinion*, memorialized for over three years. Mr. Al Torressen's *Letter of Corporate Intent* ("LCI") may come as close as anything from IBM admitting custody, merit, and extant terms, from the beginning, for IBM's use of BEST2000. Mr. Palmisano's AGC declared, on Mr. Palmisano's behalf, that IBM desires to disregard its own performance stipulation set forth in the LCI to this Caltech alum. I have no interest in parsing "wrong-doing," which is a subjective exercise. I trust IBM to help honor its own LCI and the advice of three exceptional fellow Caltech alums, which unifies everyone. That advice, now unchanged for over three years, resists the corrupting legacy in blood and treasure of *predestined* ANTI-MERITISM. Patiently I pray, and fervently I hope, the better angels of IBM's nature may prevail through its own wherewithal.

With prayerful regards as your friend, and gratitude for your fellowship,

[signature] Anthony Sneed

   Anthony Sneed   P.S. Whatever I do will be to minimize adverse effects on *everyone's* fellowships, because eternity is a long, long time. [Proverbs 6:16-19]

                                       Ex1 p84

49. Sneed filed a complaint listing Wilzbach as an unnamed VP, just as Sneed listed Palmisano as an unnamed defendant, leaving opportunity for settlement as Wilzbach proposed.

   49.1 Additionally if IBM used its legal sophistication to dismiss the complaint on a technicality, there was no need to expose Wilzbach and his family to the "social consequences" IBM could visit on them at future IBM functions.

50. In his d2/January 23, 2012 ruling, the Honorable George H. Wu gave leave to amend the complaint.

   50.1 After court, on that day, Sneed contacted Wilzbach by phone. Charlotte answered and welcomed Sneed's call. Charlotte updated Sneed on how she and Wilzbach were getting along. Sneed reciprocated the goodwill Charlotte always extends; Sneed deferred to follow-up at another time since their realtor was stopping by regarding a property matter.

   50.2 Sneed sent a written summary by Express Mail to Wilzbach of the Honorable George H. Wu's ruling regarding an affirmative statement of use rather than the factually neutral "refusal to deny use." of Attorney Spitz. Wilzbach sent Sneed a d1/January 29, 2012 letter. In it:

      50.2.1 Wilzbach assents for Sneed to continue with the proposal to sue IBM and Palmisano if Sneed chooses to do so, adhering to provisions in Wilzbach's d1/June 5, 2011 letter.

      50.2.2 Wilzbach affirms, again, that the dispute between Sneed and IBM does not involve Wilzbach in any way, and that Wilzbach does not want to become involved in any way.

      50.2.3 Wilzbach surmises that his proposal for Sneed to sue IBM and Palmisano, if Sneed chooses to continue, is nothing more than friendly, informal, practical advice.

      50.2.4 Wilzbach compares this to the same practical advice he would give if Sneed were in a **bad** car accident, or were in trouble with the IRS. With this analogy, Wilzbach's **bad** car accident may be a hit-and-run, with Palmisano driving a 10 miles per gallon "Y2K SUV" and Wilzbach his passenger; both are executives at the IBM "insurance" company. The SUV takes a tight turn too fast and hits Sneed driving a 200 miles per gallon "Y2K hybrid subcompact," knocking Sneed unconscious as the hybrid crashes off the road. Palmisano says, "Leave him. There's no one around. Let's go." But Wilzbach attends to Sneed, bandages Sneed's bleeding head, and calls 911 anonymously describing the location of the crash. An ambulance, driven by Lidow and Prouty, arrives and takes Sneed to the hospital and ICU. After a few weeks Sneed recovers and later contacts Wilzbach. He tells Wilzbach about the accident, but can't remember what happened. Instead of just filing a claim with the claims department, Wilzbach proposes Sneed seek settlement directly with the IBM "insurance" company and its current CEO Palmisano. Palmisano became CEO after Wilzbach left the company after the accident. Sneed does so and asks Palmisano if he was driving a Y2K SUV a month ago near the accident site. Palmisano says no, refuses settlement, and says he won't let Sneed see his Y2K SUV, which has a dinted front fender, with paint on it the color of Sneed's Y2K hybrid. Wilzbach then proposes that Sneed sue Palmisano and the IBM "insurance" company, though he won't tell Sneed why he proposes this course of action. When Sneed does, Wilzbach later declares he had nothing to do with the accident, which is true: Wilzbach is completely innocent and acted responsibly after the accident, and may have saved Sneed's life. The IBM "insurance" company's confidentiality agreement limits what Wilzbach, as an officer, can say about any "accident." Palmisano knows this.

      50.2.5 Wilzbach notes that using his proposal to sue IBM and Palmisano as the basis of Sneed's case would be a great mistake and directed Sneed to show Wilzbach's letter to Sneed's legal counsel. This led Attorney Spitz to later explore Caltech's Honor Code.

Ex1 p85

a *7ʰ-Gift* page

Anthony Sneed   January 29, 2012
2058 North Mills Avenue
Claremont, CA 91711

Dear Tony,

I am not sure why you wrote me the letter dated January 23, 2012. If it was just to tell me that you have acquired legal counsel, that is fine. But there are a number of statements in that letter that cause me great concern. First of all, I hope you are not using me in your court proceedings as the unnamed IBM VP, and the "refusal to deny use". It is not for me to confirm or deny use – I had nothing to do with it. As I told you before, I left IBM before the Y2K event, and my last three years preceding the Y2K event were spent entirely in the support of two outsourcing customers, Ameritech in Chicago and GE in Atlanta. Both companies had hundreds of programmers providing their own Y2K application remediation, and IBM was not involved in that aspect. So, to repeat, I had no involvement or knowledge of anything to do with your efforts.

Just because I suggested you hire legal counsel should not cause you to construe anything. I would suggest you hire legal counsel if you were in a bad accident, or were in trouble with the IRS. This is nothing more than practical advice. I would ask that you show this letter to your legal counsel and make sure they are not using me as any basis for your case. That would be a great mistake.

As always, I wish you the best in life and your pursuits,

[signature]

Peter Wilzbach

Ex/ p86

50.3 Wilzbach's positions from 1997 to 2012 may seem contradictory, particularly if his two recent letters are isolated from four prior letters between Wilzbach and Sneed.

 50.3.1 However, in Sneed's estimation, Wilzbach is more like an ethical government official under house arrest by an authoritarian regime, trying to communicate without upsetting his captors; his every communication is subject to monitoring--deposition and discovery in this analogy. Wilzbach has done his best to be forthcoming in these circumstances without harming himself or his family. Sneed disclosed to him the case is still in the pleading stage, without formal deposition and the safety it brings to override any confidentiality agreement. It is amazing Wilzbach is communicative in any fashion.

 50.3.2 Wilzbach's disclaimers in his d1/January 29, 2012 letter are fitting and proper for a person in such circumstances, now that his proposal to sue Palmisano and IBM is underway. And as for his "captors," Wilzbach's recent disclaimers would, in the future, appear to appease the requirements of any IBM secrecy agreement, while not running afoul of any future obstruction of justice charge if the case continues to move forward in unexpected ways, including DOJ inquiry of a federal contractor and Section 242 of the U.S. Criminal Code:

> [W]hoever, under color of any law...willfully subjects any person in any state to the deprivation of any rights, privileges or immunities secured or protected by the Constitution or laws of the United States" is guilty of a federal crime.
> Section 242 of U.S. Criminal Code

 50.3.3 Coded within Wilzbach's "message" is direction for Sneed to proceed with the litigation proposal; Wilzbach knows that all past correspondence would see the light of day.

 50.3.4 Wilzbach's disclaimer that he has no involvement with the dispute between Sneed and IBM is fitting and proper.

  50.3.4.1 Wilzbach has never participated in a cover-up of use of Sneed's Specification.

  50.3.4.2 He did not support Torressen's ruse call to subvert the 1956 Consent Decree, but rather exposed IBM Corporate's unnecessary perfidy in compelling Torressen's conduct under color of authority.

  50.3.4.3 Wilzbach proactively proposed settlement with Palmisano, as informal advice, even knowing, from Sneed, of IBM's denials of production-use for over a decade; per Wilzbach's proposal, Sneed attempted settlement for $1,194,349.20, respecting Wilzbach's wisdom to avoid litigation with IBM if at all possible.

  50.3.4.4 After Palmisano disregarded settlement, Wilzbach proactively proposed Sneed sue Palmisano and IBM, not withstanding 10 years of admissions by IBM of only evaluative-use that Sneed underscored to Wilzbach; Wilzbach still proposed Sneed sue Palmisano and IBM as informal, friendly advice.

  50.3.4.5 Wilzbach let his name stand with two other Caltech alums regarding a declaration of use to Palmisano; Wilzbach then memorialized that Wilzbach would say nothing to the contrary to Palmisano, even while under threat of IBM's secrecy provision that could materially effect Wilzbach. No rational person would do such for a matter that lacked merit. Wilzbach has memorialized forever that he adheres to a pattern of conduct, reflecting the highest ideals of ethical conduct, while not breaching and exposing himself and his family to the sword of IBM's confidentiality agreement. Only a person excelling through years of management, litigation, and operational training and experience, which IBM provides, could accomplish such, and inspire the efforts now making their way through the Court. Wilzbach has the "top-level view" of the technologies, parties and proceedings surrounding this matter. The Motion attempts, in as much as possible, to impart that same "top-level" view to the Court. Through this the Court will, in this matter, determine the reputations of Wilzbach, Lidow, Prouty, and Sneed for the rest of their lives, however long that may be, regarding the Caltech Honor Code, good faith adherence to Rule 11, both referenced in the last footnote, page 6, of Tentative Ruling on Motion to Dismiss Plaintiff's Amended Complaint; and respect for precedence of the U.S. Constitution and its perpetuation, while preserving a presumption of innocence for IBM that the Court, Sneed and other parties with goodwill adhere during proceedings, to the credit of this honorable Court and its magistrates, administrators and staff.

50.3.4.6 It has taken courage and patient resolve for Wilzbach to do what he has, along with Lidow and Prouty, as others within IBM have advanced their own interests using this matter however they may deign, in the past, the present, and the future.

50.3.4.7 It should be noted that Torressen and IBM separated, at or about the time the federal government issued Sneed's patent, d3/May 22, 2001; Torressen was forced to seek employment elsewhere. Torressen refused to cooperate in any way for purposes of the Court's proceedings, once Sneed found him in 2012; Torressen is non-responsive to phone or fax. Just as Specification was taken from IBM Global Services against its leadership's will, according to Wilzbach, so Torressen may have been compelled to act against his own values regarding what he said to Sneed by phone, that differs from what he later memorialized in writing.

50.3.4.8 Still, Torressen, out of respect for integrity he did manifest, is the first person at IBM to acknowledge any form of agreement between IBM and Sneed, under a legal IBM signature, before Torressen and IBM later separated and Torressen was forced to seek employment elsewhere.

50.3.4.9 Finally, just as Wilzbach left IBM of his own freewill, on the eve of Y2K, d6/December 31, 1999, so did Sneed leave IBM of his own freewill as noted in Discovery production. IBM's first written statement to the Court is defective regarding this minor, passing item. Both Wilzbach and Sneed left IBM of their own freewill, at different times. For Sneed, it was after he became the top sales performer for the entire Western Region of the United States, made 400% of quota, and graduated with honors from IBM's national training center in New York.

50.4 IBM, being a large corporation, was given ample time between d2/April 4, 2011 and d6/October 28, 2011 to amicably resolve the matter, as correspondence in July 2011 and September 2011 memorialize.

50.5 Following Wilzbach's proposal, Sneed then filed complaint d6/October 28, 2011.

50.6 The President and Attorney General were notified, in the event a national interest warranted special consideration for IBM. Neither they nor Wilzbach vetoed litigation going forward.

Exl p88

FAX TO: Mr. Sam Palmisano c/o James Brown, fax 914-945-3281   PAGES: 2 (including this one)   **CONFIDENTIAL**
FROM: Mr. Anthony Sneed ISB::Institute Fellow, *Emeritus* ("SNEED")   DATE: d6/July 15, 2011   **ISSUED DRAFT**
RE: With gratitude and respect for all of your past consideration.

Happy 6th-day, Sam Palmisano.

1. Please accept my congratulations on your firm's 100th anniversary, from its founding on or about d6/June 16th, 1911. Mr. Hollerith, by 1911, a year after the 1910 census contract went to his competitor Powers Accounting, began divesting himself of TMC, his Tabulating Machine Company. TMC benefited from the 1890 and 1900 census contracts with the U.S. government. John R. Flint acquired TMC to create an "Information Services Trust" (his term at the time) by combining four geographically disparate, apparently unrelated businesses, two being prior trusts Mr. Flint formed.

2. And that trust perseveres. I called Mr. James Brown last week, but he was on holiday for the week and unavailable. I did fall into the hands of a very capable IBM professional named Esther. Mr. Thomas Watson Sr. gave special instruction to act without favoritism, for IBM personnel handling public inquiries, when 590 Madison was headquarters. Over the past several years, Esther has far surpassed in this regard, with a clear focus on the best interests of IBM while thoughtfully considering this Caltech alum's needs. I've *never* been sent to her voice mail; each call is promptly picked up. Even with a lapse in contact of over 40 months, she readily engaged with an effortless élan and a confident wherewithal as though it had only been 40 hours. Esther adheres to the Jeffersonian notion, "Why use two words when one will do?" Her elocution is concise.

3. After the call, I remembered direct correspondence from Mr. Brown, wherein he suggested I presume his indefinite unavailability ("INTENTION") regarding any matter at hand. So being on holiday was a natural extension thereof. I respect Providence's wisdom that Esther took the call, she preserving the best interests of IBM while leaving me favorably disposed when our call amicably concluded. It is written in the book of Esther, "...And Esther won the favor of everyone who saw her." [Esther 2:15] I may never personally meet you or Esther--perhaps not in this life-time--but she has won my favor; her contact is always forthright and gracious, consistent with ideals you and Mr. Watson espouse.

4. I realize Mr. Brown has taken no oath regarding candor in any communication with me. He enjoys the presumption of innocence, while he expresses any opinion ("BROWN OPINION") regarding inquiry about BEST2000™ use at IBM. As to INTENTION, his personal entitlement to be indefinitely unavailable for some inquiries and, perhaps, less so for others, is intact. He has no oath, that I'm aware, to precepts in Proverbs:

> To show partiality is not good--yet a man will do wrong for a piece of bread. [28:21]
> Do not accuse a man for no reason--when he has done you no harm. [3:30]
> It is not good to punish an innocent man, or to flog officials for their integrity. [17:26]

Mr. Al Torresen ("TORRESSEN"), as an IBM director, sent this Caltech alum an IBM Letter of Corporate Intent ("LCI") to remunerate this Caltech alum for use of BEST2000™ by IBM. The LCI memorializes custody and confidentiality of BEST2000™ in deference to fee and terms included with BEST2000™ when IBM took custody. The LCI was, and is, within IBM policy, practice and IBM's desire to encourage productive pursuits ...d innovation by any Caltech graduate, or any student, or any computer scientist or engineer. Yet BROWN OPINION brings TORRESSEN'S ...integrity and IBM's LCI into question regarding BEST2000™'s merit and IBM's payment for use. BROWN OPINION "flogs" TORRESSEN, for an unknown "piece of bread," and effort by IBM's own Mr. Don Logan, who facilitated intake and IBM custody of BEST2000™.

5. BROWN OPINION "flogs" sworn testimony of Dr. Dale Prouty ("PROUTY"), a Caltech PhD in Applied Physics, regarding BEST2000™ use on IBM computers around the world. BROWN OPINION "flogs" a sincere IBM customer executive, Dr. Alex Lidow ("LIDOW"), a fellow Caltech alum and Caltech board member, who proved cost-effective production of BEST2000™ on IBM computers in seven countries, confirming Mr. Logan's expectation. BROWN OPINION disparages these distinguished gentlemen's integrity for an unknown "piece of bread." Mr. Palmisano, if you were to agree with these gentlemen, you too may become subject to similar "flogging" for that same unknown "piece of bread."

6. But you would be in good company. Indirectly, BROWN OPINION "flogs" Dr. Harold Brown as former president of Caltech, Chief Justice Earl Warren, John Adams, Thomas Jefferson, Abraham Lincoln and others for creating conditions that resulted in BEST2000™ as a U.S. intellectual property with acknowledged merit by TORRESSEN PROUTY, LIDOW and others inside and outside IBM, whether alums of Caltech, MIT or other schools. BROWN'S OPINION also "flogs" many others who can not speak up for themselves--including young people, who, like this Caltech alum in his youth, pursue science and technology to make meaningful contributions to society, and to the competitiveness of American endeavors.

7. In my d2/April 4, 2011, correspondence, I reviewed the practice of *predestined* ANTIMERITISM--a practice rejected by Chief Justice Earl Warren and the U.S. Supreme Court, based on foundational principles within the U.S. Constitution. IBM has resisted *predestined* ANTIMERITISM from its very founding. When Mr. Thomas Watson Sr. was candidate to be president in 1913, the board applied *predestined* ANTIMERITISM *against* him, with the *opinion* that no one *like him* could bring value to the new firm's "water-heavy" stock--because Mr. Watson had a federal conviction, even though it was *under appeal*. That instance of *predestined* ANTIMERITISM, used to disparage a man's *whole-life merit*, ignored Mr. Watson's extraordinary rise out of rural mediocrity, to selling some 1200 registers in a month for NCR, and developing competitive sales forces and strategies using the best light available to him at the time. To IBM's credit, from the beginning, it did not let *predestined* ANTIMERITISM reject its future chairman. Nor did Providence allow *predestined* ANTIMERTISIM to reject you as chairman, just because no one *like you* had ever been chairman before. Your sacrifices, in the field, in the PC division, at Global Services, and countless matters attended in your career, were not, by mere *opinion*, cast aside--especially time away from family and friends, and missed holiday--a life-time now changed forever. It is through your example, along with Mr. Watson's, that *predestined* ANTIMERTISIM *can* be resisted, even as it constantly reasserts itself against the interests of your company, this country, growth and prosperity *wherever* and through *whomever* Providence endows desire, merit, and purposed commitment. This is the *only* basis the attached warrants your attention, not withstanding the powerful legacy of *predestined* ANTIMERITISM and its teachings.

With prayerful regards and respectfully yours, and best wishes for a restful 7th-day,

Anthony Sneed

...thony Sneed   cc: Caltech; MIT; Orrick, Herington & Sutcliffe fax list;   Attachment

Ex1 p89

INVOICE 7247.2   d6/July 15, 2011

Mr. Sam Palmisano c/o Mr. Winfield James Brown
1 Orchard Road, Armonk, NY, 10504

$1,492,936.50   AMOUNT DUE d2/April 4, 2011
$6,220.57   ACCRUAL AS OF d2/May 23, 2011
$6,220.57   ACCRUAL AS OF d2/June 23, 2011

$1,505,377.64   DUE UPON RECEIPT

Payable to "ANTHONY SNEED" at 2058 North Mills Avenue, Claremont, CA, 91711, based on fee and terms acknowledged in IBM'S LETTER OF CORPORATE INTENT, sent to and received by ANTHONY SNEED. Payment within 30 days of this INVOICE, and receipt by ANTHONY SNEED, shall constitute liquidated damages and release of all claims between ANTHONY SNEED and IBM, and shall settle all matters between ANTHONY SNEED and IBM, including use of BEST2000™ described in U.S. patent #6236992 issued to ANTHONY SNEED May 22, 2001, and any derivative work thereof, applied in any way by IBM, without limit on the number of computers IBM has applied it on or before today, or applies it after today.

EX1 p90

FAX TO: Mr. Sam Palmisano c/o James Brown, 914-945-3281   PAGES: 2 (including this one)   **CONFIDENTIAL**
FROM: Mr. Anthony Sneed ISB::Institute Fellow, *Emeritus* ("SNEED")   DATE: d6/September 16, 2011   **ISSUED DRAFT**
RE: With gratitude and respect for all of your past consideration.

Happy 6th-day, Sam Palmisano.

1. There's been no payment remitted for invoice 7247.1 or 7247.2. This may mean Mr. Brown as AGC persuaded you to adopt his opinion ("OPINION") and overrule IBM's Letter of Corporate Intent ("LCI") to remunerate this Caltech alum regarding BEST2000™. BEST2000™'s specification is in US patent 6,236,992, issued to this Caltech alum. This matter is specific to the LCI, not the patent.

2. Under your leadership at IBM Global Services with Mr. Dennie Welch, IBM Global Services acquired BEST2000™ at a critical juncture in IBM's 100-year history, based on merit and need. Neither you nor Mr. Welch ever disputed the merit of this action. Similarly, Dr. Alex Lidow, as CEO, facilitated testing and installation of BEST2000™ on IBM computers at International Rectifier ("IRC"). His decision saved his firm some 70% on a two-year, and up to $10 million "best efforts" counter-proposal by Accenture (Anderson), an IBM competitor. Mr. Don Logan, as IBM Global Services Program Director, confirmed the technical merit and economy of BEST2000™ to Dr. Lidow's firm prior to IRC deploying BEST2000™ on IBM computers in seven countries. Mr. Logan did so using the same BEST2000™ specification he received from this Caltech alum, and which he tested internally at IBM under terms and conditions set forth with the specification. Mr. Al Torressen, IBM Director of Intellectual Property Licensing, distributed the BEST2000™ specification throughout IBM for evaluation, issuing the LCI after Mr. Logan's earlier findings.

3. Mr. Brown's OPINION invokes a 14th Amendment ("A14"), Section 1, entitlement ("ENTITLEMENT"), ratified in 1868:
No State shall make or enforce any law which shall abridge the privileges or immunities of citizens of the United States; nor shall any State deprive any *person* of life, liberty, or property, without due process of law; nor deny to any *person* within its jurisdiction the equal protection of the law.
A14 helped end the Coerced Uncompensated Labor Tradition ("CULT") practices which affected over a billion man-hours of production between 1789 and 1865. A14 rebukes such CULT practices and anyone attempting to perpetuate them within the US.

4. So, Mr. Brown's OPINION leverages the 1886 Supreme Court decision, *Santa Clara County v. Southern Pacific Railroad*, 118 US 394 ("394"), to subdue by OPINION any honorable motive of you, Mr. Welch, Mr. Logan, Dr. Lidow, Dr. Prouty, Mr. Torressen, or anyone else. Under 394, corporations are "*persons*" within A14's meaning, establishing two classes of *persons* within the US: One with real bodies who see, hear, talk and THINK, such as you, Mr. Welch, Mr. Logan, Dr. Lidow, Dr. Prouty, and Mr. Torressen; and another as *Immortal Bodiless Mythical Persons* ("*IBMP'S*")--*Immortal* because the life-span of an IBMP is indefinite, and limitless.

5. You and Mr. Welch as heads of IBM Global Services and as witnesses, along with Mr. Logan, Dr. Lidow, Dr. Prouty, and Mr. Torressen, reflected *reason* and *thoughtfulness* regarding the technical merit and economy of BEST2000™ when parts of IBM, like IRC and many other private and public sector institutions, were dangerously close to an abyss, adversely affecting potential growth and operations. For all such solutions, it is the _how_ as well as the _what_ that is important, since computers (the *what*) are alike, but *how* a solution is presented and implemented determines *a customer's satisfaction* and *acceptance*. Prior to BEST2000™'s specification, nothing had similarly satisfied the needs of Mr. Logan, Dr. Lidow, Dr. Prouty (a Caltech PhD) or Mr. Torressen as IBM Director of Intellectual Property Licensing. BEST2000™'s specification facilitates IBM systems operations beyond 100 years, through novelty and simplicity that Mr. Logan, Dr. Lidow and the US Patent Office found merit and economy. Dr. Lidow proved such merit and economy with BEST2000™ application on IBM computers worldwide. Mr. Brown now asserts the right to *know nothing* outside of a due process ENTITLEMENT an IBMP *may* invoke under A14, similar to asserting a 5th Amendment privilege in a criminal matter, against multiple witnesses who agree with one another. This is a strange use of A14; its original intent was to *end* CULT practices.

6. Since you, Mr. Welch, Mr. Logan, Dr. Lidow, Dr. Prouty, Mr. Torressen, this Caltech alum, and others, are *real* people, unlike some IBMP'S, our voices will fade with time, and already are, to become silent one day. And IBMP'S we serve will go on, gathering to themselves new voices, as each IBMP jealously guards its due process ENTITLEMENT and desire for immortality [Ex 20:4-6].

7. I'm grateful Providence made you, Mr. Welch, Mr. Logan, Mr. Torressen, and even Mr. Brown, *real* people, sensible to both liberty and conscience in your own life-walks. A proverb says, "If a man pays back evil for good, evil will never leave his house." [17:13]. Mr. Logan did a good thing in acquiring BEST2000™ and testing it, setting good expectations for IBM, along with, perhaps, you and Mr. Welch. They are being met, today, on IBM systems worldwide. Extant agreement is met, corroborated by Dr. Lidow, Dr. Prouty and others from Caltech, MIT and IBM. "Quota" is fulfilled for sell and install. It is entirely proper and fitting to remit the "commission" due--IBM *always* has. Approving so, now, sustains business growth, and is on the right side of history and mankind's legacy handed down to us, across generations, since the 7th-day and trust to rest on it were first established. [Gen 2:1-3]

With powerful regards for a restful 7th-day,

Anthony Sneed
Anthony Sneed   cc: Caltech; MIT; Orrick, Herrington & Sutcliffe fax list   Attachment: Invoice 7247.3
P.S. Please extend my respectful regards to Esther, who has only shown confident wherewithal and goodwill in all contact with IBM.

Ev1 p9/

a *7th-Gift* page

INVOICE 7247.3   d6/September 16, 2011

Mr. Sam Palmisano c/o Mr. Winfield James Brown
1 Orchard Road, Armonk, NY, 10504

$1,492,936.50   AMOUNT DUE d2/April 4, 2011
  $6,220.57   ACCRUAL AS OF d2/May 23, 2011
  $6,220.57   ACCRUAL AS OF d2/June 23, 2011
  $6,220.57   ACCRUAL AS OF d7/July 23, 2011
  $6,220.57   ACCRUAL AS OF d3/August 23, 2011

$1,517,818.78   DUE UPON RECEIPT

> Payable to "ANTHONY SNEED" at 2058 North Mills Avenue, Claremont, CA, 91711, based on fee and terms acknowledged by IBM, in IBM's Letter of Corporate Intent sent to, and received by ANTHONY SNEED. Payment by sunset, within 14 days of this INVOICE, and receipt by ANTHONY SNEED, shall constitute liquidated damages and release of all claims between ANTHONY SNEED and IBM, and shall settle all matters between ANTHONY SNEED and IBM, including use of BEST2000™'s specification described in U.S. patent #6236992 issued to ANTHONY SNEED May 22, 2001, and any derivative work thereof, applied in any way by IBM, without limit on the number of computers IBM has applied it, on or before today, or applies it after today.

Ex1 p92

FAX TO: 202-514-4507   DATE: d6/October 7, 2011   <u>CONFIDENTIAL</u>
FOR: The Honorable Barack H. Obama, President, United States of America
                   c/o The Honorable Eric H. Holder, Attorney General, United Sates of America
FROM: Anthony Sneed, ISB::Institute Fellow, *Emeritus*   PAGES: 7 (including this one)
RE: A witness to Caltech's purposefulness and powerful reflection of your administration's desire to recognize *and* develop merit.

1. (Executive Summary) A proverb states:
   Do not exalt yourself in the king's presence and do not claim a place among great men. It is better for him to say to you, "Come up here," than for him to
   humiliate you before a nobleman. [25:6,7]

2. Though I am a Life-member of the Caltech Alumni Association, and hold two US patents (#6236992 and #7546982), and am a former assistant professor of computer science in the California State University System, and an author, it may be a transgression of the above proverb to correspond with you without an appropriate reference. The *ONE* invoked is Proverbs[22:22,23], if you are willing.

3. My origins are modest: my father's education ended at 7th grade, my mother's at 10th. I've never owned real property, nor taken out a loan to purchase real property. My employment history has been with Caltech as a JPL intern; Hughes (Boeing) as an on-board operating system designer for the successful NASA Galileo Mission to Jupiter; assistant professor (noted above); IBM representative in the finance industry and 400% quota performer after honor graduation from IBM's two-year corporate training in New York; and founder of ISB::Institute, serving therein for the past 26 years. A year after founding ISB::Institute, my wife, Mary, allowed me to become her husband--a solemn vow of monogamy I've kept for 25 years. We have four children, evenly divided across gender.

4. I'm compelled to write to you about an invention specification (Invention), which is contained in US patent #6236992 issued to me. A friend, Dr. Alex Lidow, who is a fellow Caltech alum and member of the Caltech board of trustees, applied Invention on IBM computers in seven countries, when he was CEO of International Rectifier Corporation (IRC). This resulted in savings to his firm of some 70% on a two-year, $10 million "best efforts" counter-proposal by Accenture (Anderson Consulting at the time). IBM vetted my Invention, and confirmed to Dr. Lidow and IRC its technical merit before IRC applied it. Dr. Lidow was very pleased, and I was later awarded $236,000 under contract with his firm and, subsequently, invited by him to his parent's estate for a Caltech function.

5. IBM entered into similar agreement when it took custody of Invention, analyzed Invention's specification within IBM, and declared Invention's merit to me and, later, to IRC which deployed Invention on IBM computers in France, Germany, England, Singapore, China, the U.S. and Japan, at considerable savings over anything IBM, Accenture, or anyone else offered. IBM documented that it, too, could proceed with payment terms after d3/May 22, 2001 (Date), which is when the US Patent & Trademark Office issued a patent to me. However, after Date, and between Date and October 29, 2007, IBM demurred, in writing, that it found use that warranted any payment whatsoever to me, anywhere within IBM, worldwide, notwithstanding Dr. Lidow's and IRC's success and cost-saving application of Invention on IBM computers in France, Germany, England, Singapore, China, the U.S. and Japan.

6. This remained the case until October 2007, when I learned that Mr. Pete Wilzbach retired from IBM. Mr. Wilzbach is a fellow Caltech alum and was an IBM EVP and runner-up to become IBM CEO. He was also my hiring manager when I worked at IBM. After speaking with Mr. Wilzbach, I attempted to contact his friend, Mr. Sam Palmisano, IBM's chairman and CEO, to revisit the matter regarding use of Invention by IBM. An IBM assistant general counsel named Mr. James Brown (no relation to the recording artist) intervened, and declared IBM would no longer cooperate with any discovery requests, as IBM had done between Date and October 2007, as stipulated by IBM's *own terms* in IBM's own Letter of Corporate Intent (LCI) regarding payment. In doing so, Mr. Brown leverages a 14th Amendment reconstruction that allows IBM to consider itself a *person*, not obligated to *know or do anything* outside of a due process entitlement. Thus, it will not cooperate with any discovery outside of a due process proceeding. Multiple law firms have been contacted, each confirming the extreme hardship IBM can impose upon counsel and client in such a matter. One, Orrick, Herrington & Sutcliffe, showed the wherewithal and willingness to proceed, but later notified me that IBM had retained the firm, making it a conflict of interest for the firm to also represent me, disappointing both me and my primary contact within the firm.

7. Dr. Lidow continues to encourage a favorable outcome to the matter, based on merit and economy he and his firm achieved on IBM systems, as does Dr. Dale Prouty, a Caltech PhD who provided sworn testimony of Invention's use on IBM computers worldwide. Separately, please accept my gratitude for your nanotechnology initiative through the University of California Riverside; I greatly benefited from your 2010 funding of it, and received a nanotechnology certification, through 90 hours of training that included a comprehensive review of nanotechnology and in-lab IC fabrication techniques, first explored at Caltech. The knowledge gained is helping me develop a nanotechnology battery, which may add to my patent filings, including an unrelated revenue-enhancing invention for USPS. Resolving payment with IBM will greatly aid these efforts, consistent with free-market ideals nurtured at Caltech.

With gratitude and prayerful regards, for a restful 7th-day, and Day of Atonement,

Anthony Sneed
Anthony Sneed, 4058 North Mills Avenue, Claremont, CA, 91711; 909-753-7358
cc: Dr. Jean-lou Chameau, President Caltech; Dr. Alex Lidow, Caltech Trustee; Dr. Dale Prouty, Caltech PhD; Attachments

Exl P93

FAX TO: 202-514-4507   DATE: d6/December 16, 2011   **CONFIDENTIAL**
FOR: The Honorable Barack H. Obama, President, United States of America   **FINAL DRAFT**
          c/o The Honorable Eric H. Holder, Attorney General, United Sates of America
FROM: Anthony Sneed (Sneed), ISB.:Institute Fellow, *Emeritus*   PAGES: 1 (including this one)
RE: IBM's claim to be a $100 billion *disadvantaged person* to flee a court summons.   RE: Oct 7, 2011 Prior Correspondence

Happy 6th-day, Honorable Barack Obama. Grace and peace to you and yours.

1. Prior Correspondence reviews interest in IBM's use, if any, of this Caltech alum's specification (Specification) called BEST2000™, contained in US patent #6,236,992 issued to me. Two, fellow Caltech alums--a venture capitalist and a Caltech board member--confirmed dispositive, commercial production-use of Specification on purchased IBM AS/400 computers in seven countries.

2. IBM admits taking custody of Specification from Sneed. Mr Don Logan, an ethical IBM Global Services (IBM Global) program director, confirmed Specification's merit and economy, under terms (Terms) set forth by Sneed for no-charge testing, and a $49,764.55 per site production-use fee. Mr Logan inquired if $49,764.55 or $149,293.65 should be remitted for an IBM conversion site with three computers. Sneed confirmed the $49,764.55 per site fee, under written Terms set forth when IBM took custody of Specification.

3. Mr Logan and an unnamed, ethical IBM VP (Ethical VP), helped Sneed modify Terms with 14 cents per line of code, across an aggregate number of lines of code, and no charge after a billion lines, respecting Ethical VP's caution about IBM's counting methods. Later, one IBM client applied Specification to 1,200,000 lines of code, verified by Caltech alums Dr Dale Prouty and Dr Alex Lidow.

4. Subsequent to Mr Logan's and Ethical VP's contact, IBM Corporate Licensing Program Manager Mr Al Torressen took custody of Specification from Mr Logan and IBM Global, memorializing to Sneed the written Terms Mr Logan and Ethical VP helped establish. This demurred a Department of Justice "firewall" between IBM Corporate and the IBM Global subsidiary, obviating a markup of up to 1,400% IBM Global could legally bill IBM Corporate, or the same up to $2 per line of code IBM Global charges government or private sector clients. This "firewall" demur may have saved IBM Corporate up to $1.86 billion IBM Global could bill.

5. Mr Torressen assured Sneed, in writing, that IBM Corporate would not "expropriate" Specification, and acknowledged the Terms Mr Logan and Ethical VP helped establish in the written agreement Mr Torressen also took custody. Mr Torressen noted Specification's potential worldwide impact, and arbitrarily set d3/May 22, 2001 (Date) for Sneed to initiate contact for payment.

6. Between Date and October 29, 2007, multiple IBM legal personnel were designated to replace Mr Torressen, including personnel directed by IBM CEO Mr Sam Palmisano. Each confirmed, in writing, no benefit or discovery of use that merited payment to Sneed. This remained so until Sneed learned that an unnamed VP (Unnamed VP) retired from IBM. Unnamed VP and Sneed worked together at IBM in the 80s. After contact with Unnamed VP, Sneed contacted Mr Palmisano regarding payment for use, if any, of Specification, encouraged by Unnamed VP. Mr Palmisano directed legal counsel to contact Sneed, whereupon legal counsel declared IBM would no longer respond to inquiries about IBM's use of Specification, but later recanted its declaration of non-responsiveness.

7. Sneed filed instant action d6/October 28, 2011, with Los Angeles County court case KC062442, to establish formal inquiry notice of IBM within the statutory limit for inquiry notice and delayed discovery under California's Code of Civil Procedure. The California court can *inexpensively* resolve such matters in as little as six to eight months, including hearings, discovery, deposition, and trial. IBM did not respond to summons, and now flees state court seeking removal to federal court under 28 USC 1332 ("1332"), alleging due process apprehension by a California court. IBM uses the 1886 Supreme Court decision *Santa Clara County v. Southern Pacific Railroad*, 118 USC 394 ("394"), to construe itself as a *disadvantaged person* under section 1 (Section-1) of the 14th Amendment. Section-1's original intent is to protect *persons* subjected to Coerced Uncompensated Labor Tradition (CULT) practices--practices made a federal crime with the 13th Amendment ratified in 1865. Sneed recalls while working at IBM in the 80s, when he was elected president of the 6,000 member Western Region IBM Club, that IBM's Western Region headquarters were located in California, accounting for some 50% of IBM's $80 billion domestic revenues. IBM ranks among the top businesses licensed in California by the state's government. During the 80s and since then, IBM exercised considerable wherewithal within California courts to *economically* perpetuate its business interests, readily engaging California *citizens* and businesses through due process within California courts. IBM maintains expertise with personnel licensed through the California bar, who are some of the top practitioners in the legal field, maintaining contract-due-process wherewithal through the California courts, to this very day. IBM is NOT a *disadvantaged person* of limited means, unfamiliar with or alien to the economy of California courts. But it does attempt to claim being a $100 billion, Section-1 *disadvantaged person* for purposes of fleeing a California court. It is illegal for any such wealthy *person* to OBSTRUCT JUSTICE for a California *citizen*, to the detriment of equal protection of any *citizen's* access to a California court, especially if IBM previously waived apprehension of California courts' due process competence. Sneed disclosed *a priori*, and IBM acknowledged, that agreement between the parties is subject to California law. With hostility toward Section-1's history to *lift oppression*, IBM now claims entitlement under 1332 to construe itself a Section-1 *disadvantaged person*, a duplicity which abridges Section-1 equal protection guarantees to California *citizens* and businesses, whom IBM readily summons to California courts it now attempts to flee. In a previous case, a law firm attempted to OBSTRUCT JUSTICE regarding Specification. The California appellate court ruled in Sneed's favor and put the prominent, international law firm on notice; that law firm later disappeared from the face of the earth, closing its offices in London, Moscow, Tokyo, New York, Los Angeles, Singapore, Munich and other global locations. Its lead counsel was an MIT grad. IBM now hires a firm with similar global reach. Its lead counsel is a Harvard grad.

With prayerful regards, and best wishes for a restful 7th-day, and restful celebrations now upon us,

Anthony Sneed
Anthony Sneed

cc: Dr Jean-Lou Chameau, Caltech President; Dr Alex Lidow, Caltech board member; Dr Dale Prouty, Caltech PhD Applied Physics; Mr Sam Palmisano, IBM CEO; Mr Al Torressen, IBM Licensing Program Manager (retired), Manhattan College Graduate, Physics

Ex1 p94

FAX FAX FAX FAX FAX FAX FAX

FAX TO: 213-443-3100   10 pages (including this one)   CONFIDENTIAL   Copyright© 2012 Anthony Sneed
FOR: Mr. Shon Morgan, Esq., Distinguished Counsel, Quinn, Emanuel Urquhart & Sullivan, LLP
FROM: Anthony Sneed (Sneed), ISB: Institute Fellow, *Emeritus*   DATE: d6/February 17, 2012
RE: Gratitude for your and Ms. Bridget Anne Hauler's, Esq., efforts on behalf of me and IBM

Happy 6th-day, Shon Morgan. Grace and peace to you and yours.

1. Please accept my gratitude, that you, Bridget Hauler, and your firm are helping the parties avail themselves to peaceful means within the U.S. Constitution and due process provisions thereof, as we were unable to amicably resolve the present matter by ourselves. It was and is reassuring that you are a Harvard grad *summa cum laude*, and that your firm's reputation is well established in the four corners of the earth. Had you or your firm been second-tier, third-tier, or worse, IBM's entitlement to a perfect presumption of innocence, which in and of itself may be unattainable absent divine counsel beyond this world, would perhaps be less robust, and therefore inferior to that which it deserves from a civilized society. We all share responsibility to preserve and perpetuate precedent, in as much as circumstances and evidence permits such presumption to exist under heaven. Any doubts are now minimized that IBM's and parties' conduct therein shall be seen in other than the most favorable and honorable light possible though American jurisprudence, within the limits of earthly tribunals and human agencies to effect outcomes thereof.

2. In as much as you, Bridgette Hauler, and your firm are purposed, you are innocents in the midst of events, deeds, actions and circumstances you did not create or foment, nor desired to create or foment, nor sought out. Rather, this matter sought you out, and you now graciously give of your life, blood, time-treasure, and wholesomeness to attend it the best way you know how, within means available to you. The time away from family, the toll upon your mind and faculties, your very life, can not be restored to you--I'm already aware you and/or Bridgette Hauler have worked as late as 9:30pm on at least one occasion, and some of your court filings have approached midnight as to submittal tags, in as much as/what has been relayed to me.

3. I am a founding member of the Society for the Prevention of Cruelty to Lawyers (SPCL). I'm always aware that you are people, first, with souls, sensibilities, and lives unrelated to those who now compel you to say and do things on their behalf, even if you disagree, or even if your freedom of conscience is violated. Therefore, law practice may be one of the most intrusive forms of servitude known to mankind; its effects on the *spirit of a person* can be as profoundly disfiguring as physical servitude on the body of the boardsman or bondswoman. The only comfort I might find in this matter is that you, Bridgette Hauler, Robert Spitz, and others may be generously compensated for the time you are now dedicating on behalf of others. I shall do all I can to maximize such outcome, in this regard, for all of you, to ensure at least this meager comfort. However, be careful what you do agree to countenance on behalf of your client, either out of expediency, coercion, or avarice at another's behest, for it remains written in Proverbs:

Do not exploit the poor because they are poor and do not crush the needy in court, for the [I-SHALL-BE] will take up their case and plunder those who plunder them. [22:22,23].

If you say, "But we knew nothing about this," does not he who weighs the heart perceive it? Does not who holds your life in his hand know it? And will he not repay each person according to what he has done? [24:12]

Like a muddied spring or a polluted well is a righteous man who gives way to the wicked. [25:26]

If anyone closes a deaf ear to the law, even his *prayers* are detestable. [28:9]

4. You need not worry about me; I've already accepted an outcome to be broke and destitute the rest of my life--for however long that may be. One of my ancestors, William Jefferson, was killed when bringing a proper claim to court, after the Civil War, to take legal custody of a considerable portion of property in Louisiana--it encompassed an entire township. He was murdered while approaching the courthouse to present his claim. In a previous case, regarding my specification in the matter with IBM, on the day before trial, the fire department suddenly arrived, *with multiple engines*, and swarmed my lawyer's offices to find the conflagration at the site, reported by an anonymous tip. They evacuated us from the building, searching office-to-office for any device that could be an ignition source, allowing us to return and complete trial preparation when they were done. That case is substantially the same as the current matter, with multiple Caltech alumni helping the court ascertain the facts of the matter. My family is already in a safe location, in preparation for the duration of what your client may or may not do. So why go forward? Others, for the sake of constitutional principles, have done *much more* than be called names, or ridiculed, or mocked, or denigrated. Some dressed in uniforms and went to places and were never seen on earth again. The issue at hand involves *property*, which is at the heart of one of the *Blessings of Liberty* (there are seven) in the Constitution. It is my responsibility, as a *citizen*, to resist what your client is attempting to do, which may not be to my benefit, but to the benefit of others, much as when a wolf attacks a flock, one sheep, by engaging a wolf, may buy time for the rest of the flock until the shepherd comes--even if the one sheep doesn't survive.

5. I am aware of your client's history, which gives background on past contract practices, as referenced from my patent specification found at "Google us patent 12381901," or so I have been told it is accessible, since I've never logged on to the Internet in my life. It includes correspondence from an IBM VP to me on behalf of IBM. Prior to the coming proceedings, I committed to memory (NIV) the entire Book of John, along with the Book of Revelation, and the Book of Proverbs, since these have bearing on cites you've made in the past, are preparing to make now, and shall make in the future. They also inform 12381901 and proceedings that are pending with you and your client. Additionally, I've committed the first half of the Book of Genesis to memory, since therein is precedent for the first deposition on earth and how it was conducted, reviewed in my book *From Jupiter to Genesis*, a worldwide first-hit title on Google®, or so I'm told. This letter and other related material shall be included in an updated appendix for 2012.

6. In my last conversation with Pete Wilzbach, a retired, highly regarded IBM EVP and fellow Caltech alum, he gave permission to talk about Scripture, while limiting, at this time, discussion of other matters. He told me he had committed all of Matthew to memory, which I asked him to recite while I checked him (Pete was candidate to be a Jesuit, after Caltech and before IBM). He got every verse and name right (including Zerubbabel). I have always trusted following his directions, including keeping to a single page for correspondence like this (it probably won't be read, so the font-size doesn't matter).

7. Attached is a sample template used over the past few weeks. It may provide background on me, including my final visit to IBM, by invitation, over a quarter century ago. I've not set foot on IBM *property* since then, even when Al Torressen suggested I be "brought in," out of respect for the principle of *property*, since many could not do so, starting at IBM's founding in 1911, when the teachings of *predestined* ANTIMERITISM were first enshrined in the corporate culture.

With prayerful regard, and best wishes for a restful 7th-day,

Anthony Sneed
Attachment: Template

Ex1 p95

FAX FAX FAX FAX FAX FAX FAX

FAX TO: 610-712-3774   8 pages (including this one)
FOR: [IBMer name]
FROM: Anthony Sneed (Sneed), ISB::Institute Fellow, *Emeritus* DATE: dw/mmddyy
RE: Gratitude for your candor and adherence to what is right, and just, and fair in all your communications

FINAL DRAFT

Happy [day of week]-day, [IBMer name]. Grace and peace to you and yours.

May this reach you and yours, in the reciprocating goodwill you have extended to others throughout your life. It is fitting and proper to help you recall me, as I imagine you've extended the same dignity and grace to many others as you have with me, so this review is in the third person. It may inform subsequent contact, if any, and perpetuate the goodwill you established not so long ago.

1. At age 10, Sneed's parents gave him a computer simulator called DIGICOMP as a gift. It used marbles for electrons, released from the top of its inclined plane of about a yard in length and a foot and a half wide, tilted about 25° along its length. The tilt provided "EMF," while ridges on the molded surface acted as "wires" to guide the "marble-electrons" through "circuit" paths. When a marble reached the bottom, it passed through a trigger-gate, which pulled a wire connected to a mechanism at the top of the plane, releasing another marble. Three "registers" populated the surface of the plane. The first was the "accumulator" with its seven plastic flip-flops that, when a marble encountered one, would swing left or right, "displaying" a one or zero, depending on the previous "state" of the flip-flop, then passing on to the next flip-flop, with other components above and below the plane working to simulate binary functionality. The second was the "memory" register, as a series of four "switches" acting as pre-set bits, aligning with the first four flip-flops in the accumulator, allowing addition or subtraction. A marble passed through a switch set to one, or dropped to a sub-plane board and came out of a tunnel and went through the trigger-gate, starting the next marble down the surface of the inclined plane. The third was a "multiplier/quotient" register, containing three flip-flops, allowing the memory register to be repeatedly added or subtracted to the accumulator, simulating multiplication or division.

[Also at age 10, Sneed attempted to read the entire Scripture, but couldn't since the one he found in his father's library was in Shakespearean English. However, the 31 chapters of Proverbs were less daunting, with their packet-based verses. During two 24-mile workouts in 2010, one in January and one in February, Sneed completed committing all 31 chapters to memory, with random chapter and verse recall, or by content to identify chapter and verse. The circuit he walked was three miles, eight laps making 24 miles. There seems to be similarity between an instruction-set of a computer to determine its architecture and Proverbs as an "instruction-set" for all of Scripture.]

Let it be noted that the IBM logo was in front of Sneed every day of elementary school, populated with IBM time-clocks in every classroom. This may have been a brilliant branding decision by Thomas Watson Sr.–Sneed spent time looking at those clocks as recess or end-of-school neared. At age 12, Sneed decided to work for the company whose precision was reflected in the synchronism and tracking of time with all those clocks throughout the school. Once that decision was made, the foundation of our future meeting was set, and it, too, would involve "time."

A couple years after mastering the simulator, Sneed bought RTL (Resistor-Transistor-Logic) NAND, NOR, inverter, and multi-flip-flop chips, made by Motorola. These IC's allowed Sneed to replicate the mechanical computer simulator and experiment with much higher clock speeds. The DIGICOMP manual described a language called FORTRAN as a parallel to DIGITRAN, used to configure problems on the simulator. Intrigued, Sneed learned FORTRAN at age 13 on a 360/40 at the Claremont Colleges. When Sneed entered boarding school in Carpinteria, California, as a freshman the following year, the math department allowed him to take computer programming courses for seniors. Bill Gates, who was also a freshman that year, used the same timesharing system at his school in Seattle as Sneed did in Carpinteria. During the summer, Sneed was accepted to Granite Computer Institute with a $2000 scholarship. Granite certified programmer/analysts on IBM systems architecture, operations, and programming in three languages, including assembly. Sneed's classmates were all adults. He went in on weekends and stayed all day, since the 360/30 was not used then, so he had the entire computing complex to himself. "Debbi" console commands were mastered, along with loading and unloading disk and tape drives, running assembly and

Ex1 p96

other programs that made the 1403 spew out reports, IPLing, single stepping to watch programs execute from the front panel lights, and opening the 360 front panel and exploring its internals right down to its memory core. Also, familiarity was gained with unit record equipment, sorters, duplicators, and 400-series tabulators with their plug-boards. Sneed completed Granite's summer course and returned to boarding school in the fall. In his junior year, Sneed was moved to notify his headmaster that Sneed would be attending the California *Institute* of Technology. Having enjoyed Granite Computer *Institute*, Sneed felt a school with *Institute* in its name would be a natural next step. Sneed was accepted to Caltech in November of his senior year at boarding school.

2. The digital systems knowledge gained from ages 10 to 14 allowed Sneed to ace Caltech's freshman digital design courses, including implementing a calendar with 31 IC's. By the following year, during the summer, Sneed became a Caltech/NASA/JPL intern in Dr. Richard Goldstein's group. Just 14 years earlier, in March of 1961, Dr. Goldstein became the first person, in history, to accurately determine the earth-sun distance, or 92,955,807 miles averaged over a year. Dr. Goldstein did so by accurately computing the earth-Venus distance with radar pulses. Kepler's third law permitted the earth-Sun distance to be derived from the earth-Venus distance since it's still hard to bounce pulses off the sun, which actually contains a "sun" spinning inside the sun, first discovered in 1989; the *tachocline* is the boundary between the two. The previous best estimate was 92,880,000 miles, based on the 1874 and 1882 Venus transits, after less than satisfying results from prior worldwide efforts with the transits of 1761 and 1769.

Sneed was assigned to develop the Very Long Base Interferometry correlator, in machine language, for measuring the distances between NASA's 64-meter, Deep Space Network antennas in Canberra Australia, Madrid Spain, and Goldstone California, using digitized noise wave fronts from a quasar. The machine to process the time-marked data was a Xerox Sigma V--a perfect clone of IBM's mainframe architecture, giving Sneed ready fluency in its machine language from certification with Granite Computer Institute course work five years before. The code developed became one of the top projects for Caltech's IBM systems architecture course, resulting in straight A's for all three quarters. In Sneed's junior year, he disassembled his HP25 programmable calculator, modified its 49 steps of memory with 1,000 steps of paging memory, using three dozen IC's, and redesigned its case to have a docking port. It was a successful digital lab project. For the fun of it, Sneed took an IC design course Dr. Carver Mead taught, became team leader, and designed the masks for a four-bit shift register, successfully implemented as a MOS IC.

Later, Sneed became a summer intern at Hughes (Boeing) Space and Communications Group and was assigned to the Pioneer/Venus project. Left to himself with no specific assignment, Sneed developed his own project to simulate the entire command sequence that controlled the spacecraft from launch to orbit insertion at Venus. Sneed found a single-point, catastrophic failure-mode that could have occurred in earth orbit, after launch. This gave Sneed standing at a NASA design review of the mission, to provide a reengineering solution and remedy the failure-mode. An offer was made to Sneed to be a full-time MTS, or member of the technical staff, after graduating from Caltech.

As an MTS, Sneed was asked to trouble-shoot a probe subsystem, a key part of the NASA Galileo mission to Jupiter. The probe was to be jettisoned from an orbiter near Jupiter, reach speeds of up to 106,000 mph, and conduct a science mission below the cloud-tops of Jupiter. The relay computer on the orbiter collected and relayed science data back to earth from the probe's seven science instruments. It required a new operating system and software to be designed in the microcode of the relay computer, along with optimization of a discrete Fourier transform routine (DFT) to maintain frequency- and phase-lock on the probe's 1.4 GHZ signal affected by a 15 hz/second Doppler. Additionally, the operating system had to coordinate with other computers on the orbiter and prepare science and system maintenance data packets for down-link to the Deep Space Network on earth, and anticipate scheduled "brown-outs" due to power limitations across all subsystems on the orbiter. There was no solar power due to the distance from the sun, so radioisotope thermal electric generators were used on the orbiter.

Sneed was assigned to bring the project back on schedule and address cost overruns. So, on his own, he developed a 4,000 line simulator, in four weeks, that included modeling the RF/IF, numerical controlled oscillator, and analog-to-digital circuits receiving the probe's signal, the 7 million state instruction-set of the on-board computer, and the digital interface to other computers on the orbiter. This cut some $250,000 from the project budget, eliminating one of three hardware simulators, and advanced the microcode development schedule by six months. The Sneed Simulator allowed microcode engineers to collaborate with communication system analysts on the same computer. Sneed received the highest salary increase of second year MTS's. This led to interview and hiring as an assistant professor in the California State University system. With 90 students and a project lab, Sneed received the highest

a *7ʰ-Gift* page

department teaching quality feedback report. From that assistant professorship, Sneed was interviewed by IBM and later accepted a position offered by IBM's Data Processing Division in the finance industry. Sneed's hiring manager was a fellow Caltech alum, who had also been a Caltech/NASA/JPL summer intern, seven years before Sneed, and he, too, worked at Hughes (Boeing). Much later, he became an IBM E-5 and runner-up for IBM CEO.

3. Though beyond any job description of an IBM marketing trainee, a few months after starting work at IBM Sneed was asked to look at a technical problem that stumped account SE's (System Engineers) for some six months. It involved designing a workflow system that interfaced a bank's mainframe database with a branch-based office system, through a telecommunications access method. After analyzing the problem, Sneed developed the 1,000 lines of needed assembly code in about two weeks. Though directed by *marketing management* to trouble-shoot the problem, an SE manager accused Sneed of "violating" IBM policy, since a marketing trainee wasn't supposed to do what, apparently, had been successfully done. Since "policy" had been "violated," credit went to the account SE's for the effort. This was Sneed's introduction to the world of IBM's belief and merit system. That little project had impacted Sneed's four-week branch office assignments and preparation for Sneed's second IBM training class in Dallas. So, Sneed finished the class project on the three-hour flight to Dallas and completed the two week class.

During the next between-classes branch office period, Sneed was asked to look at a different problem which had also stumped account SE's for weeks at one of the largest accounts in the branch: Why was a bank's statewide ATM network availability only in the 80% range? The problem had stalled the sales of new IBM ATM's to the bank. Accepting the assignment, Sneed understood there was sufficient mainframe computing capacity at the bank, and the individual ATM's were operating normally. So to Sneed this seemed an operations glitch. He surmised that no single person had spent 24 hours at the control center that managed the ATM network. So Sneed assigned himself to monitor the control center workflow and the network management personnel for 24 hours. At about 2am, he found the problem and, the following day, proposed the solution. Network availability leaped into the 90s. The branch manager (BM) asked for a report on how the problem was solved, which Sneed provided. Nothing was ever heard about this effort again, perhaps becoming another "SE manager success" since this, too, may some how have "violated" IBM "policy" (marketing trainees weren't supposed to be out at an account after midnight?). By this point, Sneed was becoming a little wary of branch office projects: they somehow ended up "violating policy," there was usually no recognition, and they impacted the IBM corporate training schedule and time to prepare for classes.

At the end of a subsequent training class, the instruction manager told trainees that BM's might create hypothetical scenarios to see how trainees would respond. So, after returning to his branch, the BM called Sneed into his office and presented, what Sneed believed, was a hypothetical scenario: How to reestablish operations in one of the branch's largest accounts after a disaster. It was a bank with four IBM mainframes handling $32 billion in federal reserve float every four days, that suddenly had its entire datacenter destroyed by a three-standard-deviation municipal power surge. This knocked out all of its check processors, statewide ATM network, and statewide branch terminal network--the worst disaster of its kind in all of IBM's Western Region history. "What are YOU going to do about it??" the BM demanded. Sneed requested the newest mainframe, that IBM had just introduced and was manufacturing in Poughkeepsie New York, be shipped to the bank in Glendale California. "YOU'VE GOT IT!" Sneed requested double shifts of FE's (Field Engineers who install and do *actual* technical support) for 72 hours. "YOU'VE GOT IT!" Sneed requested to be relieved of marketing training responsibilities for one week. "YOU'VE GOT IT!" At that point, Sneed noted the only thing left to do was live at the account until it was back up. "GO DO IT!!" the BM boomed. "Yes sir," Sneed replied and promptly exited his office, quite pleased with how well the *hypothetical scenario* went.

Just for completeness, Sneed called an SE, a friend he knew at the bank. And that's when Sneed learned the bank really had flat-lined, was on its back, out cold, with *no IBM marketing personnel there* whatsoever the SE noted-- they were all leaving for the Thanksgiving holiday that started the next day. This was no hypothetical scenario: Sneed understood he had been *made* the IBM marketing principal to coordinate the bank's recovery, or he could ignore the BM's directive. So Sneed arrived at the bank in white shirt, conservative tie, Brooks Brothers suite, wing-tips and lived at the bank for four days. There were regular briefings with senior customer execs, to keep them away from the SEs and FEs. The new mainframe arrived which Sneed inspected before installation, including directing some minor reengineering of its cooling system to remedy a defective thermal coupling. The execs were impressed after their bank was brought back into business. When asked how long the new *emergency* IBM mainframe could be kept, Sneed quoted 90 days from the IBM Business Ethics Guidelines that *all* marketing personnel signed. The BM, later, told the bank it could keep the mainframe and pay for it, securing a huge year-end commission. The bank had previously paid a backlog-bookie $800,000 to get an earlier ship-date rather than wait 24 months. It was considering selling its position to someone else.

This precipitated a controversy between the BM and Sneed, though Sneed wasn't aware of the commission implication for the BM, just the IBM Legal requirement for the 90-day disclosure to avoid DOJ problems. The matter was escalated to the IBM CEO; he decided against the branch manager ("early retirement") who could not explain why he ignored the IBM Legal guideline, why he tried to fire Sneed, and why he sent a *trainee* to be the marketing principal for something of this magnitude. Sneed's marketing manager (then different from Sneed's hiring manager) was sent to Japan; the regional manager was reassigned; the division manager was overruled when he tried to back the regional manager, and later left the business. The bank tried to hire Sneed.

Sneed stayed on at IBM, chastened by the episode, yet graduated with honors from IBM's New York marketing training center. He went on to win the Western Region fast start contest with 400% of quota by the end of January. By April Sneed had enough business to make quota for the year, along with identifying and contesting with six IBM mid-range systems, six computers DEC tried to upgrade for the National Reconnaissance Organization (NRO), which processed national security photo intelligence from satellites. None of this was easy, since Sneed was assigned to a new branch, the BM thereof a 15+ year friend of the prior BM. Sneed was given a 125-mile-by-60-mile territory, of many abandoned accounts, so Sneed called on everybody and did quite well. The NRO DEC threat was disclosed by a tight-lipped CFO, after Sneed personally repaired the CFO's PC to complete a spreadsheet for a board meeting the next day. The IBM PC was coming into its own, and Sneed desired to become a branch PC specialist, but was not considered the best choice. (Two years later, the plane carrying the designated branch PC specialists from IBM PC headquarters in Florida, along with the president of the PC division and his wife, crashed in Dallas, a plane Sneed almost certainly might have been on had his wish been granted [Proverbs 14:12].)

By April of his first year on quota Sneed surmised that all the "policy violations," and ones that awaited him in the future, meant that his dream of being IBM president was not possible. So in April, Sneed left IBM and did not return. When the branch informed him he had a $7000 commission check owed him, he asked that it be mailed, along with some Spaulding Executive golf clubs Sneed won. But the new BM insisted Sneed come into the branch and pick up the check. Sneed realized that setting foot on IBM property, at that point, could probably be construed as a "violation of policy," but the BM refused to mail the check. So Sneed arrived in IBM business attire, reported to the BM's secretary, and was shown into the BM's office. The BM previously notified the IBM CEO that Sneed had been fired for "violating policy," and the CEO let Sneed know this in a letter. The BM motioned for Sneed to have a seat, possibly to talk about IBM, perhaps in violation of the CEO's disclosure of what the BM had decided. When Sneed silently remained standing, the BM, after a brief pause, handed Sneed the check and Sneed departed, still wondering why the check wasn't just mailed. The objective Sneed had set as a 12 year-old was then completed after a 17 year journey from boarding school, to Caltech, to NASA/JPL, to Hughes (Boeing), to a California State University assistant professorship, and to IBM. "Respect for the Individual" had been IBM's motto. Sneed realized if IBM was one of the most admired companies in the industry, then it seemed best to serve as an individual, or with founders of their own firms, rather than those where the founders died long ago.

4. So, Sneed started his own firm in 1985, which became ISB::Institute. He developed multiple lines of business to survive the first 24 months for a new startup. The first 24 months are when most new enterprises fail:

1. PC manufacturing: Sneed introduced the "TASCOR" brand of 100% compatible IBM computers, manufacturing PC's under this name. Motherboards were ordered directly from IBM's Boca Raton facility. The IBM BIOS and other proprietary IBM ROM were on the motherboards. Doing this allowed Sneed to build "clones" that were in fact true IBM PC's, yet priced 30% less. The largest order was 15 systems from Hughes Aircraft.
2. Contract programming: Software development was done for Gourmet Concessions across the Broadway Store chain. This database application automated workflow and inventory management. It included the sale of a TASCOR computer, database software and contract programming which helped pay law school tuition for the patent attorney who later filed Sneed's first patent, called BEST2000A™.
3. Initial development of a multitasking PC operating system: Sneed started developing one himself, but discovered that a product by Quarterdeck, called DESQview, implemented the idea. Sneed became a

Ex1 p99

distributor of Quarterdeck products, placing the largest single order for the product in the U.S. at the time, saving the company from default. For six years, Sneed acted as a marketing consultant for the firm. Its annual revenues rose from $4 million (avoiding bankruptcy), to $8 million, to $12 million, to $26 million, to $48 million, whereupon Sneed was offered a vice presidency (declined) and stock options (accepted).

4. Southern California digital radio station: Sneed, on his own, designed a digital broadcast system to link a million personal computers by FM sideband. Sneed negotiated a contract with the station manager of FM 88 KXLU, to modify their transmitter with a $1500 FM sideband modulator. A three-mile long ATT leased-line linked the KXLU FM modulator with a computer in Sneed's West LA office. Sneed negotiated the manufacture and test-model of a digital demodulator which attached to a PC like a modem, tuned to FM 88. A software and network transmission protocol was designed, allowing compressed digital transmission of news, weather, sports, stock quotes, software and other consumer content. A successful beta-test transmission between West LA and Long Beach proved that a million personal computers could be linked. Preliminary effort was made to file a patent. That effort, some 20 years later, was reviewed as evidence to prove prior art for a suit involving transmitting digital information to a handheld device like a Blackberry.

ISB.:Institute survived its founding in 1985 through 1995. During this period, Sneed started a family: Alexandria (1987), Dorion (1989) and Dorius (1993) born at Cedars Sinai in Beverly Hills; and Chante (1995) born at Pomona Valley Hospital after Sneed moved his wife and family to the Inland Empire. Diversifying his lines of business, Sneed also worked under contract with two search firms to learn the art of recruiting and placement. Sneed internalized this knowledge and applied it through ISB.:Institute to fill over $1 million in job orders, including multiple positions with a $0.5 billion HMO, a $7 billion corporate credit union, a defense contractor, a consumer products manufacturer, and an international semiconductor manufacturer. In 1996, Sneed received recognition from the president of Caltech for Sneed's work on the NASA Galileo mission, which arrived at Jupiter in December of 1995. The operating system Sneed designed, along with the optimized Discrete Fourier Transform, and all the microcode developed with the Sneed Simulator worked flawlessly. The president set expectation to hear about future successes in Sneed's career.

5. In 1995, Sneed did preliminary research on a problem that appeared already solved--Y2K. So he left it alone. Two years later, in February 1997, Sneed read in the *Los Angeles Times* that a DOD computer complex in Ohio had failed due to the Y2K problem. This startled Sneed, since DOD had more money than any agency on earth for solving this problem, yet there DOD was with a major, public failure, and less than 24 months before systemic problems could become paralyzing. Solving the Y2K problem--the problem of the century--might be a way Sneed could answer the expectation of the president of Caltech, "I look forward to hearing of future milestones in your career." So, Sneed sat down in February 1997, with a blank sheet of paper, and pondered this problem. For about a decade Sneed referred to the days of the week by their ordinal designations in Scripture (1st, 2nd, 3rd, 4th, 5th, 6th, 7th) rather than by nominative, deity-based traditions that vary country to country (*dies solis* or *sunnandaeg* or *sun-day*; *lunae dies* or *monandaeg* or *moon-day*; *Marties dies* or *Tiwesdaeg* or *god-of-war-day*; *Mercurii dies* or *Wodendaeg* or *king-of-the-gods-day*; *Jovis dies* or *Thorsdaeg* or *god-of-thunder-day*; *Venus dies* or *Frigedaeg* or *day-of-the-goddess*; *Saturnus dies* or *god-of-agriculture-day*). The ordination of days allowed Sneed to think of time as an infinite vector segmented into ordinal-based weeks, months, years and centuries. The 100 symbols of "00" to "99" were assigned on to centuries, like 1900 to 1999. In a computer, there's nothing to prevent a computer from reassigning these symbols to different 100 year periods. Due to a one day difference between the 365 days in a year, and the 364 days in the largest integral number of weeks in a year, the first day of a new year, and a particular day of the week are coincident, again, after seven years. Unless there is a leap year; this skips the coincidence four times due to leap days once every four years. So once every 4x7, or 28 years, the coincidence recurs, or 4 times any other integral multiple of 7, for a product of 91 or less. The simplicity to think about this in terms of Scripture and ordinal-based time resulted in a specification (Specification) titled "Serial Encryption System By-Pass of Year 2000 Date Malfunctions" or BEST2000™, and later BEST2000A™ as Sneed went on to develop BEST2000B™ as a vehicle thermal integrity detection system. Both were issued patents, BEST2000B™ after a successful one year national security review during patent examination, since it is a strategic technology for the aerospace industry.

6. A friend, Dr. Alex Lidow, who is a fellow Caltech alum and member of the Caltech board of trustees, applied Specification on IBM computers in seven countries, when he was CEO of International Rectifier Corporation (IRC). This resulted in savings to his firm of some 70% on a two-year, $10 million "best efforts" counter-proposal by Accenture (Anderson Consulting at the time). IBM had vetted Sneed's Specification by then, and confirmed to Dr.

Lidow and IRC its technical merit before IRC applied it. Dr. Lidow was very pleased, and Sneed was later awarded $236,000 under an IRC contract. Dr. Lidow also invited Sneed to his parent's estate for a Caltech function.

Four months prior to working with Dr. Lidow and his firm, IBM entered into similar agreement when it took custody of Specification, analyzed Specification within IBM, and declared Specification's merit to Sneed and, later, to IRC which deployed Specification on IBM computers in France, Germany, England, Singapore, China, the U.S. and Japan, at considerable savings over anything IBM, Accenture, or anyone else offered. IBM documented that it, too, could proceed with payment terms after d3/May 22, 2001 (Date), which is when a patent was issued to Sneed.

IBM admits taking custody of Specification from Sneed. Mr. Don Logan, an ethical IBM Global Services (IBM Global) program director, confirmed Specification's merit and economy, under terms (Terms) set forth by Sneed for no-charge testing, and a $49,764.55 per site production-use fee. Mr. Logan and an unnamed, ethical IBM VP (Ethical VP), helped Sneed modify Terms with 14 cents per line of code, across an aggregate number of lines of code, and no charge after a billion lines, respecting Ethical VP's caution about IBM's counting methods. As noted, one IBM client, IRC, applied Specification to 1,200,000 lines of code, verified by Caltech alums Dr. Dale Prouty and Dr. Alex Lidow and an MIT expert in IBM mainframe and midrange systems.

Subsequent to Mr. Logan's and Ethical VP's contact, IBM Corporate Licensing Program Manager Mr. Al Torressen took custody of Specification from Mr. Logan and IBM Global, memorializing to Sneed the written Terms Mr. Logan and Ethical VP helped establish. This demurred a Department of Justice "firewall" between IBM Corporate and the IBM Global subsidiary, obviating a markup of up to 1,400% IBM Global could legally bill IBM Corporate, or the same up to $2 per line of code IBM Global charges government or private sector clients. This "firewall" demur may have saved IBM Corporate up to $1.86 billion IBM Global could bill.

Mr. Torressen assured Sneed, in writing, that IBM Corporate would not "expropriate" Specification, and acknowledged the Terms Mr. Logan and Ethical VP helped establish in the written agreement Mr. Torressen also took custody. Mr. Torressen noted Specification's potential worldwide impact, and arbitrarily set d3/May 22, 2001 (Date) for Sneed to initiate contact for payment. Between Date and October 29, 2007, multiple IBM legal personnel were designated to replace Mr. Torressen, including personnel directed by IBM CEO Mr. Sam Palmisano. Each confirmed, in writing, no benefit or discovery of use that merited payment to Sneed. This remained so until Sneed learned that an unnamed VP (Unnamed VP) retired from IBM. Unnamed VP and Sneed worked together at IBM in the 80s. After contact with Unnamed VP, Sneed contacted Mr. Palmisano regarding payment for use, if any, of Specification, encouraged by Unnamed VP. Mr Palmisano directed legal counsel to contact Sneed, whereupon legal counsel declared, in writing, IBM would no longer respond to inquiries about IBM's use of Specification, but later recanted, in writing, this declaration of non-responsiveness.

7. Three Caltech alums are in agreement to move forward with this matter: Dr. Alex Lidow Caltech board member, Dr. Dale Prouty Caltech PhD Applied Physics, and an unnamed VP from within IBM.

First, in every communication with you, in your capacity as IBM [position], whether oral or written, you spoke with candor. While keeping a vow of loyalty to IBM, *simultaneously* you were informed by doing what is right and just and fair with me. When I am reminded of the highest ideals of IBM, "Respect for the Individual," it is yours and IBM CEO Mr. John Opel's example that come to mind. It is at this time I wish to express my gratitude. I also wish to express regret in not fully appreciating that your perspective goes beyond merely corporate policies, practices, beliefs, or even societal codes and regulations. There are more transcendent precepts which allow you to discern and act with confidence, when others may vacillate, equivocate, obfuscate, or denigrate efforts that manifest desire, merit, and purposed commitment. We might all do better to follow your example in life, let alone in business matters.

It is because of your and [named parties'] candor, and attempts thereof, that I gained a new respect for your firm, perhaps renewing ideals I had long ago. And out of respect for you and [named parties], I've been considerate of IBM's actions and decisions, informed by what was set forth in writing and properly acknowledged after patient and deliberative vetting. I deeply regret that IBM would not let me speak with you. I deeply regret this. Rather, instead, others were inserted in a process that seemed less committed to open communication as in the early contact with a firm you helped redefine. My comfort was that because of your integrity and commitment to excellence, you would enjoy a long and prosperous career. This comfort is injured as I learn more of your leaving the firm, since I consider you deserving the best life can still offer.

The Specification has performed flawlessly on IBM computers around the world, just as others predicted was possible, and which Dr. Lidow and Dr. Prouty and others have verified. So it is fitting and proper that your imprimatur be given weight, independent of any IBM chairman, IBM CEO, IBM executive, IBM manager or IBM employee. No one may *demand* of you anything now, except that by which you assent of your own freewill. And your name carries a legacy, through the centuries, whether in war or times of peace, that hews to doing what is right and just and fair, all the way to this very day.

It remains written in Proverbs:
    The memory of the righteous will be a blessing, but the name of the wicked will rot. [10:7]
    The faithful man will be richly rewarded, but one eager to get rich will not go unpunished. [28:20]
    A truthful witness does not deceive, but a false witness pours out lies. [14:5]

Hence forth, I consider you an Industry Intellectual Property Expert and, as circumstances may warrant, someone who may be respected in such capacity. Below, you may simply check one of two boxes, sign and date it, and return this 7ᵗʰ page by fax, as is, to 909-395-9535, before sunset on [dw/mmddyy]. If there is no response, it shall be construed that you are asserting no knowledge of this matter in any way whatsoever.

[ ] IBM did not use your Specification or anything similar to it to remediate Y2K problems on any internal IBM computer or software application.

[ ] IBM applied a two-digit date shift method, similar to or the same as all or part of your Specification, on at least one internal IBM computer or software application.

_____   _____
           [IBMer name]   Date

With prayerful regards and respectfully yours,

Anthony Sneed
Life-member Caltech Alumni Association, U.S. Patent Holder (#7546982 and #6236992),
author (*From Jupiter to Genesis*, a worldwide first-hit title on Google®)
Attachment: Correspondence with Caltech and other officials

Ex 1 p/02

6 of 6 DOCUMENTS

ANTHONY SNEED, Plaintiff and Cross-Respondent, v. INTERNATIONAL REC-
TIFIER CORPORATION, Defendant and Cross-Appellant.

B176278

COURT OF APPEAL OF CALIFORNIA, SECOND APPELLATE DISTRICT, DI-
VISION FIVE

2005 Cal. App. Unpub. LEXIS 6314

July 20, 2005, Filed

NOTICE: [*1] NOT TO BE PUBLISHED IN OFFI-
CIAL REPORTS. CALIFORNIA RULES OF COURT,
RULE 977(a), PROHIBIT COURTS AND PARTIES
FROM CITING OR RELYING ON OPINIONS NOT
CERTIFIED FOR PUBLICATION OR ORDERED
PUBLISHED, EXCEPT AS SPECIFIED BY RULE
977(B). THIS OPINION HAS NOT BEEN CERTIFIED
FOR PUBLICATION OR ORDERED PUBLISHED
FOR THE PURPOSES OF RULE 977.

PRIOR HISTORY: APPEAL from a judgment of the
Superior Court of Los Angeles County, No.
BC296142. Aurelio Munoz, Judge.

DISPOSITION: Affirmed.

COUNSEL: Coudert Brothers LLP, Glenn W. Trost,
Nancy C. Morgan, for Defendant and Cross-Appellant.

Robert J. Spitz, for Plaintiff and Cross-Respondent.

JUDGES: MOSK, J.; ARMSTRONG, Acting P.J.,
KRIEGLER, J. concurred.

OPINION BY: MOSK

OPINION

Defendant and cross-appellant International Rectifier Corporation (International Rectifier) appeals from a
judgment, following a court trial, awarding plaintiff and
cross-respondent Anthony Sneed (Sneed) $ 149,293.65
in damages and $ 54,850.07 in interest for breach of an
agreement governing the use of a method developed by
Sneed (known as Best 2000) for avoiding date-related
computer problems associated with moving from the
year 1999 to the year 2000 (commonly known as the
Y2K problem). The terms of the agreement [*2] required International Rectifier to pay Sneed $ 49,764.55

for use of Best 2000 at each of International Rectifier's
facilities.

International Rectifier contends that the trial court
erred as a matter of law by interpreting the agreement to
cover all information relating to Best 2000 provided by
Sneed, including information that was otherwise publicly
available or that was disclosed by Sneed without restriction to third parties. It claims that it used only such excluded information and therefore had no payment obligation under the agreement. International Rectifier further
contends that substantial evidence does not support the
trial court's finding that International Rectifier used Best
2000; that the statement of decision was inadequate; and
that the statute of limitations bars certain of Sneed's
claims.

We need not address International Rectifier's argument concerning the interpretation of the parties' agreement, because that agreement, even as interpreted by
International Rectifier, required it to pay Sneed for use of
proprietary information concerning Best 2000. The trial
court found that International Rectifier breached the
agreement by using such proprietary information without
[*3] payment, and substantial evidence supports that
finding. Substantial evidence also supports the findings
that Sneed did not disclose Best 2000 to any third party
without restriction on disclosure, and that before February 18, 1999, Sneed had no notice or information to put
him on inquiry notice of International Rectifier's breach
resulting from use of Best 2000 at its El Segundo facility.
Sneed's breach of contract claim with regard to the El
Segundo facility accordingly did not accrue until after
February 18, 1999 and is not barred by the statute of
limitations. The trial court's statement of decision is sufficient, and the court did not abuse its discretion by refusing to amend the statement of decision as requested
by International Rectifier. We therefore affirm the judgment.

a *7ᵗʰ-Gift* page

2005 Cal. App. Unpub. LEXIS 6314, *

BACKGROUND [1]

> 1   We recite the relevant facts "in the manner most favorable to the judgment, resolving all conflicts and drawing all inferences in favor of respondent. [Citation.]" (*Principal Mutual Life Ins. Co. v. Vars, Pave, McCord & Freedman* (1998) 65 Cal.App.4th 1469, 1475, fn. 1.)

[*4] **A. The Parties**

Sneed is a computer scientist and consultant who developed Best 2000. International Rectifier is a semiconductor manufacturer with facilities worldwide, including California, Mexico, Singapore, China, Japan, India, Italy, Germany, and England. International Rectifier relies primarily on the IBM AS/400 computer system to manage its central data processing requirements. In 1997, International Rectifier also used an accounting software program known as MACPAC Version 8, which had been modified substantially to accommodate specific needs at each of its international production sites.

B. The Y2K Problem

The Y2K problem arose because many computer programs developed before the 1990s used two numerical digits to record the year of a transaction stored in a computer database. The two numerical digits were used to represent the last two digits of a calendar or fiscal year, and the first two digits of that year were assumed to be "19." The numbers "00" would thus represent the year 1900. Although storing data in this manner could accommodate all 100 years in the twentieth century, from 1900 to 1999, the Y2K problem arose when information from the year 2000 [*5] or some later year needed to be stored. For example, if information from the year 2005 were stored as "05," an error would occur because the computer program would recognize "05" as the year 1905, not 2005.

In late 1996 through 1997, International Rectifier was investigating methods for solving its Y2K problems. One method under consideration was to upgrade International Rectifier's existing software from MACPAC Version 8 to MACPAC Version 10. The MACPAC Version 10 software was Y2K compliant because it used four digits to represent years stored in its databases.

C. Best 2000

Sneed developed Best 2000 in 1997 as a method for addressing the Y2K problem. Best 2000 is a mathematical model for encoding and decoding date-specific information and computer programs so that the information can be stored and retrieved both before and after the year

2000. Reduced to its most basic elements, Best 2000 involves two steps. The first step involves changing the data stored within a computer system's database. In this step, data is altered or encrypted so that all two-digit year fields to be stored within a database are reduced by 28 years [2]. Thus, the year 1999, which in its unencrypted [*6] state would be represented as "99," would be represented after encryption as "71." The second step involves changing a computer program itself by subtracting 28 years from all date constants within the program and by resetting the computer's internal system clock back 28 years. Encrypted data stored within a database would have to be decrypted, which involves adding back 28 years, at the time such data is retrieved. Best 2000 also included an element that permitted the use of two-digit year fields to store more than 100 years of data. A detailed explanation of how to implement the date-shifting technique described by Best 2000 is set forth in a document entitled "A Serial Encryption System-Bypass Of Year 2000 Date Malfunctions," which was the basis for a May 4, 1997 patent application by Sneed. A patent was issued to Sneed on May 22, 2001.

> 2   The number 28 is used because it is a factor of four and thereby accounts for leap years, which occur every four years. The number 28 is also a factor of seven, and thereby accounts for a seven day week, ensuring that a given date both before and after encryption falls on the same day of the week. By way of example, the date March 17, 1996, which falls on a Sunday in a leap year, would be represented in an unencrypted database as 03/17/96. That same date, after encryption, would be represented as 03/17/68, also a Sunday in a leap year.

[*7] In 1997, another computer scientist and consultant named Don Estes published a working paper entitled "Encapsulation Solutions for Year 2000 Compliance," that discussed a time-shifting strategy for resolving the Y2K problem similar to that described in Best 2000. Neither Estes nor Sneed relied upon or consulted the other's work at the time they developed their respective Y2K strategies.

D. The Parties' Agreement

In mid-1997, Sneed approached International Rectifier about possibly using Best 2000. His initial conversation was with Alex Lido, International Rectifier's CEO, and subsequently with Niel Armstrong (Armstrong), International Rectifier's Vice President of Global Information Technology and Processes. At Armstrong's request, Sneed participated in a conference call in the summer of 1997 with International Rectifier's technical staff to discuss Best 2000. Thereafter, Sneed submitted a

2005 Cal. App. Unpub. LEXIS 6314, *

proposed written contract requiring International Rectifier to pay $ 49,764.55 to Sneed for "use" of Best 2000 and preventing International Rectifier from disclosing to third parties any of the information provided by Sneed. The proposed agreement provided in relevant part as follows: [*8] [P] "International Rectifier Corporation ('Disclosee' herein) hereby agrees that all information disclosed in briefings, documents, drawings, and otherwise received from Anthony Sneed ('Discloser' or 'Inventor' herein) . . . relating to his cross-platform invention that solves the year 2000 date malfunction called *BEST 2000* ('Invention' herein) is received in confidence, and is clearly marked as being confidential, and is the proprietary right and property of Discloser, and that such information will not intentionally be reproduced, transmitted or disclosed to any other organization or entity, or to any person not in the employ of the Disclosee, or used by Disclosee, without the express prior written consent of Discloser."

International Rectifier requested that the following provisions be added to the proposed agreement: "If the information is orally or visually disclosed, it shall be identified as being confidential in writing submitted to the Disclosee within thirty (30) days after such oral or visual disclosure. Review of disclosed material for the purpose of evaluating Invention's usefulness to Disclosee does not constitute use and does not justify or require payment [*9] of $ 49,764.55 to Discloser. [P] The foregoing shall not apply to information which: [P] 1. Can be documented as within the prior knowledge of Disclosee prior to receipt of the information disclosed by Discloser; [P] 2. Was obtainable by Disclosee from sources other than Discloser who themselves obtained it from sources other than Discloser; [P] 3. Was within or becomes part of the general public knowledge through no fault of Disclosee as of the time Disclosee makes independent use of the information; [P] 4. Is provided by Discloser to any third party without restriction on disclosure; [P] 5. Is rightfully provided to the Disclosee by a third party without restriction on disclosure; or [P] 6. Is independently developed by the Disclosee." The agreement signed by the parties included the language requested by International Rectifier as well as an additional provision requested by Sneed requiring International Rectifier to make an initial non-refundable payment of $ 5,000.

E. The Parties' Subsequent Dealings

In November 1997, after the agreement was signed, Sneed delivered to International Rectifier a copy of the provisional patent filing for Best 2000. Sneed [*10] also participated in several telephone calls with Armstrong about how to implement the date-shifting remedy discussed in Best 2000. At the same time, International Rectifier's technical staff was evaluating the relative pros and cons of implementing Best 2000 versus updating the software used by International Rectifier from MACPAC Version 8 to MACPAC Version 10. In a report dated December 17, 1997, the technical staff concluded that Best 2000 was "a very common solution" and that Best 2000's ability to process date-related information which spans more than 100 years was "the only aspect of Best/2000 which could be considered proprietary or intellectual property!" In the report, the staff recommended against implementing Best 2000 and in favor of MACPAC Version 10.

Notwithstanding the recommendations of its technical staff, International Rectifier continued to pursue a date-shifting remedy for its Y2K computer problems. In December 1997, International Rectifier began testing and evaluating a date-shifting remedy at its El Segundo, California facility. In March 1998, Armstrong spoke with Sneed about implementing a 28-year date-shifting remedy at International Rectifier's facility in [*11] Singapore. After the conversation with Armstrong, Sneed sent to International Rectifier an invoice dated March 9, 1998 covering the Singapore facility. That invoice was for $ 49,764.55, but provided for an "Early Payment Discount" to $ 29,764.55, if payment was received on or before March 16, 1998. International Rectifier did not pay the invoice.

In April 1998, Sneed had two telephone conversations with Armstrong about the Singapore invoice. In the first call, Armstrong told Sneed that he had received the invoice and had forwarded it to International Rectifier's accounting department. In the second call, Armstrong said that funds for payment had not yet been released by International Rectifier's chief financial officer. At no time during these telephone conversations did Armstrong mention to Sneed that International Rectifier had developed its own solution to the Y2K problem using other publicly available information, that it had decided against implementing Best 2000 as a viable option for solving its Y2K problems, or that International Rectifier believed that it owed no payment to Sneed.

On April 21, 1998, Armstrong and Sneed's wife, Mary Stine (Stine), spoke by telephone about [*12] payment of the Singapore invoice. International Rectifier claims that in this telephone call, Armstrong told Stine that International Rectifier was using a publicly available date-shifting solution, and that it owed no payment to Sneed under the contract, but was willing to pay him a nominal amount for his consulting services. Stine thereafter sent Armstrong a memorandum dated April 21, 1998 and a letter dated April 22, 1998. Both the letter and the memorandum discussed the past due Singapore invoice and asked International Rectifier to specify an amount for the proposed consulting fee. The memorandum requested, by April 22, 1998, a letter from Interna-

Ex1   105

tional Rectifier stating its intent to pay Sneed only for his consulting services, and "[a] written declaration of intent not to honor the October 24, 1998 agreement for BEST 2000, specific to Mr. Armstrong's statement, 'there is no need to pay $ 49,764.55' for BEST 2000 and the research product therein, based on the amount paragraph # 3 of the agreement." International Rectifier did not respond to the memorandum or the letter.

Sometime in the latter part of February or the early part of March 1999, Dale Prouty, a business associate [*13] of Sneed's who was contemplating a possible joint venture involving Best 2000, met with Armstrong to discuss International Rectifier's Y2K remediation efforts and its experience with Best 2000. After the meeting, Prouty informed Sneed that International Rectifier had implemented Best 2000 at six or seven of International Rectifier's international facilities. International Rectifier in fact completed Y2K remediation of its Singapore facility in May 1998, its Shanghai facility in July 1998, Italy in October 1998, Germany in November 1998, El Segundo in February 1999, Japan in March 1999, and Great Britain in May 1999. After receiving this information from Prouty, Sneed sent International Rectifier additional invoices for use of Best 2000 at facilities other than Singapore.

In early April 1999, Sneed telephoned International Rectifier's CEO Lido about payment for Best 2000. During the call, Lido seemed surprised that Sneed had not been paid and told Sneed that he would take care of the matter the following day. After the call, Sneed believed he would be paid and sent an additional invoice to International Rectifier for use of Best 2000 at seven facilities. That invoice was not paid. [*14] Sneed subsequently received a letter from Armstrong dated April 14, 1999 claiming that International Rectifier had used only publicly available information to remediate its Y2K problems and refusing to pay for Best 2000 [3]. That letter was the first written communication Sneed received from International Rectifier stating its refusal to pay under the parties' agreement. In response to a request by Sneed for copies of the publicly available information International Rectifier purportedly used, International Rectifier sent certain documents to Sneed in August 1999. These documents included a working paper copyrighted by Don Estes in 1997 entitled "Encapsulation Solutions for Year 2000 Compliance."

> 3   The April 14, 1999 letter is not part of the record on appeal.

F. The Instant Action

On February 18, 2003, Sneed commenced the instant action for breach of contract and common count of money due and owing. The matter proceeded to a court

trial. At the conclusion of the trial, on March 30, 2004, the trial [*15] court ruled in favor of Sneed, finding that International Rectifier had used Best 2000 and that Sneed was entitled to payment under the parties' agreement for use of Best 2000 at three of International Rectifier's facilities, including the El Segundo facility, but that the statute of limitations had run as to the other four facilities. International Rectifier requested a statement of decision, and the parties thereafter filed separate proposed statements of decision. International Rectifier filed objections to Sneed's proposed statement of decision. On April 21, 2004, the trial court signed Sneed's proposed statement of decision, with a one sentence modification. Judgment was entered on April 21, 2004, awarding Sneed a total of $ 204,372.48.

International Rectifier filed a motion for a new trial, which the trial court denied. Sneed filed the instant appeal, and International Rectifier filed a cross-appeal. Sneed's appeal was subsequently dismissed, so that the only issues on appeal are those raised by International Rectifier.

DISCUSSION

A. Contract Interpretation

The parties agree that under the terms of the agreement, International Rectifier was obligated to pay [*16] Sneed $ 49,764.55 for use of the "Invention." They disagree, however, as to the meaning of the term "Invention" and a related term, "information," as used in the first two paragraphs of the agreement.

Paragraph one of the agreement provides: "International Rectifier Corporation ('Disclosee' herein) hereby agrees that all information disclosed in briefings, documents, drawings, and otherwise received from Anthony Sneed ('Discloser' or 'Inventor' herein) . . . relating to his cross-platform invention that solves the year 2000 date malfunction called *BEST 2000* ('Invention' herein) is received in confidence, and is clearly marked as being confidential, and is the proprietary right and property of Discloser, and that such information will not intentionally be reproduced, transmitted or disclosed to any other organization or entity, or to any person not in the employ of the Disclosee, or used by Disclosee, without the express prior written consent of Discloser. If the information is orally or visually disclosed, it shall be identified as being confidential in writing submitted to the Disclosee within thirty (30) days after such oral or visual disclosure. Review of disclosed material [*17] for the purpose of evaluating Invention's usefulness to Disclosee does not constitute 'use' and does not justify or require payment of $ 49,764.55 to Discloser."

Paragraph two states: The foregoing shall not apply to information which: [P] 1. Can be documented as

Ex1 106

Page 5

2005 Cal. App. Unpub. LEXIS 6314, *

within the prior knowledge of Disclosee prior to receipt of the information disclosed by Discloser; [P] 2. Was obtainable by Disclosee from sources other than Discloser who themselves obtained it from sources other than Discloser; [P] 3. Was within or becomes part of the general public knowledge through no fault of Disclosee as of the time Disclosee makes independent use of the information; [P] 4. Is provided by Discloser to a third party without restriction on disclosure; [P] 5. Is rightfully provided to the Disclosee by a third party without restriction on disclosure; or [P] 6. Is independently developed by the Disclosee."

International Rectifier contends that paragraph two of the agreement excludes from the definition of "Invention," six enumerated categories of information, including publicly available information and information disclosed without restriction by Sneed to third parties. International Rectifier [*18] claims that the exclusionary language in paragraph two was intended to cover all "information" discussed in paragraph one and that the trial court erred by construing the term "Invention" to include information previously known by International Rectifier or disclosed by Sneed to third parties without restriction on disclosure.

Sneed, on the other hand, argues that the contract defines "Invention" to include all "information," whether or not otherwise available, relating to Best 2000 and provided by him. Accordingly, any use of that information would also constitute use of the "Invention" and trigger a payment obligation under the agreement. Sneed further argues that the exclusionary language in paragraph 2 applies only to restrictions on disclosure and not restrictions on use.

In its statement of decision, the trial court made the following finding: "Extrinsic evidence regarding the interpretation of the Agreement was received by the Court, without actually admitting it, in order to determine whether the language of the Agreement was reasonably susceptible to the interpretations urged by Plaintiff and Defendants. The Court resolved the ambiguities of the Agreement in favor of the [*19] Plaintiff's interpretation, which was supported [sic] the weight of the evidence, including the extrinsic evidence admitted by the Court. Plaintiff's interpretation of the contract was most reasonable in light of the language of the Agreement and the extrinsic evidence admitted by the Court." The trial court did not specify, however, what "ambiguities" existed in the agreement, what extrinsic evidence it considered and admitted, or what its interpretation of the contract was.

We need not address any issue concerning interpretation of the parties' agreement, however, because that agreement, even as interpreted by International Rectifier, obligated it to pay for use of information disclosed by

Sneed that International Rectifier could not document as within its prior knowledge; was not part of the general public knowledge or obtainable from other sources; was not provided by Sneed to a third party without restriction on disclosure; and was not independently developed by International Rectifier. The trial court, in effect, found that International Rectifier used such information. As we discuss, substantial evidence supports that finding.

B. Breach of Contract

The trial court [*20] found that International Rectifier used information disclosed in Best 2000 that was not publicly available at the time and did not become publicly available until Sneed's patent was published in 2001. The trial court further found that Sneed did not publicly disclose Best 2000 to any third party without restriction.

We review the trial court's factual findings under the substantial evidence standard. (*Hambarian v. Superior Court (2002) 27 Cal.4th 826, 834.*) Under this standard, "the power of the appellate court begins and ends with a determination as to whether there is any substantial evidence in the record, contradicted or uncontradicted, that will support the finding." (*Associated Builders & Contractors, Inc. v. San Francisco Airports Com. (1999) 21 Cal.4th 352, 374.*) Substantial evidence means evidence that a reasonable trier of fact might accept as adequate to support a conclusion; evidence that is reasonable, credible, and of solid value. (*Guntert v. City of Stockton (1976) 55 Cal. App. 3d 131, 142, 126 Cal. Rptr. 690.*) Moreover, when the trial court's findings are set forth in a statement of decision, "any conflict in the evidence [*21] or reasonable inferences to be drawn from the facts will be resolved in support of the determination of the trial court decision." (*In re Marriage of Hoffmeister (1987) 191 Cal. App. 3d 351, 358, 236 Cal. Rptr. 543.*)

1. Use of Proprietary Information

There is substantial evidence to support the finding that International Rectifier used information disclosed by Sneed concerning Best 2000 that was not within the prior knowledge of International Rectifier, was not within the general public knowledge or obtainable from other sources, and was not independently developed by International Rectifier. That evidence consists of the following: Don Estes, the author of a publicly available working paper on Y2K solutions on which International Rectifier claims to have relied in solving its Y2K problems [1], testified as an expert at trial and concluded that International Rectifier used elements of Best 2000 that were not publicly available from other sources. Estes reviewed source code printouts provided by International Rectifier showing what it had done to solve its Y2K problems and discussed a detailed written analysis he had prepared

61 107

comparing Best 2000, publicly [*22] known solutions to the Y2K problem, and the method used by International Rectifier. The results of this analysis showed that the method used by International Rectifier was identical to Best 2000 with respect to nine distinguishing algorithm [5] elements. Estes further opined that International Rectifier's claim that it used only publicly available sources of information to implement its date-shifting remedy was implausible: "In summary, the Don Estes working paper describes 'what' but not 'how,' and TOCS [6] describes 'what' with a very distinctive 'how' that does not match with the IRC implementation. By contrast, BEST/2000 describes both 'what' and a very specific 'how' that matches the IRC implementation on 9 out of 9 measures. The IRC implementation implausibly claims to have taken the 'what' from the two sources in the public record, but it defies all probability that an identical, proprietary 'how' was implemented at the same time out of serendipity, particularly after having been tainted by in-depth exposure to the proprietary method."

4 International Rectifier refers in its brief to an-other Y2K solution that involves date-shifting, described in *United States patent number 5,600,836* issued to Harvey Alter and published on February 4, 1997 (the Alter patent, also known as the Turn of the Century or TOCS method); however, Niel Armstrong testified at trial that In-ternational Rectifier did not consider or rely upon any information contained in the Alter patent.

[*23]

5 An algorithm is defined as "[a] logical step-by-step procedure for solving a mathematical problem in a finite number of steps, often involv-ing repetition of the same basic operation" or "[a] logical sequence of steps for solving a problem, often written out as a flow chart, that can be translated into a computer program." (Encarta Dictionary: English (North America) (Microsoft 2005).)

6 As discussed, TOCS, also known as the Turn of the Century or the Alter patent, also discussed a 28-year date shift solution. Any discussion or analysis concerning TOCS is not relevant be-cause International Rectifier stated that it did not consider or rely upon TOCs when implementing its date-shifting remedy.

International Rectifier's claim that it relied on pub-licly available information in Estes' working paper is also contradicted by the testimony of its own technical staff. Christopher Casey, a manager at International Rectifier with responsibility for solving the company's Y2K prob-lems, testified that he did not recall seeing the Estes working paper at the time that International Rectifier was

[*24] exploring different options for solving its Y2K problems, but that he did recall seeing and discussing it after Sneed's lawsuit against International Rectifier had commenced.

Sneed also testified at trial, and said that at the time he developed Best 2000, the publicly available informa-tion on date-shifting discussed either data encapsulation or program encapsulation, but not both; and that Best 2000 was the first method to use a hybrid approach, in-corporating both data and program encapsulation. There was evidence that after receiving Best 2000 in November 1997 and obtaining further information about Best 2000 in subsequent telephone calls with Sneed, International Rectifier began implementing in December 1997 the same hybrid approach to solve its Y2K problems.

Although International Rectifier presented some conflicting evidence to support its claim that it did not use Best 2000, and that it had sufficient information and resources to implement its own date-shift solution, we must, under the applicable standard of review, resolve all conflicts in the evidence in favor of Sneed, and draw all legitimate and reasonable inferences to support the trial court's decision. (*In re Marriage of Okum (1987) 195 Cal. App. 3d 176, 181-182, 240 Cal. Rptr. 458;* [*25] *In re Marriage of Hoffmeister, supra, 191 Cal. App. 3d at p. 358.*) Substantial evidence supports the trial court's find-ings that International Rectifier used proprietary infor-mation disclosed in Best 2000.

2. Disclosure to Third Parties

International Rectifier argues that the terms of its agreement with Sneed allowed it to use Best 2000 with-out payment or restriction if Sneed disclosed Best 2000 to a third party without restriction on disclosure, and that Sneed's disclosure of Best 2000 to several third parties, including IBM, President Clinton, and Vice President Gore, relieved International Rectifier from any payment obligation. The trial court found, however, that there was no unrestricted "disclosure by Sneed of his BEST 2000 methodology that was available to the general public prior to the publication of his patent." Substantial evi-dence supports this finding. Sneed testified at trial that all disclosures of Best 2000 to third parties were made subject to the stipulation that the information was to be kept confidential, and that this stipulation was communi-cated in correspondence and in conversations he had with these third parties.

C. [*26] Adequacy of Statement of Decision

International Rectifier argues that the trial court failed to provide an adequate statement of decision ex-plaining the factual bases of its decision, as required by *Code of Civil Procedure section 632*. International Recti-fier contends that the statement of decision fails to identify the specific information in Best 2000 used by International Rectifier that was not publicly available, and that the court erred by refusing to amend the statement of decision after International Rectifier requested it to do so. We review this claim of error under an abuse of discretion standard. (*Hernandez v. City of Encinitas (1994) 28 Cal.App.4th 1048, 1077.*)

A statement of decision need not address every legal and factual issue raised by the parties. It must only state the grounds upon which the judgment rests, and need not specify the particular evidence considered by the trial court in reaching its decision. (*Haight v. Handweiler (1988) 199 Cal. App. 3d 85, 89-90, 244 Cal. Rptr. 488.*) "[A] trial court rendering a statement of decision under ... *section 632* is required to state only ultimate rather [*27] than evidentiary facts because findings of ultimate facts necessarily include findings on all intermediate evidentiary facts necessary to sustain them. [Citation.]" (*In re Cheryl E. (1984) 161 Cal. App. 3d 587, 599, 207 Cal. Rptr. 728.*)

The statement of decision rendered in the instant case includes the necessary findings of ultimate fact that the publicly available information at the relevant time did not show how to implement the Best 2000 encryption methodology; that International Rectifier did not have adequate information on how to implement a date-shift solution to the Y2K problem until it received the Best 2000 documentation; and that International Rectifier used Best 2000 at its various production facilities. The trial court did not abuse its discretion by not including International Rectifier's requested findings.

D. Statute of Limitations

International Rectifier claims the judgment must be reduced because the statute of limitations bars Sneed's breach of contract claim with respect to the El Segundo facility. The trial court found that International Rectifier's breach in connection with the El Segundo facility occurred after February 18, 1999. The [*28] trial court further found that Sneed did not learn of the El Segundo breach until after that date. Sneed filed his complaint on February 18, 2003.

Sneed argues that the matter should be remanded to the trial court with instructions that the limitations period did not commence running until April 14, 1999 and awarding him breach of contract damages for use of Best 2000 at six other International Rectifier facilities. Sneed's argument is not appropriate, however, as he elected not to appeal the trial court's findings in this regard.

Sneed's cause of action for breach of contract is governed by *Code of Civil Procedure section 337*, which establishes a four year limitation period for "an action upon any contract, obligation or liability founded upon

an instrument in writing." "A limitation period does not begin until a cause of action accrues, i.e., all essential elements are present and a claim becomes legally actionable." (*Glue-Fold, Inc. v. Slautterback Corp. (2000) 82 Cal.App.4th 1018, 1029 (Glue-Fold).*) "The cause of action for breach of contract ordinarily accrues at the time of *breach,* and the statute begins to run at that time [*29] regardless of whether any damage is apparent or whether the injured party is aware of his right to sue." (3 Witkin, Cal. Procedure (4th ed. 1996) Actions, § 486, p. 611.)

The delayed discovery doctrine modifies the general principles governing accrual of a cause of action. It was "developed to mitigate the harsh results produced by strict definitions of accrual," and "postpones accrual until a plaintiff discovers or has reason to discover the cause of action." (*Glue-Fold, supra, 82 Cal.App.4th at p. 1029.*) The delayed discovery doctrine may be applied to certain breach of contract causes of action. (*April Enterprises, Inc. v. KTTV (1983) 147 Cal. App. 3d 805, 832, 195 Cal. Rptr. 421 (April);* 3 Witkin, Cal. Procedure, *supra,* Actions, § 493, p. 624.)

"A common thread seems to run through all the types of actions where courts have applied the discovery rule. The injury or the act causing the injury, or both, have been difficult for the plaintiff to detect. In most instances, in fact, the defendant has been in a far superior position to comprehend the act and the injury. And in many, the defendant has had reason [*30] to believe the plaintiff remained ignorant he had been wronged. Thus, there is an underlying notion that plaintiffs should not suffer where circumstances prevent them from knowing they have been harmed. And often this is accompanied by the corollary notion that defendants should not be allowed to knowingly profit from their injuree's ignorance." (*April, supra, 147 Cal. App. 3d at p. 831.*) "Applying the discovery rule to unusual breach of contract actions poses no more burden for the courts than the date-of-injury rule does. The discovery rule contains the procedural safeguard of placing the burden on the plaintiff to plead and prove that the delayed discovery was justified." (3 Witkin, Cal. Procedure, *supra,* Actions, § 493, p. 624; *Glue-Fold, supra, 82 Cal.App.4th at p. 1030; April, supra, 147 Cal. App. 3d at p. 833.*) Whether the plaintiff has successfully met that burden is a question of fact (*April, supra, 147 Cal. App. 3d at p. 833*), that we review for substantial evidence.

Here, there were circumstances justifying application of the delayed discovery doctrine. International Rectifier knew that it was implementing [*31] the date-shifting remedy disclosed by Sneed at each of its international facilities and it controlled the timetable for that implementation. Sneed, on the other hand, knew only about International Rectifier's use of Best 2000 at its

a *7ʰ-Gift* page

Singapore facility, and he would not have known about such use had International Rectifier not disclosed this fact to him in January 1998. The terms of the parties' agreement required International Rectifier to obtain Sneed's prior written consent before using Best 2000 at any of its sites. International Rectifier did not, however, obtain Sneed's consent with respect to any facility other than Singapore. Sneed accordingly remained ignorant of International Rectifier's breaches with respect to the other facilities, including El Segundo. Thus, the breach was "difficult to detect," defendant was in "a far superior position to comprehend the act and the injury," and defendant had to have known that "plaintiff remained ignorant he had been wronged." (3 Witkin, Cal. Procedure, *supra,* Actions, § 493, p. 624.)

There was also reasonable, credible evidence to support the trial court's finding that Sneed did not know or have reason to know that International [*32] Rectifier used Best 2000 at its El Segundo facility until after February 18, 1999. Sneed testified that he learned about International Rectifier's use of Best 2000 at facilities other than the Singapore facility when Dale Prouty informed him of such use after meeting with Armstrong in February 1999. Prouty testified that he met with Armstrong "in the second half of February or the first week of March" 1999.

International Rectifier contends that there is no substantial evidence that the meeting occurred after February 18, 1999 because Prouty's own testimony concerning the date of this meeting is conflicting. Prouty did not say that he could not recall the exact date that he met with Armstrong, and admitted that the meeting could have occurred as early as February 10, 1999. International Recti-

fier also argues that in light of the conflicting evidence, Sneed did not sustain his burden of proving that the delayed discovery doctrine applies, and the trial court's finding must therefore be reversed. This argument confuses burden of proof, which is determined by the trial court, with the applicable standard of appellate review. Under the applicable standard, our analysis "'begins and ends with [*33] a determination as to whether there is *any* substantial evidence to support [the factual findings]; [we have] no power to judge of the effect or value of the evidence, to weigh the evidence, to consider the credibility of the witnesses, or to resolve conflicts in the evidence or *in the reasonable inferences that may be drawn therefrom.*' [Citation.]" (*Orange County Employees Assn. v. County of Orange* (1988) 205 Cal. App. 3d 1289, 1293, 253 Cal. Rptr. 584, original italics.) There was substantial evidence to support the trial court's finding that Sneed did not learn of International Rectifier's breach with respect to the El Segundo facility until after February 18, 1999. Sneed's breach of contract claim with respect to that facility accordingly is not barred by the statute of limitations.

DISPOSITION

The judgment is affirmed. Sneed is awarded his costs on appeal.

MOSK, J.

We concur.

ARMSTRONG, Acting P.J.

KRIEGLER, J.

A SERIAL ENCRYPTION SYSTEM-BYPASS
OF
YEAR 2000 DATE MALFUNCTIONS

Inventor: Anthony Sneed, 2058 North Mills Avenue, Claremont, CA 91711

Application: David O'Reilly, Patent Attorney, 1800 Bridgegate St. Ste 200, Westlake, CA 91361

Filed After: d7/April 19th, 1997

Abstract

The invention is an encryption system that regresses internal date representations used in stored-program computers and their associated databases. The invention removes the Year 2000 Malfunction from large-scale, and generally used computer systems without restructuring or expanding the field widths for storing date information.

The arithmetic encryption and re-engineering tools of the invention encode both databases and programs, using a unique interface to synchronize information exchanges between peripheral devices and the central processor in a computing complex. A series of two encoder and two decoder protocols are applied to both databases and programs to encrypt the internal date representations of a computing system while maintaining synchrony with extant calendar precedence for computation, external transfer, display or printing of date-sensitive information.

Ex 1   111

1.0 Field of the Invention

The invention herein is specific to the Year 2000 Malfunction on large-scale, and widely used computer systems. This particular malfunction is isolated to systems programmed to use two digits to represent years in calendar dates. For example, such systems abbreviate "1955" to "55." Because of this, calculations beyond the year 1999 cause system malfunctions from programs that maintain a two-digit precedence to represent years throughout their programs and database files.

In such systems, "00" represents the year "2000" since only two digits are used to record the year in a calendar date. Computations that use a value of "00" to denote the year 2000 violate the chronological sequence of actual calendar dates (..., 97, 98, 99, 00, 01,...). The truncation, therefore, makes "2000" chronologically less than the two-digit representations of "1901" to "1999," which distorts any time relationships for events, calculations or stored dates projecting beyond 1999.

The two-digit year precedence helped save space and reduce costs when converting to newer storage technology at mid-century. Such technology enabled database files to be accessed randomly instead of only sequentially. Current solutions to retro-actively address the Year 2000 Malfunction involve restructuring such databases, and rewriting support programs to accommodate full four-digit representations of calendar years instead of two-digit abbreviations.

This re-engineering task involves *multiple billions* of lines of code, and *multiple trillions* of bytes of data spread throughout government, academic and private sector databases. All of this is running on a host of different computing platforms built during the last 40 years, some of which may no longer be commercially produced, if the manufacturers remain in business at all. Also, due to age, some of the source code for programs in many production systems may no longer be available. Conventional tools to address the Year 2000 Malfunction may not produce useful results sooner than 18 to 36 months. In many cases, this may make a plethora of systems that electronically handle multiple trillions of dollars of resources--including airline reservations, banking transactions, accounting systems, utilities, government tax collections, social services, and others--inoperable as the frequency of Year 2000 Malfunctions increases over the next 36 months.

The invention within this claim is an encryption system for any manner of computing hardware and storage technology subject to the Year 2000 Malfunction. It achieves this without altering the two-digit year precedence of existing databases, and with only minor parametric changes to software supporting such databases. The simplicity of this encryption and re-engineering invention allows the Year 2000 Malfunction to be addressed in a matter of weeks, instead of months or years for most systems. This may permit a large number of organizations to obviate the malfunction in as little as seven weeks, and no more than seven months for even the most complex systems. The invention enables each organization to address the malfunction with existing staff, avoiding the expense and uncertainty of trying to find and hire specialized

EX 1 112

consultants as the supply of such personnel continues to shrink. As such, this invention helps increase the useful life by 20 to 80 years of already-purchased systems through in-place staff.

When applied, the invention herein is a direct replacement for additional hardware storage devices and additional central computers that may otherwise be needed and purchased in mass to allow large-scale information systems to be modified and operate beyond the year 1999. In this sense, the encryption and re-engineering invention acts as a direct substitute for tangible and costly hardware-intensive alternatives now being considered to avoid the Year 2000 Malfunction.

2.0 Background of the Invention

Since 1900, and for census purposes shortly before, the year portion of calendar dates has been truncated to two digits ("55" instead of "1955," for example). The precedence helped maximize the amount of information encoded on 80-column punch cards in use throughout the first half of the 20th century.

Such cards were indispensable for the decennial census (1890, 1900, 1910, 1920, 1930, 1940), significantly speeding census counts and post-census count analysis. Without such punch card technology, according to the U.S. government, the 1890 census would not have been completed before the 1900 census began. The machines that processed the cards were called tabulators, later having plug-in boards that could be wired to perform various computational and printed report functions. The same technology gained wide-scale commercial acceptance in business (in addition to census applications) after 1913, maintaining the same two-digit census precedence for the year portion of dates.

By the 1950's, electronic computers began replacing tabulators to support census and commercial data processing. Data storage advances included magnetic tape (herein "mag tape") and electromagnetic disk drives (herein "disk drives"). Vast amounts of data originally stored on punch cards were transferred to the new media of mag tape and disk drives. These new storage devices added the ability to randomly and electronically access files through stored program machines, a dramatic improvement over the manual handling of millions of punch cards. The new storage technology also used less physical space to store data than punch cards, as well as obviating much of the labor expense to manually process punch-card-based data files.

The two-digit precedence for year portions of calendar dates (instead of four digits) was also used with the newer electronic computers and their storage devices. This simplified the transfer of volumes of punch card data while avoiding additional expense to store four-digit year fields. In doing so, the punch card images were transferred unaltered to the new storage devices, speeding conversion and removing up to an additional 2.5% storage-expense per record.

To have done otherwise, such as converting to four-digit years, may have inflated the already high cost of newer storage offerings as compared to punch card technology and its media. No manufacturer desired to include in a product proposal a 2.5% higher cost, which might give competitors an instant 2.5% price advantage who stayed with two-digit years already on punch cards. Thus, the remedy (of staying with two-digit years) was technologically expedient and helped reduce a price objection by customers, making the storage devices *appear* as seamless extensions of proven punch card protocols. This expediency, and its widespread perpetuation during the last four decades, helped set the worldwide conditions for the Year 2000 Malfunction.

None of the computing technology advances since then have reversed the long-term effects of maintaining the two-digit year precedence. Though the cost of disk drive storage has dropped dramatically in the last few decades, the reprogramming expense to reverse-out the two-digit date limitation has reciprocally increased each year. New functions were incrementally added to existing programs decade after decade. Such software changes were stimulated by regulatory requirements, new tax reporting and accounting methods, corporate acquisitions, new telecommunications links between multiple computer sites, new hardware and other reasons.

With each and every change, the two-digit year precedence spread like yeast within a batch of dough. Every year, it seemed easier to justify avoiding radical changes to production software that had taken years to develop and stabilize. Leaving software the same as it was the year before, even if the hardware was changed, seemed operationally and fiscally prudent. In numerous cases, this reasoning was followed until the Year 2000 Malfunction began to tangibly manifest itself in economically damaging ways over the past 18 to 24 months.

3.0 Brief Description of the Drawings (Invention Components in Italics)

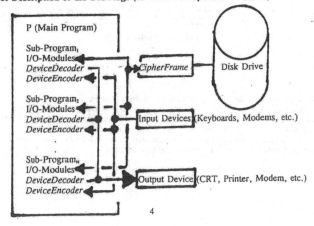

4

3.1 *CipherFrame* Definition Across Actual Calendar Time

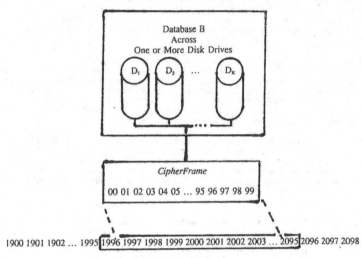

1900 1901 1902 ... 1995 1996 1997 1998 1999 2000 2001 2002 2003 ... 2095 2096 2097 2098

(Actual Calendar Time)

Diagram 3.1

The *CipherFrame* is the interface between encoded dates of database B stored on disk drives D_1, D_2,..., D_K (which may be a single drive), and internal computations done within the central processor by any program P. It permits all year-specific fields on database B to remain two-digit, arithmetically sequential symbols (but no longer *actual* dates after encoding). It is only when a specific symbol in the *CipherFrame* is decoded, as through the *DeviceDecoder* in diagram 3.4, that the symbol gains actual calendar and time meaning.

5

exl 115

a *7ʰ-Gift* page

3.2 DiskEncoder

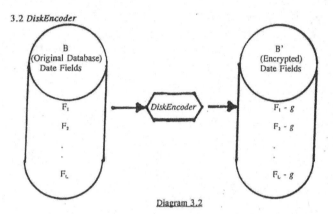

Diagram 3.2

The *DiskEncoder* takes all or part of the data image of any database B on a disk drive (D) and encrypts all year-specific fields through the *CipherFrame* that contains 100 two-digit symbols ("00" to "99"). The *DiskEncoder* allows the *CipherFrame* to be recursively encrypted, if needed, in subsequent years, as proscribed by a value g, an integer multiple of four selected for the *DiskEncoder*. If $g=0$, the *CipherFrame* will encompass calendar years 1900 to 1999, the default for virtually all systems, based on the historical 2-digit year precedence used during the 20th century (see "2.0 Background of Invention"). Table 3.2.1 shows a range of g values and their encoding effect through the *CipherFrame*.

Table 3.2.1

g Value	Database Encoded Years through *CipherFrame*
0	1900-1999
4	1904-2003
.	.
.	.
92	1992-2091
96	1996-2095
etc.	

6

<u>This makes *g* the encryption key for the entire system.</u>

The *DiskEncoder* subtracts *g* from every two-digit year-specific field of database B on the disk drive (or drives). This encodes database B through the *CipherFrame*, making it the encoded database B'. A corresponding encryption through the *ProcessorEncoder* (diagram 3.3) ensures that any program P that alters or uses database B' in any way whatsoever recognizes and correctly interfaces to database B through the *CipherFrame*.

The year 2000 is a regular leap year (unlike the years 1900 and 2100, since these centennial years do not have a February 29th), based on the quadricennial adjustment initiated in 1582 for most of Europe (1752 for England and its possessions) to correct an error in Julius Caesar's reckoning of the length of a year in 46 B.C. The fact that *g* is an <u>integer factor of four</u> ensures consistency with quadrennial (leap) years that a *CipherFrame* may span. With 2000 being a regular leap year, the *CipherFrame* may transparently encompass it as well.

Note: If *y* contains an encrypted symbol for a year within the *CipherFrame* and *g*=20, then

$$y + g + 1900$$

is the full four-digit calendar year encoded within *y*.

Finally, as the *DiskEncoder* detects year dates in B' that can not be encoded into the *CipherFrame*, the *DiskEncoder* produces a series of year-specific sub-databases, $B_1,...,B_r$, for each year that is outside the *CipherFrame*. For example, if *g*=20, then sub-databases $B_{1919},...,B_{1880}$ would contain all records outside of the *CipherFrame* if B spanned a period from 1880 to 2019. Years 1920 to 2019 would be encoded within the *CipherFrame*, and records with dates in this range would remain encoded within B'.

According to U.S. census data as of 1994, less than 5% of all demographic data lies outside of the *CipherFrame* spanning 1920 to 2019. On the U.S. Social Security database, individuals born before 1920 are 78 years and older. If all such persons fit a uniform pattern of benefit payments, then $P_{1919},...,P_{1880}$ may be identical to each other to the point that they may collapse into a single program, $P_{1919-1880}$. This pattern may be consistent across many private and public sector databases. Again, B' would contain the most volatile demographics (77 years and year), which represents 95% of the entire database population—and it would all be in the *CipherFrame*.

With extant policy regarding each demographic group, a 100 year *CipherFrame* may be more than sufficient if the expense of all other Year 2000 Malfunction alternatives become prohibitive or untimely.

3.3 ProcessorEncoder

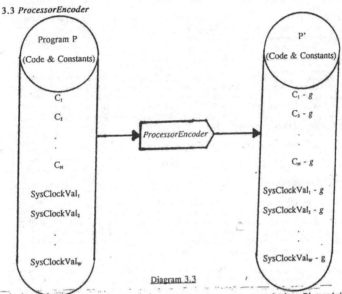

Diagram 3.3

The *ProcessorEncoder* modifies any program P that updates or uses any database B' encoded through the *DiskEncoder* (Diagram 3.2). The modified program, P', becomes chronologically aligned with any database B' through the *CipherFrame* interface for any date-related computation or any data retrieval of a date-related field. The *ProcessorEncoder* takes all chronological constants in P and subtracts g, where g is the same integer multiple of four used with the *DiskEncoder* for database B'. It also subtracts g from the year portion of all system clock values ("SysClockVal's") resulting from system clock calls ("SysClockCalls"). A system clock call provides the year, month, day, and hours, minutes and seconds that a computer maintains internally. g is subtracted from the year portion of every $SysClockVal_i$ before it is transferred to any variable within a program or used for any other type of computation within program P.

8

3.4 *DeviceDecoder*

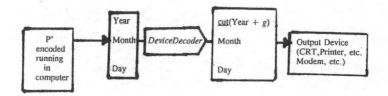

Diagram 3.4

Note: <u>cut</u>(*value*) truncates *value* to its two right-most digits; <u>cut</u>(112) = 12

The *DeviceDecoder* converts any encoded two-digit symbol in the *CipherFrame* into its two-digit year equivalent. This is done by adding *g* to the two-digit encoded symbol and truncating the sum to two digits. The *ProcessorEncoder* and *DiskEncoder* align programs and data through the *CipherFrame* interface, maintaining the two-digit year precedence and avoiding a conversion to four digit year representations.

The *DeviceDecoder* is required before any encoded date may be transmitted for printing, displaying or interchanging with any other device.

9

3.5 *DeviceEncoder*

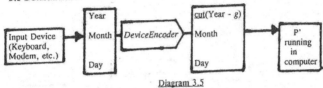

Diagram 3.5

The *DeviceEncoder* takes any un-encoded four-digit year value received from a keyboard or other device, and encodes it through the *CipherFrame*. The encoded value may then be used within any program P' modified by the *ProcessorEncoder*, and any associated database B' encrypted with the *DiskEncoder*.

After subtracting g, the *DeviceEncoder* then decides if the four-digit year can be encoded into a two-digit symbol within the *CipherFrame*, or if it represents a year outside of the *CipherFrame*. For example, if $g = 20$ then the *DeviceEncoder* logic is:

```
If (Year - g) > 1899 then Year := cut(Year - g) else
begin
  Case Year of begin
    1919: P₁₉₁₉
    1918: P₁₉₁₈
    1917: P₁₉₁₇
          .
          .
          .
    1882: P₁₈₈₂      (if needed)
    1881: P₁₈₈₁      (if needed)
    1880: P₁₈₈₀      (if needed)
  end
end
```

Each program $P_{1919}, \ldots, P_{1880}$ is a clone of P' with fixed year constants encoded into each clone program for its particular year. When a year outside of the *CipherFrame* is detected by the *DeviceEncoder*, it invokes the year-specific P_r to handle whatever data is received with the year. Otherwise, the regular program path within P' handles the data entered along with the year. Each P_r has its own database B_r which is partitioned from B' by the *DiskEncoder* based on each particular year outside of the *CipherFrame*.

Years outside of the *CipherFrame* represent at most 5% of all demographic information in a database. This is based on government census information as of 1994.

10

4.0 Detailed Description of the Preferred Embodiment

The serial encryption system is for the two-digit year precedence used in *multiple billions* of lines of code and *multiple trillions* of bytes of databases developed over the last four decades through a host of computer languages, hardware platforms and storage technologies. The invention permits the rapid re-engineering and proto-typing of large-scale and generally used computer systems such that their operation and date integrity is unaffected by the Year 2000 Malfunction. The malfunction is triggered by the two-digit year precedence within extant programs and data structures.

As such, the encryption system, herein, displaces processor hardware and additional storage that might otherwise be required to convert to a four-digit year protocol throughout multiple billions of lines of code and multiple trillions of bytes stored on large-scale and generally used computer systems.

The central component of this invention is the *CipherFrame* which synchronizes the interchange of data between a processor, its storage and other connected devices, while removing the Year 2000 Malfunction. As such, the *CipherFrame* and its 100 symbols ("00" to "99") are detached from any specific group of years. Through related decoders and encoders, the *CipherFrame* may be serially aligned across multiple 100 year periods, including any period that may trigger the Year 2000 Malfunction. An integer multiple of four, called g, arithmetically encrypts the *CipherFrame* to represent a specific period, as needed, through the *DiskEncoder* and *ProcessorEncoder*.

Dates Outside of the *CipherFrame*

For example, if $g=20$, the *CipherFrame* encompasses the years 1920 to 2019 (represented by the *CipherFrame* symbols "00" to "99"). If there are records with dates on the main database, B_{MAIN}, that precede 1920 (representing information 78 years or older relative to 1997), then these records are assigned to sub-databases, $B_{1919},...,B_{1900}$. A worst case involves setting $g=20$ for a very large database, like the U.S. Social Security database. For this database, 95% of the database in B_{MAIN} would contain the records of all people 77 years and younger, and only 5% would be spread across $B_{1919},...,B_{1880}$ (1880 is the lower limit because no known person is over 117 years old).

In this example, $B_{1919},...,B_{1880}$ are handled by $P_{1919},...,P_{1880}$. $P_{1919},...,P_{1880}$ are identical to P_{MAIN} (the main program) except that each has fixed date constants programmed into them, unique to their sub-databases' year-specific contents. P_{MAIN}, along with $P_{1919},...,P_{1880}$, are treated as a single object since 95% to 99% of their code is identical. This part of the encryption architecture supports the few years outside of the *CipherFrame* and is called <u>*Chronological Encapsulation*</u>, using earlier mentioned chronologically specific programs with each time-segmented portion of the main database, B_{MAIN}.

11

EX 1   121

a *7ʰ-Gift* page

The Social Security database is a worst case example, since most databases do not encompass a 117-year span. Yet even with this, for one of the most complex databases in the world, over 95% of the entire database is within the *CipherFrame*, with *Chronological Encapsulation* accommodating the remaining 5%.

Re-engineering Efficiency of the Invention

The invention reduces the time to re-engineer virtually any large-scale, generally used computer system and its database, compared to extant methods that require two-digit to four-digit conversions of year representations. Restructuring databases holding multiple trillions of bytes of data creates a very dramatic and costly change in the multiple billions of lines of code supporting such databases. These multiple billions of lines of code have evolved over a four-decade period, across a range of computer architectures and manufacturers, not all of which are still in business.

If the multiple billions of lines of code are looked at as a single software object, then the two-digit to four-digit conversion may be prohibitive to complete, in aggregate, within the next 12 to 36 months. Even if accurate source code were available (which it may not always be for a significant percentage of code in both the private and public sectors), there may remain a shortage of trained programmers and consultants with sufficient expertise to handle the size of the task within available budgets and time.

The *CipherFrame* simplifies the task by focusing only on an arithmetic encryption that preserves existing software and hardware investments. The *DiskEncoder* can readily encrypt virtually any type of database overnight, without changing the record architecture within a target database. By leaving data structures intact, this helps obviate the huge expense of reprogramming multiple billions of lines of program code.

Output routines are usually few in number and can be readily identified by their standard system calls. Furthermore, the output images of virtually every printed report are available from active production systems, which provides a well-documented road-map of how each output segment within a program works. The *ProcessorEncoder*, that converts a program P to P' and aligns it with the *CipherFrame*, is even simpler. It takes each two-digit, year constant (used for comparison or computational purposes as fixed values) within P and subtracts *g* (the integer multiple of four used in all decoders and encoders of the encryption system).

Even for the most massively complex programs, involving hundreds of thousands of lines of code, there are fewer than several dozen year-constants pre-programmed for computational and comparison purposes. In source code, such constants may be readily found by doing routine text searches for "00" to "99" values. If only executable code is available, a reverse compiler still forces the dates to standout through identified instructions with constant fields. In either event, the task may be trivial compared to tracing every usage of a two-digit field that needs to be expanded to four digits across multiple billions of lines of code.

12

The Integrity and Fault-tolerance of the Invention

The invention herein permits serial encryption of an entire computer processor and time-sensitive data on its storage devices to a known state that existed g years ago, where g is the integer multiple of four used as the encryption key for the *ProcessorEncoder*, *DiskEncoder*, *DeviceDecoder* and *DeviceEncoder*. The regressed state of a program P is designated P', created by the *ProcessorEncoder* design through manual or automatic means.

As such, the dates outputed from an encrypted computer system appear externally consistent with "today's" date through this invention. Yet, internally the machine thinks it's "1977," if $g = 20$ for example. The invention maintains all relative time lengths (between internally encrypted dates) which ensures the chronological integrity of subsequent computations for any year encoded through the *CipherFrame*.

The *DeviceDecoder* converts encrypted-symbols to years before output or transfer to other devices or systems. However, if for some reason a single output module or section of code does not decode an encrypted symbol before output, then the fail-safe mode of the encryption system only results in a "regressed" year being printed or displayed. A regress-year-dated check of more than a year or two, for example, is usually non-negotiable. This appears to be an advantage the encryption system has over other methods that require database restructuring and reprogramming massive amounts of code.

The encryption system only requires minor changes to those few portions of the system that generate time-based values (a small fraction of most software programs, as compared to every instance within a large program that must be altered to accommodate new, four-digit year fields). Encryption only introduces incremental changes to specific parts of a computer's software, as opposed to altering the entire infrastructure of databases and programs. Encryption helps avoid many of the risks inherent in trying massive conversions with limited budgets, tight deadlines and competing priorities of available staff.

In virtually every date-aware program, there is a routine that computes the length of time between dates, usually in days. So long as the base-date used in these calculations is below the lowest calendar date encrypted through the *CipherFrame*, all relative time lengths so computed will be chronologically consistent for internal, interim computations.

13

Claims

This invention is not to be limited by the embodiment shown in the drawings, or described herein, or given by way of example and not indicative of any limitation.

What is claimed is:

1. The *CipherFrame* design to synchronize computer, disk drives and peripherals through a stored-program computer, and with date-sensitive data stored on such a system, to mitigate the Year 2000 Malfunction and its detrimental effects on such a system.

2. The encryption that relocates the *CipherFrame* to encompass any period up to 100 years, extending the useful life of software and computer hardware that uses the invention to mitigate the effects of the Year 2000 Malfunction.

3. The re-engineering tools that implement and synchronize the *CipherFrame*, designated herein as the *ProcessorEncoder, DiskEncoder, DeviceDecoder* and *DeviceEncoder.*

4. The design according to claim 1, claim 2 and claim 3 that maintains chronological integrity of a computer and database without expanding two-digit fields used to represent years stored on a database, while mitigating the effects of the Year 2000 Malfunction.

5. The design to handle years outside of the *CipherFrame* using *Chronological Encapsulation* by replicating a main program, P_{MAIN}, into a series of date-specific clone programs, $P_1,...,P_T$, and their corresponding date-specific sub-databases, $B_1,...,B_T$, from a main database B_{MAIN}.

6. An interim buffer, called I_{MAIN}, that accepts database updates or other transactions through the main program, P_{MAIN}, for database records outside of the *CipherFrame*. The interim buffer, I_{MAIN}, is later sorted into individual files, $I_1,...I_T$, which are individually applied for updating sub-databases, $B_1,...,B_T$, through each sub-databases' date-specific program, $P_1,...,P_T$.

7. The serial encryption of the *CipherFrame*. The design permits an encryption key g where:

$$g = g_1 + g_2 + g_3 + ... + g_n$$

so that each g_i may be used to encrypt successive, *overlapping*, 100-year periods through the *CipherFrame*, which permits the invention to be used serially. This requires complete fidelity with the quadricennial (400-year) precedence in subsequent centennial years (2100, 2200, etc. which <u>do not</u> contain a quadrennial leap-day), as it is consistently reflected in the year 2000 (which <u>does</u> contain a quadrennial leap-day) for extant programs, systems and databases subject to the Year 2000 Malfunction.

14

FAX FAX FAX FAX FAX FAX FAX

FAX TO: 610-712-3774   8 pages (including this one)
FOR: Mr. Al Torressen, Manhattan College graduate, physics; Industry Intellectual Property Expert
FROM: Anthony Sneed (Sneed), ISB∴Institute Fellow, *Emeritus*   DATE: d2/February 13, 2012
RE: Gratitude for your candor and adherence to what is right, and just, and fair in all your communications

FINAL DRAFT

Happy 2nd-day, Al. Grace and peace to you and yours.

May this reach you and yours, in the reciprocating goodwill you have extended to others throughout your life. It is fitting and proper to help you recall me, as I imagine you've extended the same dignity and grace to many others as you have with me, so this review is in the third person. It may inform subsequent contact, if any, and perpetuate the goodwill you established not so long ago.

1. At age 10, Sneed's parents gave him a computer simulator called DIGICOMP as a gift. It used marbles for electrons, released from the top of its inclined plane of about a yard in length and a foot and a half wide, tilted about 25° along its length. The tilt provided "EMF," while ridges on the molded surface acted as "wires" to guide the "marble-electrons" through "circuit" paths. When a marble reached the bottom, it passed through a trigger-gate, which pulled a wire connected to a mechanism at the top of the plane, releasing another marble. Three "registers" populated the surface of the plane. The first was the "accumulator" with its seven plastic flip-flops that, when a marble encountered one, would swing left or right, "displaying" a one or zero, depending on the previous "state" of the flip-flop, then passing on to the next flip-flop, with other components above and below the plane working to simulate binary functionality. The second was the "memory" register, as a series of four "switches" acting as pre-set bits, aligning with the first four flip-flops in the accumulator, allowing addition or subtraction. A marble passed through a switch set to one, or dropped to a sub-plane board and came out of a tunnel and went through the trigger-gate, starting the next marble down the surface of the inclined plane. The third was a "multiplier/quotient" register, containing three flip-flops, allowing the memory register to be repeatedly added or subtracted to the accumulator, simulating multiplication or division.

[Also at age 10, Sneed attempted to read the entire Scripture, but couldn't since the one he found in his father's library was in Shakespearean English. However, the 31 chapters of Proverbs were less daunting, with their packet-based verses. During two 24-mile workouts in 2010, one in January and one in February, Sneed completed committing all 31 chapters to memory, with random chapter and verse recall, or by content to identify chapter and verse. The circuit he walked was three miles, eight laps making 24 miles. There seems to be similarity between an instruction-set of a computer to determine its architecture and Proverbs as an "instruction-set" for all of Scripture.]

Let it be noted that the IBM logo was in front of Sneed every day of elementary school, populated with IBM time-clocks in every classroom. This may have been a brilliant branding decision by Thomas Watson Sr.–Sneed spent time looking at those clocks as recess or end-of-school neared. At age 12, Sneed decided to work for the company whose precision was reflected in the synchronism and tracking of time with all those clocks throughout the school. Once that decision was made, the foundation of our future meeting was set, and it, too, would involve "time."

A couple years after mastering the simulator, Sneed bought RTL (Resistor-Transistor-Logic) NAND, NOR, inverter, and multi-flip-flop chips, made by Motorola. These IC's allowed Sneed to replicate the mechanical computer simulator and experiment with much higher clock speeds. The DIGICOMP manual described a language called FORTRAN as a parallel to DIGITRAN, used to configure problems on the simulator. Intrigued, Sneed learned FORTRAN at age 13 on a 360/40 at the Claremont Colleges. When Sneed entered boarding school in Carpinteria, California, as a freshman the following year, the math department allowed him to take computer programming courses for seniors. Bill Gates, who was also a freshman that year, used the same timesharing system at his school in Seattle as Sneed did in Carpinteria. During the summer, Sneed was accepted to Granite Computer Institute with a $2000 scholarship. Granite certified programmer/analysts on IBM systems architecture, operations, and programming in three languages, including assembly. Sneed's classmates were all adults. He went in on weekends and stayed all day, since the 360/30 was not used then, so he had the entire computing complex to himself. "Debbi" console commands were mastered, along with loading and unloading disk and tape drives, running assembly and other programs that made the 1403 spew out reports, IPLing, single stepping to watch programs execute from the front panel lights, and opening the 360 front panel and exploring its internals right down to its memory core. Also, familiarity was gained with unit record equipment, sorters, duplicators, and 400-series tabulators with their plugboards. Sneed completed Granite's summer course and returned to boarding school in the fall. In his junior year, Sneed was moved to notify his headmaster that Sneed would be attending the California *Institute* of Technology. Having enjoyed Granite Computer *Institute*, Sneed felt a school with *Institute* in its name would be a natural next step. Sneed was accepted to Caltech in November of his senior year at boarding school.

2. The digital systems knowledge gained from ages 10 to 14 allowed Sneed to ace Caltech's freshman digital design courses, including implementing a calendar with 31 IC's. By the following year, during the summer, Sneed became a Caltech/NASA/JPL intern in Dr. Richard Goldstein's group. Just 14 years earlier, in March of 1961, Dr. Goldstein became the first person, in history, to accurately determine the earth-sun distance, or 92,955,807 miles averaged over a year. Dr. Goldstein did so by accurately computing the earth-Venus distance with radar pulses. Kepler's third law permitted the earth-Sun distance to be derived from the earth-Venus distance since it's still hard to bounce pulses off the sun, which actually contains a "sun" spinning inside the sun, first discovered in 1989; the *tachocline* is the boundary between the two. The previous best estimate was 92,880,000 miles, based on the 1874 and 1882 Venus transits, after less than satisfying results from prior worldwide efforts with the transits of 1761 and 1769.

Sneed was assigned to develop the Very Long Base Interferometry correlator, in machine language, for measuring the distances between NASA's 64-meter, Deep Space Network antennas in Canberra Australia, Madrid Spain, and Goldstone California, using digitized noise wave fronts from a quasar. The machine to process the time-marked data was a Xerox Sigma V–a perfect clone of IBM's mainframe architecture, giving Sneed ready fluency in its machine language from certification with Granite Computer Institute course work five years before. The code developed became one of the top projects for Caltech's IBM systems architecture course, resulting in straight A's for all three quarters. In Sneed's junior year, he disassembled his HP25 programmable calculator, modified its 49 steps of memory with 1,000 steps of paging memory, using three dozen IC's, and redesigned its case to have a docking port. It was a successful digital lab project. For the fun of it, Sneed took an IC design course Dr. Carver Mead taught, became team leader, and designed the masks for a four-bit shift register, successfully implemented as a MOS IC.

Later, Sneed became a summer intern at Hughes (Boeing) Space and Communications Group and was assigned to the Pioneer/Venus project. Left to himself with no specific assignment, Sneed developed his own project to simulate the entire command sequence that controlled the spacecraft from launch to orbit insertion at Venus. Sneed found a single-point, catastrophic failure-mode that could have occurred in earth orbit, after launch. This gave Sneed standing at a NASA design review of the mission, to provide a reengineering solution and remedy the failure-mode. An offer was made to Sneed to be a full-time MTS, or member of the technical staff, after graduating from Caltech.

As an MTS, Sneed was asked to trouble-shoot a probe subsystem, a key part of the NASA Galileo mission to Jupiter. The probe was to be jettisoned from an orbiter near Jupiter, reach speeds of up to 106,000 mph, and conduct a science mission below the cloud-tops of Jupiter. The relay computer on the orbiter collected and relayed science data back to earth from the probe's seven science instruments. It required a new operating system and software to be designed in the microcode of the relay computer, along with optimization of a discrete Fourier transform routine (DFT) to maintain frequency- and phase-lock on the probe's 1.4 GHZ signal affected by a 15 hz/second Doppler. Additionally, the operating system had to coordinate with other computers on the orbiter and prepare science and system maintenance data packets for down-link to the Deep Space Network on earth, and anticipate scheduled "brown-outs" due to power limitations across all subsystems on the orbiter. There was no solar power due to the distance from the sun, so radioisotope thermal electric generators were used on the orbiter.

Sneed was assigned to bring the project back on schedule and address cost overruns. So, on his own, he developed a 4,000 line simulator, in four weeks, that included modeling the RF/IF, numerical controlled oscillator, and analog-to-digital circuits receiving the probe's signal, the 7 million state instruction-set of the on-board computer, and the digital interface to other computers on the orbiter. This cut some $250,000 from the project budget, eliminating one of three hardware simulators, and advanced the microcode development schedule by six months. The Sneed Simulator allowed microcode engineers to collaborate with communication system analysts on the same computer. Sneed received the highest salary increase of second year MTS's. This led to interview and hiring as an assistant professor in the California State University system. With 90 students and a project lab, Sneed received the highest department teaching quality feedback report. From that assistant professorship, Sneed was interviewed by IBM and later accepted a position offered by IBM's Data Processing Division in the finance industry. Sneed's hiring manager was a fellow Caltech alum, who had also been a Caltech/NASA/JPL summer intern, seven years before Sneed, and he, too, worked at Hughes (Boeing). Much later, he became an IBM E-5 and runner-up for IBM CEO.

3. Though beyond any job description of an IBM marketing trainee, a few months after starting work at IBM Sneed was asked to look at a technical problem that stumped account SE's (System Engineers) for some six months. It involved designing a workflow system that interfaced a bank's mainframe database with a branch-based office system, through a telecommunications access method. After analyzing the problem, Sneed developed the 1,000 lines of needed assembly code in about two weeks. Though directed by *marketing management* to trouble-shoot the problem, an SE manager accused Sneed of "violating" IBM policy, since a marketing trainee wasn't supposed to do what, apparently, had been successfully done. Since "policy" had been "violated," credit went to the account SE's for the effort. This was Sneed's introduction to the world of IBM's belief and merit system. That little project had impacted Sneed's four-week branch office assignments and preparation for Sneed's second IBM training class in Dallas. So, Sneed finished the class project on the three-hour flight to Dallas and completed the two week class.

During the next between-classes branch office period, Sneed was asked to look at a different problem which had also stumped account SE's for weeks at one of the largest accounts in the branch: Why was a bank's statewide ATM network availability only in the 80% range? The problem had stalled the sales of new IBM ATM's to the bank. Accepting the assignment, Sneed understood there was sufficient mainframe computing capacity at the bank, and the individual ATM's were operating normally. So to Sneed this seemed an operations glitch. He surmised that no single person had spent 24 hours at the control center that managed the ATM network. So Sneed assigned himself to monitor the control center workflow and the network management personnel for 24 hours. At about 2am, he found the problem and, the following day, proposed the solution. Network availability leaped into the 90s. The branch manager (BM) asked for a report on how the problem was solved, which Sneed provided. Nothing was ever heard about this effort again, perhaps becoming another "SE manager success" since this, too, may some how have "violated" IBM "policy" (marketing trainees weren't supposed to be out at an account after midnight?). By this point, Sneed was becoming a little wary of branch office projects: they somehow ended up "violating policy," there was usually no recognition, and they impacted the IBM corporate training schedule and time to prepare for classes.

At the end of a subsequent training class, the instruction manager told trainees that BM's might create hypothetical scenarios to see how trainees would respond. So, after returning to his branch, the BM called Sneed into his office and presented, what Sneed believed, was a hypothetical scenario: How to reestablish operations in one of the branch's largest accounts after a disaster. It was a bank with four IBM mainframes handling $32 billion in federal reserve float every four days, that suddenly had its entire datacenter destroyed by a three-standard-deviation municipal power surge. This knocked out all of its check processors, statewide ATM network, and statewide branch terminal network--the worst disaster of its kind in all of IBM's Western Region history. "What are YOU going to do about it??" the BM demanded. Sneed requested the newest mainframe, that IBM had just introduced and was manufacturing in Poughkeepsie New York, to be shipped to the bank in Glendale California. "YOU'VE GOT IT!" Sneed requested double shifts of FE's (Field Engineers who install and do *actual* technical support) for 72 hours. "YOU'VE GOT IT!" Sneed requested to be relieved of marketing training responsibilities for one week. "YOU'VE GOT IT!" At that point, Sneed noted the only thing left to do was live at the account until it was back up. "GO DO IT!!" the BM boomed. "Yes sir," Sneed replied and promptly exited his office, quite pleased with how well the *hypothetical scenario* went.

Just for completeness, Sneed called an SE, a friend he knew at the bank. And that's when Sneed learned the bank really had flat-lined, was on its back, out cold, with *no IBM marketing personnel there* whatsoever the SE noted-- they were all leaving for the Thanksgiving holiday that started the next day. This was no hypothetical scenario: Sneed understood he had been *made* the IBM marketing principal to coordinate the bank's recovery, or he could ignore the BM's directive. So Sneed arrived at the bank in white shirt, conservative tie, Brooks Brothers suite, wing-tips and lived at the bank for four days. There were regular briefings with senior customer execs, to keep them away from the SEs and FEs. The new mainframe arrived which Sneed inspected before installation, including directing some minor reengineering of its cooling system to remedy a defective thermal coupling. The execs were impressed after their bank was brought back into business. When asked how long the new *emergency* IBM mainframe could be kept, Sneed quoted 90 days from the IBM Business Ethics Guidelines that *all* marketing

a *7ʰ-Gift* page

personnel signed. The BM, later, told the bank it could keep the mainframe and pay for it, securing a huge year-end commission. The bank had previously paid a backlog-bookie $800,000 to get an earlier ship-date rather than wait 24 months. It was considering selling its position to someone else.

This precipitated a controversy between the BM and Sneed, though Sneed wasn't aware of the commission implication for the BM, just the IBM Legal requirement for the 90-day disclosure to avoid DOJ problems. The matter was escalated to the IBM CEO; he decided against the branch manager ("early retirement") who could not explain why he ignored the IBM Legal guideline, why he tried to fire Sneed, and why he sent a *trainee* to be the marketing principal for something of this magnitude. Sneed's marketing manager (then different from Sneed's hiring manager) was sent to Japan; the regional manager was reassigned; the division manager was overruled when he tried to back the regional manager, and later left the business. The bank tried to hire Sneed.

Sneed stayed on at IBM, chastened by the episode, yet graduated with honors from IBM's New York marketing training center. He went on to win the Western Region fast start contest with 400% of quota by the end of January. By April Sneed had enough business to make quota for the year, along with identifying and contesting with six IBM mid-range systems, six computers DEC tried to upgrade for the National Reconnaissance Organization (NRO), which processed national security photo intelligence from satellites. None of this was easy, since Sneed was assigned to a new branch, the BM thereof a 15+ year friend of the prior BM. Sneed was given a 125-mile-by-60-mile territory, of many abandoned accounts, so Sneed called on everybody and did quite well. The NRO DEC threat was disclosed by a tight-lipped CFO, after Sneed personally repaired the CFO's PC to complete a spreadsheet for a board meeting the next day. The IBM PC was coming into its own, and Sneed desired to become a branch PC specialist, but was not considered the best choice. (Two years later, the plane carrying the designated branch PC specialists from IBM PC headquarters in Florida, along with the president of the PC division and his wife, crashed in Dallas, a plane Sneed almost certainly might have been on had his wish been granted [Proverbs 14:12].)

By April of his first year on quota Sneed surmised that all the "policy violations," and ones that awaited him in the future, meant that his dream of being IBM president was not possible. So in April, Sneed left IBM and did not return. When the branch informed him he had a $7000 commission check owed him, he asked that it be mailed, along with some Spaulding Executive golf clubs Sneed won. But the new BM insisted Sneed come into the branch and pick up the check. Sneed realized that setting foot on IBM property, at that point, could probably be construed as a "violation of policy," but the BM refused to mail the check. So Sneed arrived in IBM business attire, reported to the BM's secretary, and was shown into the BM's office. The BM previously notified the IBM CEO that Sneed had been fired for "violating policy," and the CEO let Sneed know this in a letter. The BM motioned for Sneed to have a seat, possibly to talk about IBM, perhaps in violation of the CEO's disclosure of what the BM had decided. When Sneed silently remained standing, the BM, after a brief pause, handed Sneed the check and Sneed departed, still wondering why the check wasn't just mailed. The objective Sneed had set as a 12 year-old was then completed after a 17 year journey from boarding school, to Caltech, to NASA/JPL, to Hughes (Boeing), to a California State University assistant professorship, and to IBM. "Respect for the Individual" had been IBM's motto. Sneed realized if IBM was one of the most admired companies in the industry, then it seemed best to serve as an individual, or with founders of their own firms, rather than those where the founders died long ago.

4. So, Sneed started his own firm in 1985, which became ISB::Institute. He developed multiple lines of business to survive the first 24 months for a new startup. The first 24 months are when most new enterprises fail:

1. PC manufacturing: Sneed introduced the "TASCOR" brand of 100% compatible IBM computers, manufacturing PC's under this name. Motherboards were ordered directly from IBM's Boca Raton facility. The IBM BIOS and other proprietary IBM ROM were on the motherboards. Doing this allowed Sneed to build "clones" that were in fact true IBM PC's, yet priced 30% less. The largest order was 15 systems from Hughes Aircraft.
2. Contract programming: Software development was done for Gourmet Concessions across the Broadway Store chain. This database application automated workflow and inventory management. It included the sale of a TASCOR computer, database software and contract programming which helped pay law school tuition for the patent attorney who later filed Sneed's first patent, called BEST2000A™.
3. Initial development of a multitasking PC operating system: Sneed started developing one himself, but discovered that a product by Quarterdeck, called DESQview, implemented the idea. Sneed became a

distributor of Quarterdeck products, placing the largest single order for the product in the U.S. at the time, saving the company from default. For six years, Sneed acted as a marketing consultant for the firm. Its annual revenues rose from $4 million (avoiding bankruptcy), to $8 million, to $12 million, to $26 million, to $48 million, whereupon Sneed was offered a vice presidency (declined) and stock options (accepted).

4. Southern California digital radio station: Sneed, on his own, designed a digital broadcast system to link a million personal computers by FM sideband. Sneed negotiated a contract with the station manager of FM 88 KXLU, to modify their transmitter with a $1500 FM sideband modulator. A three-mile long ATT leased-line linked the KXLU FM modulator with a computer in Sneed's West LA office. Sneed negotiated the manufacture and test-model of a digital demodulator which attached to a PC like a modem, tuned to FM 88. A software and network transmission protocol was designed, allowing compressed digital transmission of news, weather, sports, stock quotes, software and other consumer content. A successful beta-test transmission between West LA and Long Beach proved that a million personal computers could be linked. Preliminary effort was made to file a patent. That effort, some 20 years later, was reviewed as evidence to prove prior art for a suit involving transmitting digital information to a handheld device like a Blackberry.

ISB::Institute survived its founding in 1985 through 1995. During this period, Sneed started a family: Alexandria (1987), Dorion (1989) and Dorius (1993) born at Cedars Sinai in Beverly Hills; and Chante (1995) born at Pomona Valley Hospital after Sneed moved his wife and family to the Inland Empire. Diversifying his lines of business, Sneed also worked under contract with two search firms to learn the art of recruiting and placement. Sneed internalized this knowledge and applied it through ISB::Institute to fill over $1 million in job orders, including multiple positions with a $0.5 billion HMO, a $7 billion corporate credit union, a defense contractor, a consumer products manufacturer, and an international semiconductor manufacturer. In 1996, Sneed received recognition from the president of Caltech for Sneed's work on the NASA Galileo mission, which arrived at Jupiter in December of 1995. The operating system Sneed designed, along with the optimized Discrete Fourier Transform, and all the microcode developed with the Sneed Simulator worked flawlessly. The president set expectation to hear about future successes in Sneed's career.

5. In 1995, Sneed did preliminary research on a problem that appeared already solved--Y2K. So he left it alone. Two years later, in February 1997, Sneed read in the *Los Angeles Times* that a DOD computer complex in Ohio had failed due to the Y2K problem. This startled Sneed, since DOD had more money than any agency on earth for solving this problem, yet there DOD was with a major, public failure, and less than 24 months before systemic problems could become paralyzing. Solving the Y2K problem--the problem of the century--might be a way Sneed could answer the expectation of the president of Caltech, "I look forward to hearing of future milestones in your career." So, Sneed sat down in February 1997, with a blank sheet of paper, and pondered this problem. For about a decade Sneed referred to the days of the week by their ordinal designations in Scripture (1st, 2nd, 3rd, 4th, 5th, 6th, 7th) rather than by nominative, deity-based traditions that vary country to country (*dies solis* or *sunnandaeg* or *sun-day*; *lunae dies* or *monandaeg* or *moon-day*; *Marties dies* or *Tiwesdaeg* or *god-of-war-day*; *Mercurii dies* or *Wodendaeg* or *king-of-the-gods-day*; *Jovis dies* or *Thorsdaeg* or *god-of-thunder-day*; *Venus dies* or *Frigedaeg* or *day-of-the-goddess*; *Saturnus dies* or *god-of-agriculture-day*). The ordination of days allowed Sneed to think of time as an infinite vector segmented into ordinal-based weeks, months, years and centuries. The 100 symbols of "00" to "99" were assigned on to centuries, like 1900 to 1999. In a computer, there's nothing to prevent a computer from reassigning these symbols to different 100 year periods. Due to a one day difference between the 365 days in a year, and the 364 days in the largest integral number of weeks in a year, the first day of a new year, and a particular day of the week are coincident, again, after seven years. Unless there is a leap year; this skips the coincidence four times due to leap days once every four years. So once every 4x7, or 28 years, the coincidence recurs, or 4 times any other integral multiple of 7, for a product of 91 or less. The simplicity to think about this in terms of Scripture and ordinal-based time resulted in a specification (Specification) titled "Serial Encryption System By-Pass of Year 2000 Date Malfunctions" or BEST2000™, and later BEST2000A™ as Sneed went on to develop BEST2000B™ as a vehicle thermal integrity detection system. Both were issued patents, BEST2000B™ after a successful one year national security review during patent examination, since it is a strategic technology for the aerospace industry.

6. A friend, Dr. Alex Lidow, who is a fellow Caltech alum and member of the Caltech board of trustees, applied Specification on IBM computers in seven countries, when he was CEO of International Rectifier Corporation (IRC). This resulted in savings to his firm of some 70% on a two-year, $10 million "best efforts" counter-proposal by Accenture (Anderson Consulting at the time). IBM had vetted Sneed's Specification by then, and confirmed to Dr.

Lidow and IRC its technical merit before IRC applied it. Dr. Lidow was very pleased, and Sneed was later awarded $236,000 under an IRC contract. Dr. Lidow also invited Sneed to his parent's estate for a Caltech function.

Four months prior to working with Dr. Lidow and his firm, IBM entered into similar agreement when it took custody of Specification, analyzed Specification within IBM, and declared Specification's merit to Sneed and, later, to IRC which deployed Specification on IBM computers in France, Germany, England, Singapore, China, the U.S. and Japan, at considerable savings over anything IBM, Accenture, or anyone else offered. IBM documented that it, too, could proceed with payment terms after d3/May 22, 2001 (Date), which is when a patent was issued to Sneed.

IBM admits taking custody of Specification from Sneed. Mr. Don Logan, an ethical IBM Global Services (IBM Global) program director, confirmed Specification's merit and economy, under terms (Terms) set forth by Sneed for no-charge testing, and a $49,764.55 per site production-use fee. Mr. Logan and an unnamed, ethical IBM VP (Ethical VP), helped Sneed modify Terms with 14 cents per line of code, across an aggregate number of lines of code, and no charge after a billion lines, respecting Ethical VP's caution about IBM's counting methods. As noted, one IBM client, IRC, applied Specification to 1,200,000 lines of code, verified by Caltech alums Dr. Dale Prouty and Dr. Alex Lidow and an MIT expert in IBM mainframe and midrange systems.

Subsequent to Mr. Logan's and Ethical VP's contact, IBM Corporate Licensing Program Manager Mr. Al Torressen took custody of Specification from Mr. Logan and IBM Global, memorializing to Sneed the written Terms Mr. Logan and Ethical VP helped establish. This demurred a Department of Justice "firewall" between IBM Corporate and the IBM Global subsidiary, obviating a markup of up to 1,400% IBM Global could legally bill IBM Corporate, or the same up to $2 per line of code IBM Global charges government or private sector clients. This "firewall" demur may have saved IBM Corporate up to $1.86 billion IBM Global could bill.

Mr. Torressen assured Sneed, in writing, that IBM Corporate would not "expropriate" Specification, and acknowledged the Terms Mr. Logan and Ethical VP helped establish in the written agreement Mr. Torressen also took custody. Mr. Torressen noted Specification's potential worldwide impact, and arbitrarily set d3/May 22, 2001 (Date) for Sneed to initiate contact for payment. Between Date and October 29, 2007, multiple IBM legal personnel were designated to replace Mr. Torressen, including personnel directed by IBM CEO Mr. Sam Palmisano. Each confirmed, in writing, no benefit or discovery of use that merited payment to Sneed. This remained so until Sneed learned that an unnamed VP (Unnamed VP) retired from IBM. Unnamed VP and Sneed worked together at IBM in the 80s. After contact with Unnamed VP, Sneed contacted Mr. Palmisano regarding payment for use, if any, of Specification, encouraged by Unnamed VP. Mr Palmisano directed legal counsel to contact Sneed, whereupon legal counsel declared, in writing, IBM would no longer respond to inquiries about IBM's use of Specification, but later recanted, in writing, this declaration of non-responsiveness.

7. Three Caltech alums are in agreement to move forward with this matter: Dr. Alex Lidow Caltech board member, Dr. Dale Prouty Caltech PhD Applied Physics, and an unnamed VP from within IBM.

First, Al, in every communication with you, in your capacity as IBM Licensing Program Manger, whether oral or written, you spoke with candor. While keeping a vow of loyalty to IBM, *simultaneously* you were informed by doing what is right and just and fair with me. When I am reminded of the highest ideals of IBM, "Respect for the Individual," it is yours and IBM CEO Mr. John Opel's example that come to mind. It is at this time I wish to express my gratitude. I also wish to express regret in not fully appreciating that your perspective goes beyond merely corporate policies, practices, beliefs, or even societal codes and regulations. There are more transcendent precepts which allow you to discern and act with confidence, when others may vacillate, equivocate, obfuscate, or denigrate efforts that manifest desire, merit, and purposed commitment. We might all do better to follow your example in life, let alone in business matters.

It is because of your and Mr. Logan's candor, and attempts thereof, that I gained a new respect for your firm, perhaps renewing ideals I had long ago. And out of respect for you and Mr. Logan, I've been considerate of IBM's actions and decisions, informed by what was set forth in writing and properly acknowledged after patient and deliberative vetting. I deeply regret that IBM would not let me speak with you. I deeply regret this. Rather, instead, others were inserted in the process who seemed less committed to open communication as in the early contact with a firm you helped redefine. My comfort was that because of your integrity and commitment to excellence, you would enjoy a

Ex1 p130

long and prosperous career. This comfort is injured as I learn more of your leaving in 2003, since I consider you deserving the best life can still offer.

The Specification has performed flawlessly on IBM computers around the world, just as you predicted was possible, and which Dr. Lidow and Dr. Prouty and others have verified. So it is fitting and proper that your imprimatur be given weight, independent of any IBM chairman, IBM CEO, IBM executive, IBM manager or IBM employee. No one may *demand* of you anything now, accept that by which you assent of your own freewill. And the name Torressen carries a legacy, through the centuries, whether in war or times of peace, that hews to doing what is right and just and fair, all the way to this very day.

It remains written in Proverbs:
> The memory of the righteous will be a blessing, but the name of the wicked will rot. [10:7]
> The faithful man will be richly rewarded, but one eager to get rich will not go unpunished. [28:20]
> A truthful witness does not deceive, but a false witness pours out lies. [14:5]

Hence forth, I consider you an Industry Intellectual Property Expert and, as circumstances may warrant, someone who may be respected in such capacity. Below, you may simply check one of two boxes, sign and date it, and return this 7ᵗʰ page by fax, as is, to 909-395-9535, before sunset on d4/February 15, 2012. If there is no response, it shall be construed that you are asserting no knowledge of this matter in any way whatsoever.

[ ] IBM did not use your Specification or anything similar to it to remediate Y2K problems on any internal IBM computer or software application.

[ ] IBM applied a two-digit date shift method, similar to or the same as all or part of your Specification, on at least one internal IBM computer or software application.

_____   _____
Mr. Al Torressen   Date

With prayerful regards and respectfully yours,

Anthony Sneed
Anthony Sneed
Life-member Caltech Alumni Association, U.S. Patent Holder (#7546982 and #6236992),
author (*From Jupiter to Genesis*, a worldwide first-hit title on Google®)
Attachment: Correspondence with Caltech and other officials

EX1 p131

1

2

ANTHONY SNEED, PRO SE
3   2058 NORTH MILLS AVENUE
CLAREMONT, CA 91711
4

5

6   Pro Se Plaintiff,
ANTHONY SNEED
7

8

9

10   UNITED STATES DISTRICT COURT

11   CENTRAL DISTRICT OF CALIFORNIA - WESTERN DIVISION

12

13

14   ANTHONY SNEED, An Individual,   )   CASE NO. CV 11-010217GW (PJWx)
)
15   Plaintiff,   )   **SECOND AMENDED COMPLAINT FOR:**
)
16   v.   )   **1. BREACH OF WRITTEN CONTRACT**
)
17   )   **2. BREACH OF ORAL CONTRACT**
)
18   )   **3. BREACH OF QUASI CONTRACT**
)
19   )   **4. MISAPPROPRIATION OF TRADE**
20   )   **SECRETS**
)
21   )
22   )   2nd Amend. Complaint: Jul. 10, 2012
)   Post-Mediation Status Conf.   Jul. 12, 2012
23   )   Discovery Cutoff   Aug. 15, 2012
)   Expert Discovery Cutoff   Sep. 7, 2012
24   )   Motion Hearing Cutoff   Oct. 15, 2012
   INTERNATIONAL BUSINESS   )   Pretrial Conference   Nov. 19, 2012
25   MACHINES CORPORATION, A New   )   Court Trial: Dec. 4, 2012
   York Corporation, and DOES 1   )
26   THROUGH 10, Inclusive.   )
)
27   Defendants   )   Judge: Hon. Judge George H. Wu
28   )   Ctrm: 10

SECOND AMENDED COMPLAINT: RESPONSIVE TO COURT NOTICE TO PLAINTIFF JUNE 29, 2012
EXHIBITS RESPONSIVE TO *MISHLER V. CLIFT*, 191 F.3d 998, 1008 n.7 (9th Cir. 1999)

1

2

3 **GENERAL ALLEGATIONS**

4   1.   Plaintiff Anthony Sneed (Sneed) is an individual who resides and has been

5 doing business in the County of Los Angeles, State of California.

6

7   2.   Plaintiff is informed and believes and thereon alleges that Defendants

8 International Business Machines Corporation (IBM), is a corporation duly organized and existing

9 under the laws of the State of New York, and doing business within the State of California.

10   3.   Plaintiff is ignorant of the true names and capacities of Defendants sued herein

11 as DOES 1 through 10, inclusive, and therefore sues said Defendants by such fictitious names.

12 Plaintiff will amend this Complaint to allege their true names and capacities when ascertained.

13 Plaintiff is informed and believes and thereon alleges, that each of said fictitiously named

14 Defendant is liable and responsible in some manner for the occurrences herein alleged, and that

15 Plaintiff's damages as herein alleged were proximately caused by said conduct.

16

17   4.   Plaintiff is informed and believes and thereon allege, that at all times mentioned

18 in this Complaint, each Defendant was the employee and/or agent and/or representative of each

19 Co-Defendant, and in doing the things alleged in this Complaint, were acting within the course

20 and scope of said employment, agency and representation.

21

22   5.   All acts, occurrences and transactions hereinafter mentioned occurred and

23 transpired in Los Angeles County, State of California.

24   6.   The obligations and claims sued upon herein were made and entered into and are

25 due and payable in the above mentioned county, State of California, and are not subject to the

26 provisions of California Civil Code Section 395(b).

27

28

2

SECOND AMENDED COMPLAINT: RESPONSIVE TO COURT NOTICE TO PLAINTIFF JUNE 29, 2012
EXHIBITS RESPONSIVE TO *MISHLER V. CLIFT*, 191 F.3d 998, 1008 n.7 (9ᵗʰ Cir. 1999)

a *7th-Gift* page

FOR A FIRST, SEPARATE, AND DISTINCT CAUSE OF ACTION FOR BREACH OF CONTRACT AGAINST DEFENDANTS IBM AND DOES 1 THROUGH 10, PLAINTIFF ALLEGES AS FOLLOWS:

7.   Plaintiff incorporates by reference paragraphs 1 through 6, as though fully set forth herein.

8.   By 1997 it became widely known that various models of IBM computers utilizing only two digits to represent year information would have calculation errors when the Year 2000, represented by the digits 00, was used in year calculations. For example, certain calculations that subtracted the year 2000 from the year 1999 would generate the incorrect answer of either -99 or an error signal, rather than the correct answer of the number 1. In some cases, the computer software program would not function after generating such an error. These errors would be unavoidable in most accounting programs using year information.

9.   At all times relevant herein, Plaintiff was the inventor, creator and owner of all rights to a certain invention and trade secret (hereinafter referred to as the "Specification") that had the ability to remediate the computation problems associated with computer errors caused by calculations involving the year 2000, or as it is more commonly referred to -- the "Y2K" computer problem.

10.   The Specification and description of the invention was completed by Plaintiff in or about February 1997. Plaintiff refined his Specification and description for submission of a Patent Application to the United States Patent and Trademark Office (PTO) in April of 1997.

11.   In the year 1997, the Specification in its entirety and the methodology for the implementation of this solution as described and explained by Plaintiff was not common

3

SECOND AMENDED COMPLAINT: RESPONSIVE TO COURT NOTICE TO PLAINTIFF JUNE 29, 2012
EXHIBITS RESPONSIVE TO *MISHLER V. CLIFT*, 191 F.3d 998, 1008 n.7 (9ᵗʰ Cir. 1999)

knowledge as a viable solution to the Y2K computer problem for the computers that were manufactured by IBM. Plaintiff invented the methodology by which the computer database could be encrypted and input and output modules and devices could be modified in order to generate accurate year computations, without the typical Y2K errors.

12.   The Specification did provide a viable solution to the Y2K problem for many applications and software environments that was superior to alternative solutions in terms of time and effort required to remediate the computation errors. Other solutions to the Y2K problem could be more time-consuming and labor-intensive for certain software applications.

13.   In the normal business use of computers such as the IBM AS/400 it was common for each computer to contain one million lines of code or more for the various software programs that were being operated. These computers had the capacity to analyze and perform computations on large amount of data and required sophisticated and complex computer programs.

14.   It is believed and thereupon alleged that under special circumstances that required the evaluation of the entire software programming for these IBM computers, IBM would charge its customers between $1 an $2 per line of code in order to remediate software problems. Since the Y2K problem had the potential to affect virtually all lines of code for these computers, the remediation of the Y2K problem could generate revenues and/or costs of more than $1,000,000 per computer.

15.   It is believed and thereupon alleged that the labor cost for IBM to remediate the Y2K problem, on its computers that were used internally, by means other than Specification, would cost IBM approximately $1 per line of code. Since application of Specification to

remediate the Y2K problem did not require the review of all of the software lines of code, it could be the most cost-effective methodology for the remediation of the Y2K computation errors in many software applications.

16.   On or about June, 1997, Plaintiff submitted a proposal to Don Logan, one of the executives at IBM with whom Plaintiff developed a bond of trust and a confidential relationship. Having been a former IBM employee, Plaintiff was well aware of the strict culture of secrecy and confidentiality that was imbued in all of IBM's employees and especially its executives. One of the reasons that IBM has maintained its status as the world's leading manufacturer of large business computers is the secrecy and confidentiality of its operations.

17.   As a result of the foregoing, Plaintiff knew that he could trust the verbal representations of maintaining the confidentiality and secrecy of the Specification with the executives of IBM that were presented with the knowledge of the Specification. The executives at IBM are people of the highest levels of business integrity, as it is an integral part of the culture and custom of that esteemed organization.

18.   At all times herein, Plaintiff has relied upon the confidential relationship that he has with the executives of IBM. Plaintiff reasonably believed that his oral agreement of confidentiality for his Specification would be honored.

19.   During 1997, Plaintiff submitted a written offer and contract (hereinafter referred to as the "Agreement") to IBM and DOES 1-10 regarding its use of the Specification.

20.   The essential terms of that written agreement were that IBM would pay Plaintiff the amount of $49,764.55 for each site or location at which the Specification was used to remediate or mitigate the Y2K problem on any IBM computers that were used internally or within the corporation.   In addition, Plaintiff's offer stated that if IBM used the Specification to remediate the Y2K problem for its customers, it would pay Plaintiff the amount of 14 cents per line of code contained on each computer that was remediated using the Specification. As part of the Agreement, Plaintiff offered to IBM that upon acceptance of the Agreement, IBM would not be obligated to pay for any use of the Specification that exceeded the remediation of computers with an aggregate of more than a total of one billion lines of code.   The Agreement that was offered limited IBM's payment for use of the Specification in remediating lines of code to $140,000,000.

21.   During 1997, after undertaking review of the utility of the Specification, Al Torressen an authorized representative of IBM agreed that it would pay Plaintiff pursuant to the terms of the Agreement, if IBM used the Specification to remediate the Y2K problem on its computers or for its customers, on the condition that Plaintiff obtained approval for his Specification from the PTO with the issuance of a Patent. Defendant, IBM agreed that it would pay Plaintiff pursuant to the terms of the Agreement, if the Specification was used to remediate the Y2K problem on the condition that Plaintiff obtained a Patent.

22.   On May 22, 2001, the United States Patent and Trademark Office issued Plaintiff, Sneed the US Patent 6,236,992, meeting the condition of performance set by Torressen during contract formation with IBM's Letter of Corporate Intent, dated June 20, 1997.

23.   Plaintiff has performed all of his obligations, conditions and requirements for this performance of the terms of the Agreement, except those he has been excused from performing by the actions of Defendants.

24.   After receipt of the US Patent Number 6,236,992, Plaintiff notified IBM that he expected payment for use of the Specification pursuant to Agreement. IBM notified Plaintiff that it had not used the Specification and on that basis was not obligated to pay Plaintiff any amount pursuant to the Agreement.

25.   Over the years, Plaintiff has continued to make reasonable and repeated inquiries to IBM regarding its use of the Specification. Plaintiff has placed several phone calls to the designated representative of IBM with whom he was allowed to speak, inquiring about its use of the Specification. Plaintiff has been instructed by officials at IBM that he should only correspond to the executive designated by IBM's CEO with regard to his Specification and that he should have no communications with any other employees of IBM.

26.   Plaintiff has periodically sent letters to IBM over the years in which he has inquired about IBM's use of the Specification. At all times and in all conversations and responses by IBM, Plaintiff has been informed that IBM has not used and did not use the Specification to remediate the Y2K problem on any computer or for any customer.

27.   Upon information and belief, Plaintiff is informed and believes and thereupon alleges that IBM did use the Specification for the remediation of the Y2K problem. Plaintiff could not have reasonably discovered the use of the Specification by IBM due to the strong culture of secrecy and confidentiality at IBM, and the use of confidentiality agreements with its employees that contain severe penalties for breaches of such confidential matters.

28.   Plaintiff has been unable, despite his reasonable diligence, to have discovered the use of the Specification by IBM. He has made repeated contacts with IBM representatives over the years and each time he has been informed that IBM has not made use of the Specification.

29.   Only within the past year has Plaintiff obtained written-probative-notice that is factually suggestive that IBM used Plaintiff's Specification and owes Plaintiff for that use.   The attached references are excerpted from one exhibit accepted as filed by the Court d3/June 19, 2012, in Plaintiff's REQUEST TO EXTEND DISCOVERY AND DEPOSITION CUTOFF DATE DUE TO INTERNATIONAL BUSINESS MACHINES CORPORATION'S DISREGARD OF FRCP 34.

30.   This one exhibit, excerpts noted herein, is included with respect to *Mishler v. Clift*, 191 F.3d 998, 1008 n.7 (9th Cir. 1999) ("When a complaint alleges the contents of a document, but the document is not attached to the pleading, the court may consider the document in ruling on a Rule 12(b)(6) motion.") specific to references to the Caltech Honor Code, communications thereof, and correspondence that is factually suggestive to written-probative-notice that IBM used Plaintiff's Specification and owes Plaintiff for that use.

31.   In exhibit 1, page 82 (Ex.1, p.82, first paragraph, last line), Mr. Pete Wilzbach (Wilzbach), as a retired IBM Executive Vice President and friend of Plaintiff, assures Plaintiff:

> "I won't copy [IBM CEO] Mr. Palmisano's office on this note, but if my name appears on any further correspondence, I will be forced to tell [Palmisano and IBM]."
> d1/June 5, 2011 letter from Wilzbach to Plaintiff

a *7th-Gift* page

32.   The "correspondence" Wilzbach refers to is the d2/April 4, 2011 letter (Ex.1, p.79) from Plaintiff to IBM CEO Mr. Sam Palmisano (Palmisano) that contains a statement, assented to by Wilzbach herewith, that IBM used Specification and owes Plaintiff for said use. Wilzbach affirms in his d1/June 5, 2011 letter that he will let the statement stand *in full* as written to Palmisano by Plaintiff, that IBM used Specification and owes Plaintiff for that use. Dr. Alex Lidow and Dr. Dale Prouty are cc'ed and accepted the letter without dissent:

> By 2007, [Plaintiff] learned WILZBACH retired from IBM. WILZBACH, as a retired IBM EVP, corroborated [Dr. Alex Lidow's] opinion, as an IBM customer executive, that [Specification's] use on computers *inside* of IBM merited payment to [Plaintiff]. Both WILZBACH and [Dr. Alex Lidow] are Caltech alums, [Dr. Alex Lidow] a member of the Caltech board of trustees. In addition, Dr. Dale Prouty ("PROUTY"), a third Caltech alum, vetted [Specification] with sworn testimony of its use and economy on IBM computers.   d2/April 4, 2011 letter from Plaintiff to Palmisano

33.   Wilzbach's "I will be forced to tell [Palmisano and IBM]." refers only to <u>future correspondence</u> from Plaintiff to IBM. Wilzbach stipulates that no matter what Plaintiff or IBM may or may not do in the future, the d2/April 4, 2011 letter's statement that IBM used Specification and owes Plaintiff for that use will be unaffected. This represents a *public (Public)* statement, without restriction. Wilzbach assented to its receipt by IBM's CEO and allows it to be cc'ed to Caltech's president Dr. Jean-Lou Chameau, Caltech board of trustee member Dr. Alex Lidow (Lidow), and Caltech alumnus Dr. Dale Prouty (Prouty) <u>without restriction</u>.

34.   For at least three years prior to the d2/April 4, 2011 letter, Wilzbach, Lidow, and Prouty, each 25 year+ friends of Plaintiff, were insistent that Plaintiff sue IBM. Wilzbach reminds Plaintiff in Wilzbach's d1/June 5, 2011 letter of Wilzbach's similar, past advice to get a lawyer involved. The *Public* statement becomes admissible to the very lawyer Wilzbach recommends Plaintiff hire to file a complaint.   Wilzbach's assent of the *Public* statement,

without restriction, in his d1/June 5 letter to Plaintiff, constitutes the first and only written-probative-notice Plaintiff received from any party to file a claim that IBM used Specification and owes Plaintiff for that use.

35.   The *Public* statement, without restriction, to Palmisano in the d2/April 4, 2011 letter, builds on past communication between Palmisano, Plaintiff, and Wilzbach regarding actions, deeds, statements or disclosures by Wilzbach that merits Wilzbach's d1/June 5, 2011 letter written-probative-notice to file a complaint. In a d4/December 2, 2007 phone conversation, Wilzbach directs Plaintiff how to sue IBM, after IBM rejects Wilzbach's friendly advice and effort through Plaintiff for settlement with Palmisano and IBM.

36.   A d5/December 3, 2007 letter (Ex.1, p.76) from Plaintiff to Wilzbach memorializes Wilzbach's advice and directions to Plaintiff on the mechanics of how to sue IBM. Wilzbach provided such friendly, informal advice after IBM rejected Wilzbach's good-faith initiative for settlement between Plaintiff and Palmisano and IBM. Therein is specific notice to Wilzbach regarding the Caltech Honor Code, in paragraphs numbered "4.", "6." and "7." With such references, Plaintiff reminds Wilzbach, Lidow and Prouty not to take unfair advantage of one another, Plaintiff, or of anyone else in the actions being recommended to sue IBM.

37.   At that point, for the first time, Wilzbach, Lidow, and Prouty were all in agreement that Plaintiff sue IBM. The d5/December 3, 2007 letter was accepted without dissent by Wilzbach, Lidow, and Prouty, and cc'ed to Palmisano as well.

38.   Plaintiff supplied similar notice regarding the Caltech Honor Code to Wilzbach, Lidow, and Prouty during steps of the settlement process initiated by Wilzbach and carried out by Plaintiff, from d6/October 19, 2007 to d4/December 2, 2007. The settlement

process ended d4/December 2, 2007 when Wilzbach recommended Plaintiff sue IBM, in harmony with the position Prouty and Lidow had held for some two years prior.

39.   Again, the Caltech Honor Code was a reminder by Plaintiff to Wilzbach, Lidow, and Prouty not to take unfair advantage of one another, Plaintiff, or anyone else through the course of action being recommended to Plaintiff. The Caltech Honor Code is so referenced *and underscored* five times in the d6/October 19, 2007 settlement letter (Ex.1, p.64) from Plaintiff to Palmisano and cc'ed to Wilzbach, Lidow, and Prouty. Wilzbach, Lidow, and Prouty accepted the letter without dissent.

40.   Again, the Caltech Honor code is referenced *and underscored* in the second settlement letter from Plaintiff to Palmisano, d2/November 5, 2007 (Ex.1, p.69), cc'ed to Wilzbach, Lidow, and Prouty. Wilzbach, Lidow, and Prouty accepted the letter without dissent. All parties were given repeated, written notice of Plaintiff's clear expectation that they not take unfair advantage of one another, Plaintiff, or anyone else through the course of action being recommended, whether in settlement with Palmisano and IBM or, later, suit; Wilzbach, Lidow, and Prouty accepted this notice, each time, without dissent.

41.   Wilzbach's assent, through his d1/June 5, 2011 letter, to the *Public* statement in Plaintiff's d2/April 4, 2011 letter to Palmisano, is informed by the legacy of notices given above. Plaintiff filed complaint based on that legacy and IBM's non-response to the *Public* statement supported by Wilzbach, Lidow, and Prouty. IBM's non-response broke its pattern and practice from 1997 to 2007, and the condition of performance IBM set on Plaintiff in its d6/June 20, 1997 Letter of Corporate Intent regarding inquiries to IBM. As Plaintiff performed, IBM responded to inquiries in 2004, 2006, and in October and November of 2007. IBM did not

respond after the d2/April 4, 2011 letter to Palmisano that included the *Public* statement, supported by Wilzbach, Lidow, and Prouty.

42.   Plaintiff filed complaint d6/October 28, 2011. Wilzbach had set expectation that Plaintiff sue IBM d4/December 2, 2007, after initiating a settlement effort through Plaintiff in October 2007. Absent probative notice, Plaintiff continued the *pro forma* pattern of contacting IBM approximately every 24 to 36 months to make inquiry, respecting the condition of performance on Plaintiff set forth by IBM in its d6/June 20, 1997 Letter of Corporate Intent.

43.   The d2/April 4, 2011 letter was continuation of such *pro forma* contact. By then, Prouty and Lidow were insistent that Plaintiff sue IBM. Wilzbach's prior communication supported Prouty's and Lidow's conviction that IBM had used Specification and owed Plaintiff for that use. The d2/April 4, 2011 letter expressed their collective conviction regarding the matter. Wilzbach assented to the *Public* statement contained in the d2/April 4, 2011 letter, referenced as a retired IBM Executive Vice President therein. The *Public* statement memorialized that IBM used Specification and owed Plaintiff for that use.

44.   Wilzbach, Prouty and Lidow know that Plaintiff is subordinate to them as a fellow Caltech alum and within the Caltech Honor Code. If any one of them ever directed Plaintiff to cease litigation, Plaintiff would do so forthwith. Not one ever has. All have urged the litigation move forward, and continue to do so now.

45.   This is inclusive of Wilzbach's recent letter to Plaintiff, d1/January 29, 2012 (Ex.1, p.86).) As with Wilzbach's d1/June 5, 2011 letter, a simple "cease litigation" directive from Wilzbach would suffice for Plaintiff to end the case. There is no such directive, ever, from Wilzbach. Rather, Wilzbach encourages Plaintiff regarding the complaint filed based on the

SECOND AMENDED COMPLAINT: RESPONSIVE TO COURT NOTICE TO PLAINTIFF JUNE 29, 2012
EXHIBITS RESPONSIVE TO *MISHLER V. CLIFT*, 191 F.3d 998, 1008 n.7 (9ᵗʰ Cir. 1999)

Public statement:

> I am not sure why you wrote me the letter dated d2/January 23, 2012. If it was just to tell
> me that you have acquired legal counsel, that is fine.

In other words, Wilzbach approves of acquiring legal counsel and filing a complaint. Wilzbach's "I am not sure why you wrote me" means Wilzbach needs no details on the litigation since Wilzbach's expectation set forth d4/December 2, 2007 to sue IBM is now being met by Plaintiff, consistent with Lidow's and Prouty's affirmation of the *Public* statement with Wilzbach.

46.   Wilzbach assents to all documents that reference his name being provided to the Court, or produced in discovery. It is impossible for expectation Wilzbach set to sue IBM to exclude these documents. Thus, Wilzbach's "I hope you are not using me…" acknowledges the reality of discovery, because as Wilzbach affirms, "[If] you have acquired legal counsel, that is fine." This is consistent with expectation Wilzbach set forth d4/December 2, 2007 to sue IBM which would include discovery of all relevant communication with Wilzbach.

47.   In his last paragraph, Wilzbach requests, but does not demand, the following:

> I would ask that you show this letter to your legal counsel and make sure they are not
> using me as any basis for your case. That would be a great mistake.

Wilzbach knows that written communication between Wilzbach/Sneed/Palmisano/IBM is part of any discovery.   As a retired IBM Executive Vice President, Wilzbach's actions, deeds, statements, and disclosures are probative, along with his and Plaintiff's 30+ years of amicable relations.

13

SECOND AMENDED COMPLAINT: RESPONSIVE TO COURT NOTICE TO PLAINTIFF JUNE 29, 2012
EXHIBITS RESPONSIVE TO *MISHLER V. CLIFT*, 191 F.3d 998, 1008 n.7 (9th Cir. 1999)

48.   Though Wilzbach declares referencing him in discovery and pleading for the case would be "a great mistake," he does not preclude Plaintiff from making this "mistake." Wilzbach's disclaimers are consistent with maintaining a neutral position relative to the litigation: he is neither obstructing litigation nor *willingly* aiding deliberation, outside of a subpoena or a Court order *beyond the pleading stage.*

49.   Wilzbach's disclaimer of "no involvement or knowledge" does not refer to his d3/May 13, 1997 letter (Ex.1, p.2) to Plaintiff, wherein he acknowledges custody of a copyright-restricted, evaluation-only copy of Specification; or to Wilzbach's effort to vet Specification; or Wilzbach's encouragement to Plaintiff to patent Specification.   Nor does this disclaimer refer to Wilzbach apprising Plaintiff of IBM's internal politics, in or about June 1997, when IBM Corporate "seized" Specification from IBM Global Services.   Nor does this disclaimer refer to Wilzbach's settlement effort initiative in or about October 2007. Nor does this disclaimer refer to directions from Wilzbach to Plaintiff about how to sue IBM in or about December 2007. Nor does this disclaimer refer to the *Public* statement supported by Wilzbach with Lidow and Prouty, without restriction, through Wilzbach's d1/June 5, 2011 to Plaintiff.

50.   Wilzbach's disclaimer does refer to not being involved with efforts to cover-up use, deny payment to Plaintiff, mislead Plaintiff through repeated denials, or obstruct justice during a Department of Justice inquiry on behalf of Plaintiff.   Wilzbach rejects outright any specific involvement or knowledge in any way of such conduct by other parties at IBM.

51.   IBM would have the Court believe that Wilzbach is an "empty suit," ignorant of IBM's internal Y2K politics regarding Specification, and that Wilzbach is a has-been sent out to pasture over a decade ago.   Nothing could be further from the truth since Wilzbach maintains

a *7ᵗʰ-Gift* page

extensive contacts throughout IBM and remains a highly regarded person by many at IBM. Plaintiff concurs.

52.   Wilzbach is one of the most astute executives in the history of IBM. He closed the largest outsourcing contract known in the annals of IBM. He was Plaintiff's hiring and training manager, recruiting Plaintiff from an assistant professorship at California State University Northridge. Wilzbach provided the foundation for Plaintiff to graduate with honors form IBM's New York marketing school, attain 400% of quota recognized as the top performer for the entire western region of the U.S., and be elected president of the IBM club for the entire western U.S. Wilzbach and Plaintiff are fellow Caltech alums who made notable contributions to developing interplanetary spacecraft as undergraduate interns and expanded on such efforts after graduating from Caltech.

53.   Wilzbach knows that Plaintiff is subordinate to Wilzbach's wishes regarding this case, and those wishes are expressed by Wilzbach to, (1) proceed with litigation, (2) respond to *all* discovery requests and, (3) as it may please the Court, move the case forward from the pleading stage based on the merits that IBM used Plaintiff's Specification and owes Plaintiff for that use.

54.   As noted by the US Patent Office, those merits include the Specification's ability to preserve the two-digit date precedent programmed into billions of lines of code developed during the last 40 years, across a host of different computers and languages, while allowing databases on such systems to retain and access records older than 100 years, and maintain such software's viability indefinitely into the future. Additionally, the Specification

SECOND AMENDED COMPLAINT: RESPONSIVE TO COURT NOTICE TO PLAINTIFF JUNE 29, 2012
EXHIBITS RESPONSIVE TO *MISHLER V. CLIFT*, 191 F.3d 998, 1008 n.7 (9ᵗʰ Cir. 1999)

1

2

3   allowed its application to be automated, greatly reducing the time and manual labor expense to

4   convert billions of lines of code to become Y2K complaint.

5   55.   Absent any of these features, explained through a simple encoding/decoding

6   paradigm that executives *and* technical personnel can grasp, many CEO's waited, their

7   companies drifting like boats edging closer and closer to the Y2K equivalent of Niagara Falls.

8

9   56.   The Specification's completeness compares favorably to other schemes limited

10   to records less than 100 years old, and schemes prone to Y2K-all-over-again after a period of

11   time, or that require billions of lines of code to be inspected and manually converted to a

12   different date format.

13

14   57.   Dr. Alex Lidow as CEO of NYSE-listed International Rectifier during Y2K, was

15   about to be forced to sign an up to a $10 million contract, or more, to repair his IBM AS/400

16   computers worldwide, with no guarantee the task could be completed before Y2K problems

17   began hitting his systems in less than 18 months. The Caltech Honor Code allowed Lidow and

18   Plaintiff to trust one another while addressing the Y2K problem. Both Lidow and Plaintiff

19   resided in Fleming House while undergraduates at Caltech.

20

21   58.   One site in Singapore with 1.2 million lines of code was repaired in only six

22   weeks with Specification. There were no post-production issues whatsoever. The now $1 billion

23   firm's sites in England, France, Germany, China, Japan, and the U.S. rapidly followed, all

24   completed well before Y2K, for under $3 million, with International Rectifier's existing staff

25   rather than an army of contractors under the $10 million, no-guarantee alternative.

26

27   59.   Lidow, who sits on Caltech's board of trustees *with* IBM, is convinced that

28   IBM had substantially the same problem as his firm, except on a much, much, larger, worldwide

1
2
3   scale. Like his firm, IBM waited too long to address some billion+ lines of code that needed
4   repair within the only 18 month window before early Y2K failures were triggered. This informs
5   Lidow's conviction, supported by Wilzbach, through the *Public* statement, that IBM used
6   Specification and owes Plaintiff for that use.
7
8   60.   Based upon the foregoing, Plaintiff alleges that he has been thwarted in his
9   efforts to discover the use of his Specification by Defendant, IBM. Despite Plaintiff making
10 reasonably diligent efforts to discover such use, Defendant, IBM has repeatedly represented to
11 Plaintiff that no such use of the Specification has been made and that IBM has no obligation to
12 pay Plaintiff. Plaintiff believes that the representations of IBM have been false and that there
13 has been actual use of the Specification for which payment is due.
14
15   61.   Defendant, IBM has breached Agreement by failing and refusing to pay
16 Plaintiff the compensation that is due to him pursuant to the terms of the Agreement. Due to
17 the strict culture of confidentiality within IBM, Plaintiff is unable to ascertain the scope of the
18 use of the Specification by IBM without discovery pursuant to Rule 34. Defendant, IBM has
19 exempted itself from Rule 34, while readily referencing documents as exhibits contained in
20 Plaintiff's full, discovery production provided in REQUEST TO EXTEND DISCOVERY AND
21 DEPOSITION CUTOFF DATE DUE TO INTERNATIONAL BUSINESS MACHINES
22 DISREGARD OF FRCP 34--before ever receiving such production from Plaintiff.
23
24   62.   In so doing, regarding non-confidential documents and derivatives thereof not
25 produced by IBM, specific to this case, that IBM received from Plaintiff in 1997, IBM asserts it
26 is above the law and the authority of this honorable Court regarding Discovery and FRCP
27 34(a)(1).
28

17

SECOND AMENDED COMPLAINT: RESPONSIVE TO COURT NOTICE TO PLAINTIFF JUNE 29, 2012
EXHIBITS RESPONSIVE TO *MISHLER V. CLIFT*, 191 F.3d 998, 1008 n.7 (9ᵗʰ Cir. 1999)

63.   Nevertheless, it is believed and thereupon alleged that the use of Specification was extensive within IBM and the Specification was used in no less than 500 sites, consistent with the economy to standardize and replicate a given solution across a large organization, and avoid having multiple, incompatible solutions that divide and diminish the effectiveness of technical, personnel, training and other resources.

64.   It is believed and thereupon alleged that IBM has, or will tacitly, indirectly, or surreptitiously admit it did use Specification in submittals to the Court. Such tacit, indirect, or surreptitious admission of use by IBM may be used to fabricate an *unwritten-unspoken-probative* notice that it disclosed using Specification in the 1997 to 2001 time-frame, contradicting written communications to Plaintiff and the Department of Justice during that period. Such communication is contained in non-confidential documents IBM did not produce under Rule 34 for discovery. Some such documents are included in Plaintiff's REQUEST TO EXTEND DISCOVERY AND DEPOSITION CUTOFF DATE, filed d3/June 19, 2012, with the Court.

65.   It is believed and thereupon alleged that IBM has breached the Agreement by failing to pay Plaintiff the requisite amount of $49,764.55 per site in which the Specification has been used. It is believed and thereupon alleged that IBM is obligated to pay Plaintiff the amount of no less than $25,000,000 for its use of the Specification in IBM computers at more than 500 sites. IBM has not paid any amount that is owed to Plaintiff.

66.   Plaintiff is informed and believes and thereupon alleges that IBM utilized the Specification for the remediation of the Y2K problem for its customers. IBM has never disclosed to Plaintiff the amount or extent of its use of the Specification for the remediation of the Y2K problem for customers and users of its computers. Plaintiff is informed and believes

and thereupon alleges that IBM provided services for the remediation of the Y2K problem on numerous IBM computers containing an aggregate of more than one billion lines of code. Former IBM CEO Louis Gerstner references the billion lines of code figure in his book *Who Says Elephants Can't Dance*, Harper Collins, 2002, p.292. Therein he admits IBM had Y2K remediation responsibilities *exceeding* a billion lines of code.

67.   Pursuant to the terms of the Agreement, IBM agreed to pay Plaintiff 14 cents per line of code that existed on computers that it used the Specification to remediate the Y2K problem. IBM has breached the Agreement by failing and refusing to pay Plaintiff the amounts owed for the number of lines of code in the computers that were remediated.

68.   It is believed and thereupon alleged that IBM has breached the Agreement by failing to pay Plaintiff the requisite amount of 14 cents per line of code for the use of the Specification for the remediation of the Y2K problem on computers containing an aggregate of over one billion lines of code. IBM is obligated to pay Plaintiff the aggregate amount of $140,000,000 for its use of the Specification in remediating the Y2K problem. IBM has not paid any amount that is owed to Plaintiff under the Agreement.

69.   Plaintiff is entitled to damages pursuant to the Agreement plus costs of suit and legal interests from the date the Specification was used at each site and payment under the Agreement was due and payable.

70.   As there is no provision in Agreement for attorney fees, Plaintiff does not seek attorneys fees pursuant to contractual agreement. Plaintiff does seek all attorneys fees to which he might otherwise be entitled by law.

1
2
3
4
5
6
7
8
9
10   THIS PAGE LEFT INTENTIONALLY BLANK
11
12
13
14
15
16
17
18
19
20
21
22
23
24
25
26
27
28

SECOND AMENDED COMPLAINT: RESPONSIVE TO COURT NOTICE TO PLAINTIFF JUNE 29, 2012
EXHIBITS RESPONSIVE TO *MISHLER V. CLIFT*, 191 F.3d 998, 1008 n.7 (9th Cir. 1999)

a *7ᵗʰ-Gift* page

1

2

3   **FOR A SECOND, SEPARATE, AND DISTINCT CAUSE OF ACTION FOR BREACH**

4   **OF ORAL CONTRACT AGAINST DEFENDANTS IBM AND DOES 1 THROUGH 10,**

5   **PLAINTIFF ALLEGES AS FOLLOWS:**

6   71.   Plaintiff incorporates by reference paragraphs 1 through 70, as though fully set

7   forth herein.

8   72.   Plaintiff entered into an oral agreement with authorized representatives of IBM

9

10   and DOES 1-10 in 1997 that is hereinafter referred to as the Oral Agreement. The essential

11   terms of that Oral Agreement were that IBM would pay Plaintiff the amount of $49,764.55 for

12   each site or location at which the Specification was used to remediate or mitigate the Y2K

13

14   problem on any IBM computers that were used internally or within the corporation. In addition,

15   Plaintiff's offer stated that if IBM used the Specification to remediate the Y2K problem for its

16   customers, it would pay Plaintiff the amount of 14 cents per line of code contained on each

17   computer that was remediated using the Specification. As part of the Oral Agreement, Plaintiff

18   offered to IBM that upon acceptance of the Oral Agreement, IBM would not be charged for any

19   use of the Specification that exceeded the remediation of computers with an aggregate of more

20   than a total of one billion lines of code. The Oral Agreement that was offered limited IBM's

21   payment for use of the Specification in remediating lines of code to $140,000,000. Defendant,

22

23   IBM and DOES 1 through 10 accepted Plaintiffs terms on the Oral Agreement on the condition

24   that Plaintiff obtain a patent for Specification.

25   73.   Plaintiff has been unable, despite his reasonable diligence, to have discovered

26   the use of the Specification by IBM. He has made repeated contacts with IBM representatives

27

28

over the years and each time he has been informed that IBM has not made use of the Specification.

74.   Only within the past year has Plaintiff obtained written-probative-notice that is factually suggestive that IBM used Plaintiff's Specification and owes Plaintiff for that use.

75.   Based upon the foregoing, Plaintiff alleges that he has been thwarted in his efforts to discover the use of his Specification by Defendant, IBM. Despite Plaintiff making reasonably diligent efforts to discover such use, Defendant, IBM has repeatedly represented to Plaintiff that no such use of the Specification has been made and that IBM has no obligation to pay Plaintiff. Plaintiff believes that the representations of IBM have been false and that there has been actual use of the Specification for which payment is due.

76.   Defendant, IBM has breached Oral Agreement by failing and refusing to pay Plaintiff the compensation that is due to him pursuant to the terms of the Oral Agreement.   Due to the strict culture of confidentiality within IBM, Plaintiff is unable to ascertain the scope of the use of the Specification by IBM without discovery pursuant to Rule 34. Defendant, IBM has exempted itself from Rule 34, while readily referencing documents as exhibits contained in Plaintiff's full, discovery production provided in REQUEST TO EXTEND DISCOVERY AND DEPOSITION CUTOFF DATE DUE TO INTERNATIONAL BUSINESS MACHINES DISREGARD OF FRCP 34--before Plaintiff produced such Discovery.

77.   In so doing, regarding non-confidential documents and derivatives thereof not produced by IBM, specific to this case, received by IBM from Plaintiff in 1997, IBM asserts it is above the law and the authority of this honorable Court regarding Discovery and FRCP 34(a)(1).

22

SECOND AMENDED COMPLAINT: RESPONSIVE TO COURT NOTICE TO PLAINTIFF JUNE 29, 2012
EXHIBITS RESPONSIVE TO *MISHLER V. CLIFT*, 191 F.3d 998, 1008 n.7 (9ᵗʰ Cir. 1999)

78.　　Nevertheless, it is believed and thereupon alleged that the use of Specification was extensive within IBM and the Specification was used in no less than 500 sites, consistent with the economy to standardize and replicate a given solution across a large organization, and avoid having multiple, incompatible solutions that divide and diminish the effectiveness of technical, personnel, training and other resources.

79.　　It is believed and thereupon alleged that IBM has, or will tacitly, indirectly, or surreptitiously admit it did use Specification in submittals to the Court. Such tacit, indirect, or surreptitious admission of use is to assert it provided some type of *unwritten-unspoken-probative* notice when it began using Specification in the 1997 to 2001 time-frame, contradicting written communications with Plaintiff and the Department of Justice during that period. Such communication is contained in non-confidential documents IBM did not produce under Rule 34 for discovery. Some such documents are included in Plaintiff's REQUEST TO EXTEND DISCOVERY AND DEPOSITION CUTOFF DATE, filed d3/June 19, 2012 with the Court.

80.　　It is believed and thereupon alleged that IBM has breached the Oral Agreement by failing to pay Plaintiff the requisite amount of $49,764.55 per site in which the Specification has been used. It is believed and thereupon alleged that IBM is obligated to pay Plaintiff the amount of no less than $25,000,000 for its use of the Specification in IBM computers at more than 500 sites. IBM has not paid any amount that is owed to Plaintiff.

81.　　Plaintiff is informed and believes and thereupon alleges that IBM utilized the Specification for the remediation of the Y2K problem for its customers. IBM has never disclosed to Plaintiff the amount or extent of its use of the Specification for the remediation of the Y2K problem for customers and users of its computers. Plaintiff is informed and believes and thereupon alleges that IBM provided services for the remediation of the Y2K problem on numerous IBM computers containing an aggregate of more than one billion lines of code. Former IBM CEO Louis Gerstner references the billion lines of code figure in his book *Who Says Elephants Can't Dance*, Harper Collins, 2002, p292. Therein he admits that IBM had Y2K remediation responsibilities *exceeding* a billion lines of code.

82.   Pursuant to the terms of the Oral Agreement, IBM agreed to pay Plaintiff 14 cents per line of code that existed on computers that it used the Specification to remediate the Y2K problem. IBM has breached the Oral Agreement by failing and refusing to pay Plaintiff the amounts owed for the number of lines of code in the computers that were remediated.

83.   It is believed and thereupon alleged that IBM has breached the Oral Agreement by failing to pay Plaintiff the requisite amount of 14 cents per line of code for the use of the Specification for the remediation of the Y2K problem on computers containing an aggregate of over one billion lines of code. IBM is obligated to pay Plaintiff the aggregate amount of $140,000,000 for its use of the Specification in remediating the Y2K problem. IBM has not paid any amount that is owed to Plaintiff under the Oral Agreement.

84.   Plaintiff is entitled to damages pursuant to the Oral Agreement plus costs of suit and legal interests from the date the Specification was used at each site and payment under the Agreement was due and payable.

1

2

3   **FOR A THIRD, SEPARATE, AND DISTINCT CAUSE OF ACTION FOR BREACH OF**

4   **QUASI CONTRACT AGAINST DEFENDANTS IBM AND DOES 1 THROUGH 10,**

5   **PLAINTIFF ALLEGES AS FOLLOWS:**

6   85.   Plaintiff incorporates by reference paragraphs 1 through 84, as though fully set

7   forth herein.

8

9   86.   On the basis of the representations and discussions between Plaintiff and IBM

10   and DOES 1-10 a quasi contract (Quasi Contract) was established for which IBM became

11   obligated to pay Plaintiff the amounts alleged above and on the terms of the Oral Agreement.

12   87.   It is believed and thereupon alleged that IBM has breached the Quasi Contract

13   by failing to pay Plaintiff the requisite amount of $49,764.55 per site in which the Specification

14   has been used. It is believed and thereupon alleged that IBM is obligated to pay Plaintiff the

15   amount of no less than $25,000,000 for its use of the Specification in IBM computers at more

16

17   than 500 sites. IBM has not paid any amount that is owed to Plaintiff.

18   88.   Plaintiff is informed and believes and thereupon alleges that IBM utilized the

19   Specification for the remediation of the Y2K problem for its customers. IBM has never

20   disclosed to Plaintiff the amount or extent of its use of the Specification for the remediation of

21

22   the Y2K problem for customers and users of its computers. Plaintiff is informed and believes

23   and thereupon alleges that IBM provided services for the remediation of the Y2K problem on

24   numerous IBM computers containing an aggregate of more than one billion lines of code.

25   89.   Pursuant to the terms of Quasi Contract, IBM agreed to pay plaintiff 14 cents

26   per line of code that existed on computers that it used the Specification to remediate the Y2K

27

28

problem. IBM has breached the Quasi Contract by failing and refusing to pay Plaintiff the amounts owed for the number of lines of code in the computers that were remediated.

90.   It is believed and thereupon alleged that IBM has breached the Quasi Contract by failing to pay Plaintiff the requisite amount of 14 cents per line of code for the use of the Specification for the remediation of the Y2K problem on computers.   It is believed and thereupon alleged that IBM has utilized the Specification to remediate computers containing an aggregate of over one billion lines of code. Based upon the Quasi Contract, IBM is obligated to pay Plaintiff the aggregate amount of $140,000,000 for its use of the Specification in remediating the Y2K problem on these computers.   IBM has not paid any amount that is owed to Plaintiff under Quasi Contract.

91.   Plaintiff is entitled to damages pursuant to the Quasi Contract plus costs of suit and legal interest from the date that the Specification was used at each site and payment under the Quasi Contract was due and payable.

FOR A FOURTH, CAUUSE OF ACTION FOR TRADE SECRET MISAPPROPRIATION AGAINST DEFENDANTS IBM AND DOES 1 THROUGH 10, PLAINTIFF ALLEGES AS FOLLOWS:

92.   Plaintiff incorporates by reference paragraphs 1 through 91, as though fully set forth herein.

93.   Plaintiff has treated and considered Specification as a Trade Secret Information that could only be used for the purposes that Plaintiff specifically authorized to the Defendants, IBM and DOES 1-10.

94.   Plaintiff's Specification contained Trade Secret Information that had actual

a *7th-Gift* page

economic value, as it could be used to remediate the Y2K problem in an effective and economic manner.

95.   Defendants, IBM and DOES 1-10 have misappropriated and used Plaintiff's trade secret Specification without his permission or authorization.

96.   IBM's wrongful conduct of misappropriating the trade secret information of the Specification has been deliberately used to improve its business and business opportunities.

97.   Plaintiff is entitled to damages from IBM for its misappropriation and use of the trade secrets of Plaintiff without his authorization.

98.   The amount of damages of Plaintiff is unknown at this time but believed and alleged to be in the amount of $165,000,000.

WHEREFORE, Plaintiff pray for judgment against Defendants, and each of them, as follows:

ON THE FIRST, SECOND, THIRD, AND FOURTH CAUSES OF ACTION:

1.   For General Damages in a sum in excess of the jurisdictional minimum of this Court and no less than $165,000,000;

2.   For Special Damages according to proof;

3.   For Prejudgment Interest permitted by law;

4.   For attorneys fees as allowed by law;

5.   For costs of suit incurred herein; and

6.   For such other and further as the Court deems just and proper.

I declare under penalty of perjury the foregoing is true and correct.

Executed this 4th day of the week, 4th of July, 2012, in Claremont, California.

Anthony Sneed

ANTHONY SNEED, Plaintiff

SECOND AMENDED COMPLAINT: RESPONSIVE TO COURT NOTICE TO PLAINTIFF JUNE 29, 2012
EXHIBITS RESPONSIVE TO *MISHLER V. CLIFT*, 191 F.3d 998, 1008 n.7 (9th Cir. 1999)

Addendum
Aerospace Trauma Detection Patent,
Issued Prior to *Sneed v. IBM* in 2012

After a five-year examination period, and a one-year national security review, the U.S. Patent and Trademark Office (USPTO) issues the following patent to the author.   During the one-year national security review, all documentation and materials regarding this patent "disappear" from USPTO, according to the patent examiner of record and his supervisor. As the patent neared approval, neither could find a trace of the patent's submittal, anywhere at USPTO, as though it never existed. Subsequent to the one-year national security review, when the materials "reappear" at USPTO, the patent examiner of record notifies the author by phone that the "outside agency" conducting the national security review directs USPTO to issue the patent to the author, immediately, which is promptly done.

Know-how within this patent may constitute a significant, aerospace break-through for detecting trauma or stress to flight surfaces of commercial, military, or experimental aircraft, vehicles, or vessels; and to surfaces of aircraft, vehicles, or vessels operating in conditions below the sea, on the sea, on land, in the air, or in space--that is, for conditions under the heavens. It builds on art from 1938 called strain-gauges, which, at that time, greatly enhances the efficient design of aircraft during World War II, and in other areas of manufacturing and design since then.

The Director of the United States Patent and Trademark Office

Has received an application for a patent for a new and useful invention. The title and description of the invention are enclosed. The requirements of law have been complied with, and it has been determined that a patent on the invention shall be granted under the law.

Therefore, this

United States Patent

Grants to the person(s) having title to this patent the right to exclude others from making, using, offering for sale, or selling the invention throughout the United States of America or importing the invention into the United States of America for the term set forth below, subject to the payment of maintenance fees as provided by law.

If this application was filed prior to June 8, 1995, the term of this patent is the longer of seventeen years from the date of grant of this patent or twenty years from the earliest effective U.S. filing date of the application, subject to any statutory extension.

If this application was filed on or after June 8, 1995, the term of this patent is twenty years from the U.S. filing date, subject to any statutory extension. If the application contains a specific reference to an earlier filed application or applications under 35 U.S.C. 120, 121 or 365(c), the term of the patent is twenty years from the date on which the earliest application was filed, subject to any statutory extensions.

John Doll

Acting Director of the United States Patent and Trademark Office

US007546982B2

(12) **United States Patent** (10) **Patent No.:** **US 7,546,982 B2**
Sneed (45) **Date of Patent:** **Jun. 16, 2009**

(54) **SHUTTLE THERMAL INTEGRITY DETECTION SYSTEM**

(76) Inventor: **Anthony Sneed**, 2058 N. Mills Ave., Claremont, CA (US) 91711-2812

(*) Notice: Subject to any disclaimer, the term of this patent is extended or adjusted under 35 U.S.C. 154(b) by 490 days.

(21) Appl. No.: **10/773,511**

(22) Filed: **Feb. 5, 2004**

(65) **Prior Publication Data**

US 2004/0238686 A1 Dec. 2, 2004

Related U.S. Application Data

(60) Provisional application No. 60/445,329, filed on Feb. 5, 2003.

(51) **Int. Cl.**
B64G 1/58 (2006.01)
(52) **U.S. Cl.** **244/159.1**; 244/171.7
(58) **Field of Classification Search** .... 244/158 R, 244/160, 162, 163, 158 A, 159.3, 159.1, 171.7, 244/158.1; 180/274; 280/735
See application file for complete search history.

(56) **References Cited**

U.S. PATENT DOCUMENTS

3,826,452 A * 7/1974 Little 244/160

OTHER PUBLICATIONS

http://www.answers.com/ space%20shuttle%20columbia%20disaster.*
http://science.ksc.nasa.gov/shuttle/missions/sts-107/mission-sts-107.html.*

* cited by examiner

Primary Examiner—Timothy D Collins
(74) *Attorney, Agent, or Firm*—Roberta D. German

(57) **ABSTRACT**

A detection-grid is disclosed that is part of a vehicle's thermal protection layer, such as that of a space shuttle. A hybrid digital/analog system detects electrical changes in the detection grid caused by mechanical trauma to a vehicle's external surface. The system produces timely and useful display of such events. Furthermore, with redundant verification of such real-time data, the vehicle can detach from other apparatus, such as an external fuel tank or booster rockets, to execute pre-planned glide or descent scenarios maximizing a crew's and vehicle's safe return before proceeding to orbit. The detection-grid ablates off during re-entry of a regular mission.

11 Claims, 3 Drawing Sheets

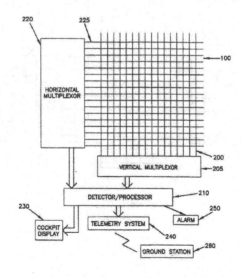

FIG. 1

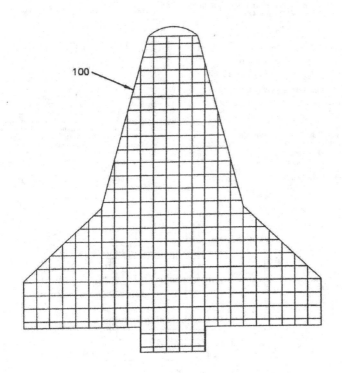

a *7th-Gift* page

484 FROM JUPITER TO GENESIS

FIG. 2

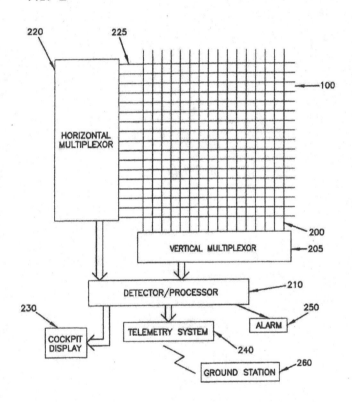

U.S. Patent   Jun. 16, 2009   Sheet 3 of 3   US 7,546,982 B2

FIG. 3

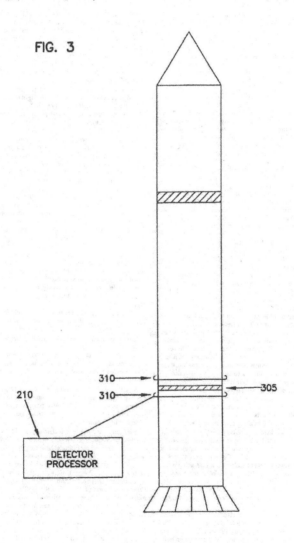

210

310

310

DETECTOR
PROCESSOR

305

US 7,546,982 B2

SHUTTLE THERMAL INTEGRITY DETECTION SYSTEM

RELATED APPLICATION DATA

The present application claims the benefit under 35 U.S.C. 119(e) of the priority date of Provisional Application Ser. No. 60/445,329 filed Feb. 5, 2003, which is hereby incorporated by reference.

FIELD OF THE INVENTION

This invention (herein "Invention") generally relates to sensor technology for detecting structural compromise of thermally-sensitive surfaces when exposed to mechanical trauma. Such thermally-sensitive surfaces may be attached to a static structure or a vehicle, including those operating on the ground, in the air, or space, such as space shuttles.

BACKGROUND OF THE INVENTION

An object of the present Invention is to help return a crew operating a vehicle, like a space shuttle, when compromise of the vehicle's thermal-protection surface could lead to catastrophic loss of crew and vehicle. Of the 113 shuttle launches during a 22 year period, the safe return of crews was achieved 98.3% of the time. In the only two exceptions, crews may have safely returned using existing operational capabilities of the shuttle. One of the two vehicles was Challenger, Jan. 28, 1986, whose crew consisted of commander Francis R. Scobee, pilot Michael J. Smith, and astronauts Judith A. Resnik, Ronald E. McNair, Ellison S. Onizuka, Greg B. Jarvis (a fellow member of the technical staff at what was then Hughes Space and Communications Group), and Sharon "Christa" McCaliffe. The other vehicle was Columbia, Feb. 1, 2003, with commander Rick D. Husband, pilot William C. McCool, and astronauts Michael P. Anderson, Kalpana Chawla ("Culp-na Chav-la"), Laurel B. Salton Clark, David M. Brown and Ilan Ramon. This Invention is motivated out of respect for the profound commitment these husbands, wives, fathers, mothers, sons and daughters made in a shared national and international purpose. And it respects the sentiment of the only U.S. president ever to be awarded a patent since the U.S. Constitution was set in motion, regarding those who purposefully give "the last full measure of devotion":

> . . . It is rather for us to be here dedicated to the great task remaining before us—that from these honored dead we take increased devotion to that cause for which they gave the last full measure of devotion; that we here highly resolve that these dead shall not have died in vain . . .
> President Abraham Lincoln, Gettysburg, Nov. 19, 1863

Providence's wisdom, in permitting the U.S. Constitution to exist in its present form, allows a citizen to present a petition, such as this one, regarding useful innovation that may benefit society as a whole. Though powerless while receiving the news reports of Columbia a year ago, Feb. 1, 2003, developing this Invention was the only means available for this writer to check the downward cycle of despair that enveloped himself, his fellow Caltech alums, the nation and the world, as occurred 17 years prior, with Challenger on Jan. 28, 1986.

It is also remembered that certain events precipitated NACA (National Advisory Committee on Aeronautics—the precursor to NASA leading up to the 1950s) to evolve, having been formed to allow the U.S. to achieve and maintain leadership in the aeronautic arts. Yet, as the level of technical excellence and precision was raised to achieve space flight,

NACA had to give way to NASA (National Aeronautics and Space Administration), respecting a fundamentally different type of leadership, philosophy and technical commitment to express the nation's new aspirations beyond aerodynamics—research stimulated and advanced by the Wright brothers, Samuel Langley and others in America, as well as internationally, near the turn of the last century.

The NACA/NASA change was and is in complete harmony with the U.S. Constitution:

> . . . [T]hat whenever any form of government becomes destructive of these ends, it is the right of the people to alter or to abolish it, and to institute new government, laying its foundation on such principles and organizing its powers in such form, as to them shall seem most likely to effect their safety and happiness.
>
> Prudence, indeed, will dictate that governments long established should not be changed for light and transient causes; and accordingly all experience hath shown, that mankind are more disposed to suffer, while evils are sufferable, than to right themselves by abolishing the forms to which they are accustomed. Congress, Jul. 4, 1776

In other words, the issues embracing the nation's attention, first with Challenger, and later with Columbia, are not specific to just a technological imperative, an institution, or even to this present generation. Because of such issues' visibility to the entire world, they evidence the most sublime meaning of American-based ideals, through a present-day people, and within a shared national purpose which, from the beginning, like so many other American endeavors, has been pregnant with risk and danger, as well as unparalleled achievement and success. The Declaration of Independence acknowledges as much, suggesting that expediency show deference to prudence, while avoiding destructive ritual that may be indifferent to new avenues of safety and happiness for any citizen, regardless of station.

Thus, the matter, at hand, transcends merely the technological arts, exposing the underlying values of those whom, today, must also, by necessity, send others into harm's way.

NACA constituted sacrifice, on an unprecedented scale, to advance the aeronautic arts. And NASA benefited from that knowledge, advancing it far beyond what the NACA charter could encompass. NASA, though formed during the Cold War, inspired a new generation to embrace an impossible challenge—and achieve it.

Now, before us is a new challenge, and it shall again require the boundless energy, enthusiasm, innovation and commitment of a new generation. As NACA provided the foundation for NASA to come into being, so NASA may become the precursor of . . . what?

The California Institute of Technology; the Massachusetts Institute of Technology; Illinois Institute of Technology; Rensselaer Polytechnic Institute; New York Institute of Technology; New England Institute of Technology; Georgia Institute of Technology; Oregon Institute of Technology and universities such as Stanford, Berkeley, Princeton, Chicago, Rice, Rutgers, Dartmouth, Ohio, Michigan, Oregon, Purdue, Colorado, Delaware, Florida, Houston, Louisville, Maryland, Massachusetts, Minnesota, Missouri, Mississippi, New Mexico, North Carolina, Oklahoma, Puerto Rico, Rhode Island, Alabama, Texas, Washington, Wisconsin and Illinois and others'represent academic institutions with a profound interest in space and technology to explore it, whether through manned-vehicles or remote probes and robots. It will be from institutions such as these that men and women will come with vision of greater achievements and successes.

US 7,546,982 B2

3

NACA was a precursor organization in the evolution that lead to NASA in the 1950s. It is proposed that NISAA continues this progress, in the first decade of the 2000s, as the National Institute for Space and Aeronautic Achievement. NISAA will draw on the best talent from America and the world to advance the cause of shared-human achievement through space exploration, vehicle innovation, sustainable system design, and aerodynamic research. Intrinsic to NISAA's mission is the recognition that all progress, in the discovery and exploratory arts, entail prudent acceptance of risk, uncertainty and danger that the unknown always presents. Yet the confidence to go forward respects the unparalleled sacrifice and accomplishments of those who always answered the call, before, that now compels excellence from a new generation, today.

NISAA goes beyond administering the exploration of space to achieving the highest levels of excellence in the discovery and exploration arts. NISAA benefits from the perspective, wisdom and practical skills NASA helped foster, just as NACA did for NASA prior to the 1950s. The level of investment through NASA was extraordinary compared to anything NACA ever attempted. At the time NASA reflected a national purpose visible to the whole world, affirming American values and the respect for every living soul who is sent into harms way to fulfill such purpose.

NISAA will express these values in new venues, protocols and future endeavors. Its very newness establishes a vibrancy and expectation unlike anything before. It will accept personnel, technology and know-how from its predecessor. However, as NASA did before it, NISAA will define itself in new ways, with new personnel, new innovations, and new achievements that will be wholly its on. Its leadership will be more attuned to the constituent technological institutions, which themselves are among the best the world can offer. With "Achievement" in NISAA's name, it will be a light from and to academe and industry, fulfilling its mission, disciplined by wherewithal inherited to carry the torch forward.

The Invention herein is to help complete the legacy of NASA technology and accomplishments with the space shuttle that are extraordinary in the history of nations. This Invention is to help ensure that NASA is honored for all that it has done for America and the world through the lives of those who served it, whether on earth or above it. And as NISAA may come into being through a new birth of commitment, let it always be remembered that such birth, when and where it is permitted to be, has been through the lives of those who gave the "last full measure of devotion."

Thus, this Invention is a declaration of honor and gratitude to NASA, its current administrator, its past administrators, and its talented explorers, scientists, engineers, technicians, management, administrative, operations and support staffs who have carried the torch through the unknown, and lighted the way for an unlimited future that awaits a new generation, with courage and dedication to embrace and advance what NACA and NASA have set before mankind. This will not be achieved within the first 100 days of NISAA, nor perhaps the first thousand days, nor within the present generation. There may yet be unforeseen set backs. But with the steady confidence and progress NACA and NASA established, let us begin.

SUMMARY OF INVENTION

The Invention herein detects anomalies at critical phases of a shuttle's operation, especially between launch and attaining orbit. In the case of Challenger, the Invention may have helped make the decision to detach the solid rocket boosters

4

before their catastrophic failure. Well-planned scenarios for crew and vehicle recovery are extant with timely information to make such a decision. In the case of Columbia, multiple abort options were available between launch and attaining orbit if thermal compromise was detectable in real-time. In both cases, real-time, telemetry information on thermal integrity regarding booster section seams and heat-sensitive surfaces on the shuttle itself may have helped permit the safe return of all 113 crews. It should be noted that in so complex a research vehicle, such as the shuttle, its 98.3% safe-return of crews speaks volumes to NASA's clear intentions regarding the well being of every astronaut.

Space shuttles, as research vehicles, using ceramic or brittle thermal protection technology on their outer surfaces, are vulnerable to catastrophic failure. The Invention herein detects compromise of such a thermal protection layer, whether ceramic tiles or other thermally protective material, within the critical window a shuttle may safely return to earth. It allows such determination before a shuttle goes into orbit, where compromise of thermal protection on the outer surface of a vehicle can lead to catastrophic failure, such as during re-entry. Furthermore, it considers the thermal integrity of "structural seams" and component parts of a shuttle's booster rockets, particularly such seams that when compromised by heat or fuel leakage of any type, could lead to the loss of such vehicles. Additionally, all such detected data are injected into a telemetry stream for real-time decisions that may aid the safe-return of crews operating such vehicles.

Finally, since any decision regarding such thermal integrity may affect some $500 million to launch and operate such a vehicle, or some $5 billion to reproduce a new vehicle, redundant verification of thermal layer integrity on such a vehicle helps reduce to acceptable levels, or prevent, false positives or negatives during operation and or flight.

If in fact there is no compromise of the integrity of the thermal layer that protects a shuttle or other vehicle or surface, a sensor grid used to verify thermal integrity ablates off during the course of re-entry.

BRIEF DESCRIPTION OF THE DRAWINGS

FIG. 1. shows the affected surface area of a vehicle and the sensor grid 100 that spans it. In this implementation, a horizontal/vertical grid is indicated, though other geometries can be accommodated depending on the shape of a vehicle, or the areas of compromise to be detected.

FIG. 2. shows vertical grid elements 200 feeding into vertical multiplexor 205 that samples each grid element at an appropriate rate. Horizontal grid elements 225 feed into horizontal multiplexor 220. Both 205 and 220 feed into detector/processor 210 which converts the signals from the grid elements to produce digital values defining an area or areas of compromise.

FIG. 3. shows a rocket having continuity loops attached above and below the seams of the sections of the rocket.

DETAILED DESCRIPTION OF INVENTION

The present invention provides a 100 sensor grid on the surface area of a vehicle FIG. 1. The preferred embodiment is described of a system and attached apparatus for detecting thermal integrity of a protective layer on a vehicle's surface, such as a shuttle, during its operation. FIG. 2. shows the vertical grid elements 200 that attach to the surface areas of interest through either mechanical or adhesive technology, or are suspended above the areas of interest with suspension points at intervals across the surface areas of interest. The

US 7,546,982 B2

5

Invention detects compromise of all or part of the grid elements. The option of multiple contact points and feeds along any grid element is not excluded, which permits other grid elements to continue operating around a compromised area, allowing horizontal multiplexor 220 and vertical multiplexor 205 to receive redundant data, or to receive localized data independent of other grid areas. Connective terminals or fasteners or sockets allow grid elements to attach to vertical multiplexor 205. The same is true also for horizontal grid elements 225 and horizontal multiplexor 220.

Both vertical multiplexor 205 and horizontal multiplexor 220 feed digitized values to detector/processor 210. Detector/processor 210 then produces further refined digital values that can be used to plot thermal integrity in a cockpit display 230 or for transfer to telemetry system 240 for down-link to ground station 260. Ground station 260 has regular processing systems for displaying all telemetry from a vehicle. Additionally, such display is available to the vehicle's crew in whole or in part.

When an external event or events compromise the detection grid, data values produced from detector/processor 210 allow rapid assessment of the area of compromise. Multiple levels of detection grids may be installed on the outside, within and/or beneath the vehicle's thermal protection layer to determine the depth of compromise, if such refinement is needed. Alarm 250 can automatically sound for extraordinary compromise, processor 210 sensing grid trauma that precedes events indicative of potential vehicle failure in different phases of its operation.

The detection grid may also be comprised of wire-strips, containing one or more wires in very close proximity. These may also be twisted-pair or topologically similar configurations to reduce noise. An alternative implementation is to use conductive paint to form all or part of the detection grid. Additionally, transmission delays along grid elements could be detected to indicate compromise, when different grid transmission times are compared. To aid in such detection, passive or active components (such as resistors, transistors, diodes) may be at the intersection of grid elements, and attach to different grid elements that physically (but not electrically) intersect. Passive or active components that do create electrical connection between different, intersecting grid elements are selected such that the electrical characteristics of the grid are sensitive to even a small subset of such elements, or an individual element, being compromised by external trauma determined to be indicative of potential vehicle failure.

Alternatively, fiber-optic, or other light sensitive grid elements, may be used to detect either breaks or attenuation of light transmittance, indicating compromise as determined from definable limits in detector/processor 210, telemetry system 240, cockpit display 230 or ground station 260. If needed, the entire grid could be fiber-optic, or a subpart of the grid in hybrid configuration, as noise and other electrical characteristics warrant.

Since the Invention detects in real-time, the decision can be made to proceed with a mission or to terminate the mission during easily recoverable phases, such as prior to reaching orbit.

The sensing gird and components are selected based on weight and the ability to readily break when exposed to definable mechanical trauma, or to provide detectable changes when exposed to such trauma. Additionally, the sensor and other components may be selected to allow ready ablation in the final phase of a successful mission.

As needed, a redundant grid or grids may co-exist with a primary grid, other grids offset vertically or horizontally or

6

laterally from one another, since the information from the Invention could determine if a multi-million-dollar mission continues or is terminated.

The integrity of the system is tested prior to launch of a vehicle. Once confirmed, launch proceeds and the system operates continually, including while in orbit to detect any events during orbital operations: space debris, stray tools or material from the vehicle, or mechanical accidents during extra vehicular activity that might compromise vehicle thermal integrity.

The digital components are kept to a minimum to meet weight requirements, perhaps allowing the system, independent of grid and connecting elements, to weigh about a kilogram, if possible, at an off-the-shelf-cost of actual circuit components of about $200 per unit. To meet military specifications, the weight might increase as would the price of a unit depending on grid technology and detection sophistication (propagation delays and multi-compromise detection).

An alternative implementation is for each grid element to detect the absence of a thermal element on the surface of a vehicle (such as a tile). In this case, the grid and its detection elements could be exclusively beneath the thermal protection layer and the elements thereof, requiring little if any reinstallation between missions. The grid detection elements could be one-to-one for the thermal elements (such as tiles) or one-to-many depending on weight and the geometries to accommodate individual thermal surface elements per grid element or multiple thermal surface elements per grid element.

An additional embodiment is to include 310 "continuity loops" above and below the 305 seams of the sections of solid rocket boosters attached to a shuttle FIG. 3. Such wire, fiber optic, conducting paint, or other lines of material would be on the circumference of the booster cylinders (or other areas of the vehicle) above and below sections that are sealed together. Each 310 loop could be within protective insulation or covering. The 310 continuity loops may be redundant and also use a twisted pair topology to reduce noise. The material and insulation used to make the 310 "continuity loops" would melt through when a localized temperature on any part of a loop is elevated beyond a pre-determined threshold.

The wires of the 310 continuity loop on the boosters or other thermally sensitive areas would feed into detector/processor 210, for display or down link with other horizontal grid elements 225 and vertical grid elements 200. This would allow a relatively complete, real-time "thermal integrity" check of all relevant vehicle surfaces and seams or other thermally sensitive areas or components. Should the integrity of any of these components be comprised, both the ground station and on-board personnel would have timely information to make critical decisions after launch and before proceeding to orbit, as well as while in orbit and before initiating re-entry. This, for example, allows on-board or ground personnel to initiate a safe-glide return alternative, after launch, and before a vehicle goes into orbit, or to choose alternatives to re-entry if any vehicle thermal-surface integrity compromise is detected before dropping out of orbit.

Finally, in the event of thermal protective layer form that may produce cavities of empty space beneath the protective layer's surface, 270 pressure sensing devices would act as a redundant detectors of compromise. That is, in addition to the protection-grid and telemetry resulting from it, pressure sensors within hallow cavities would produce separate telemetry of pressure changes. An example of this would be the hollow area between re-enforced carbon-carbon elements on the leading edge of a shuttle's wing, and the flat forward edge of the wing itself. As such, if there is a precipitous change of

US 7,546,982 B2

7

pressure within such a cavity and trauma to the detection grid, then the two independent telemetry sources would indicate crew and vehicle safety is compromised or may soon be compromised during continued operation of the vehicle.

What is claimed is:

1. A system for detecting damage to a thermal protection surface of a spacecraft, the system comprising:
   a). A grid mounted with the thermal protection surface;
   b). An apparatus connected with the grid for detecting the change in the property of the grid; and
   c). An analyzer connected with the apparatus for receiving and analyzing the change in the property of the grid.

2. The system of claim 1, wherein the grid comprises a material that ablates upon re-entry of the spacecraft into the earth's atmosphere.

3. The system of claim 1, wherein the grid mounted with the thermal protection surface is mounted on the exterior of the thermal protection surface, embedded within the thermal protection surface, or mounted beneath the thermal protection surface, or any combination of these positions.

4. The system of claim 1, wherein the spacecraft is a space shuttle.

8

5. The system of claim 1, wherein the apparatus for detecting the change in property of the grid is a multiplexer.

6. The system of claim 1, wherein the analyzer is a detector or processor.

7. The system of claim 1, further comprising a cockpit display that receives the detected change in the property of the grid from the apparatus that detects the change in property of the grid.

8. The system of claim 1, further comprising a telemetry system that receives the detected change in the property of the grid from the apparatus that detects the change in property of the grid.

9. The system of claim 1, wherein the grid mounted with the thermal protection surface is mounted on a seal that fastens parts of the spacecraft.

10. The system of claim 1, wherein the grid comprises a material that undergoes a detectable change in a property of the grid when the thermal protection surface is damaged.

11. The system of claim 10, wherein the material is metallic wire, optical fiber, conductive paint, or any combination of these materials.

* * * * *

a *7ʰ-Gift* page

Father's Limitless Library

Roaming through time's grand temple,
   Spinning wheel to newborn stars
A youthful mind seen resting,
   Beside Jupiter next to Mars

Metallic rings of Saturn,
   Bestriding with hands and feet
Heaven's orbs near home's green lawn,
   Dream Park spreads across Welcome Street

Jupiter spans two meters,
   NASA space-probe there one day?
First to Father's Library,
   Mother's chores the gateway to play

No fire's incited dance,
   Stays Love's faithful joy *foreseen*
Marbles feigning electrons,
   Father's toy computing machine

7th-*Gift* on shelf-lined wall,
   Royal *books* fill one large *Book*
Braves *10-year-old* wherewithal,
   Abideth? Believeth? Forsook?

Admission "GO!" at *14*,
   Toy's helpful ones and zeroes
IBM-Caltech-NASA,
   Other worlds and quantum heroes

Before new patent treasures,
   The 7th in humble heart
Bits sleep half-a-billion miles,
   Lips recalling the sea did part

Appendix ‫ו‬

What Angels Can Never Do

As of 1978 and thereafter, book publisher Zondervan prints as true this English translation of Genesis 5:1,2 and Genesis 6:1,2, with "man" and "men" uppercased and italicized for emphasis:

> This is the written account of Adam's line. When God created *MAN*, he made him in the likeness of God. He created them male and female and blessed them. And when they were created, he called them *MAN*. Genesis 5:1,2

> When *MEN* began to increase in number on the earth and daughters were born to them, the sons of God saw that the daughters of *MEN* were beautiful, and they married any of them they chose. Gen 6:1,2

"Chapter 6: The First Born Daughter on Earth," of the book you are now reading, delineates the difference between *MAN* in the context of Genesis 5:1,2, and *MEN* in the context of Genesis 6:1,2. "Son of God" in the Book of John is always interchangeable with "Son of *MAN*." There is *never* such equivalence between "Son of God" and "Son of *MEN*," or "Son of *MANKIND*," or "Son of *HUMAN BEINGS*," or "Son of *HUMANS*." This is because *MAN* sets precedent of the original, divine, likeness and image of God as male and female; the Son of *MAN* fulfills that precedent in an extraordinary way, being born of a virgin:

> Then God said, "let us make *MAN* in our image, in our likeness . . ." Genesis 1:26

> So God created *MAN* in his own image, in the image of God he created him, male and female he created them. Genesis 1:27

Thus "Son of *MAN*," through testimony in the Book of John, harmonizes with and fulfills this precedent:

> I tell you the truth, a time is coming and has now come when the dead will hear the voice of the *Son of God* and those who hear will live. For as the Father has life in himself, so he has granted the Son to have life in himself. And he has given him authority to judge because he is the *Son of MAN*. John 5:25-27

The Son of God in John 5:25,26, who is the Son of *MAN* in John 5:27, fulfills the testimony that *MAN*, as male and female, is made in the image of God, just as Genesis 1:26,27 and Genesis 5:1,2 declare. Later translations prefer changing *MAN* to "mankind" in Genesis 1:26,27 and Genesis 5:1,2, to improve readability for contemporary readers. However, no such substitution preference *ever* occurs for "Son of *MAN*."

"Mankind" is composed of two, distinct lines: *MAN* as the descendants of Adam who remain faithful to the Creator and his image and likeness; and *MEN* who choose their own ways, which includes rebellion against the Creator. This appendix vets the premise that an "angel," as a spirit being, may marry a human, flesh-and-blood female. The premise, and vetting it, contrasts *MEN* in Genesis 6:1,2 with "humans," that some translations prefer, to improve readability.

¶1.   Positional Declarations regarding an angel marrying.

¶1.1.   An angel can marry.
¶1.2.   An angel can never marry.

Central to this inquiry is, "Can any angel ever marry?" Matthew 22:30 precludes ¶1.1.:

At the resurrection people will neither marry nor be given in marriage; they will be like the angels in heaven. Matthew 22:30

Fallen angels—that is, angels expelled from heaven—are also given consideration regarding a marriage premise.

¶2.   Positional Declarations regarding the "daughters of men" in Genesis 6:2.

¶2.1.   A least one fallen angel marries at least one such daughter in Genesis 6:2.
¶2.2.   A fallen angel never marries a "daughter of men" in Genesis 6:2.

Hebrews 1:5 precludes ¶2.1 being true, or construing an "angel" as a "son of God" in Genesis 6:2:

> For to *which* of the *angels* did God *ever* say, "You are my *Son*; today I have become your *Father*"? [Psalm 2:7] Or again, "I will be his *Father* and he will be my *Son*"? [2 Samuel 7:14; 1Chronicles 17:13]

The answer to the rhetorical question is *never.* Hebrews 1:7 and 1:14 provide an affirming witness:

> In speaking of the angels he says, "He makes his angels winds, his servants flames of fire." [Psalms 104:4] Hebrews 1:7

> Are not all angels ministering *spirits* sent to serve those who will inherit salvation? Hebrews 1:14

Faithful angels can have physical bodies, which Scripture gives numerous witnesses, consistent with the Spirit of truth's ability to manifest himself anthropomorphically. Such manifestations can occur in multiple places, simultaneously, in full unity with the purpose of the [I-SHALL-BE]: the Father, the Son, and the Spirit of truth; that is, as the God of Abraham, the God of Isaac, and the God of Jacob.

> The [I-SHALL-BE] appeared to Abraham near the great trees of Mamre while he was sitting at the entrance to his tent in the heat of the day. Abraham looked up and *saw* three *men* standing nearby. When he saw them, he hurried from the entrance of his tent to meet them and bowed low to the ground. Genesis 18:1,2

> He [Abraham] then brought some curds and milk and the calf that had been prepared, and set these

a *7ʰ-Gift* page

before them. While *they ate*, he stood near them under a tree. Genesis 18:8

With the coming of dawn, the *angels* urged Lot, saying, "Hurry! Take your wife and your two daughters who are here, or you will be swept away when the city is punished."

When he hesitated, the *men grasped* his hand and the hands of his wife and of his two daughters and led them safely out of the city, for the [I-SHALL-BE] was merciful to them. Genesis 19:15, 16

Now when Joshua was near Jericho, he looked up and *saw a man* standing in front of him with a drawn sword in his hand. Joshua went up to him and asked, "Are you for us or for our enemies?"
    "Neither," he replied . . . Joshua 5:13

The angel of the [I-SHALL-BE] *appeared* to [Samson's mother] and said, "You are sterile and childless, but you are going to conceive and have a son." Judges 13:3

Then *the angel* of the [I-SHALL-BE] *appeared* to [Zechariah] . . . Luke 1:11

The *angel* answered, "I am Gabriel . . ." Luke 1:19

Fallen angels, as *spirit beings, never* have physical bodies, yet they can dwell within flesh-and-blood beings, including living *people* and *animals*:

Whenever the *spirit from God* came upon Saul, David would take his harp and play. Then relief would come to Saul; he would feel better, and the *evil spirit* would leave him. 1Samuel 16:23

When he arrived at the other side in the region of the Gadarenes, two demon-possessed men coming from the tombs met him. They were so violent that no one could pass that way. "What do you want with us, Son of God?" they shouted. "Have you come here to torture us before the appointed time?" Some distance from them a large herd of pigs were feeding. The demons begged [יֵשׁוּעַ], "If you drive us out, send us into the herd of pigs."

He said to them, "Go!" So they came out and went into the pigs, and the whole herd rushed down the steep bank into the lake and died in the water." Matthew 8:28-32

The next day, when they came down from the mountain, a large crowd met him. A man in the crowd called out, "Teacher, I beg you to look at my son, for he is my only child. A spirit seizes him and he suddenly screams; it throws him into convulsions so that he foams at the mouth. It scarcely ever leaves him and is destroying him. I begged your disciples to drive it out, but they could not."

"O unbelieving and perverse generation," [יֵשׁוּעַ] replied, "how long shall I stay with you and put up with you? Bring your son here."

Even while the boy was coming, the demon threw him to the ground in a convulsion. But [יֵשׁוּעַ]

rebuked the evil spirit, healed the boy and gave him back to his father. Luke 9:37-42

[ישוע] answered, "It is the one to whom I will give this piece of bread when I have dipped it in the dish." Then, dipping the piece of bread, he gave it to Judas Iscariot, son of Simon. As soon as Judas took the bread, Satan entered into him.

"What you are about to do, do quickly," [ישוע] told him[.] John 13:26,27

Once when we were going to the place of prayer, we were met by a slave girl who had a spirit by which she predicted the future. She earned a great deal of money for her owners by fortune-telling. This girl followed Paul and the rest of us, shouting, "These men are servants of the Most High God, who are telling you the way to be saved." She kept this up for many days. Finally Paul became so troubled that he turned around and said to the spirit, "In the name of [ישוע] Christ I command you to come out of her!" At that moment the spirit left her.

Acts 16:16-18

Finally, to call a *fallen angel* a "son of God" would elevate such an angel *above all* the *faithful angels* who are always obedient to my Father, his Son, and the Spirit of truth. No *faithful angel* is ever called a "son of God":

For to *which* of the *angels* did God *ever* say, "You are my *Son*; today I have become your *Father*"? [Psalm 2:7] Or again, "I will be his *Father* and he will be my *Son*"? [2Samuel 7:14; 1Chronicles 17:13]

Hebrews 1:5

ꝟ3.   Positional declarations regarding Nephilim in Genesis 6:4

ꝟ3.1.   Nephilim are angelic beings.

ꝟ3.2.   Nephilim are people made of flesh and blood.

Genesis 6:4 precludes ꝟ3.1 by declaring the Nephilim were "hereos of old, *men* of renown.".

ꝟ4.   Positional declarations on procreation resulting in Nephilim of Genesis 6:4

ꝟ4.1.   Male Nephilim are offspring of fallen angels and the "daughters of men" in Genesis 6:2.

ꝟ4.2.   Each daughter in Genesis 6:2 is a virgin, *before* she consummates her first relationship of any kind.

ꝟ4.3.   A Nephilim is never an offspring of a fallen angel and a virgin in Genesis 6:2.

ꝟ4.4.   An angel, fallen or not, is never a "son of God" in Genesis 6:1, 6:2, or 6:4.

If ꝟ4.1 and ꝟ4.2 are both true, then *one* such union between a fallen angel *as a spirit being,* and *a virgin daughter* in Genesis 6:2, results in a *son*—who would be a flesh-and-blood male-child, based on Genesis 6:4. To conceive such a son—a flesh-and-blood male-child, *before the flood,* through the union of an angel, *as a spirit being,* and a *virgin*—requires the false premise that "*son of God*" and "*angel*" are interchangeable in Genesis 6:1, 6:2, or 6:4. Thus, all references to "*sons of God*" in Genesis 6:1, 6:2 and 6:4 only refer to male, flesh-and-blood persons, and never to angels *as spirit beings.*

There is *one and only one* flesh-and-blood male-child who *is* born of a spirit being—*the Holy Spirit*—and a *virgin,* in Scripture, as it is written:

> Then Isaiah said, "Hear now, you house of David! Is it not enough to try the patience of men? Will you try the patience of my God also? Therefore the [I-SHALL-BE] himself will give you a sign: The *virgin* will be with child and will give birth to a *son,* and will call him Immanuel [God with us]."
>
> Isaiah 7:13, 14

> "How will this be," Mary asked the angel, "since I am a *virgin?*" The angel answered, "*The Holy Spirit* will come upon you, and the power of the Most High will overshadow you. So the holy one to be born will be called the Son of God."
>
> Luke 1:34,35

> But after he had considered this, an angel of the [I-SHALL-BE] appeared to him in a dream and said, "Joseph son of David, do not be afraid to take Mary home as your wife, because what is conceived in her is from *the Holy Spirit.* She will give birth to a son, and you are to give him the name [I-SHALL-BE-SALVATION], because he will save his people from their sins."

> All this took place to fulfill what the [I-SHALL-BE] had said through the prophet: "The *virgin* will be with child and will give birth to a son, and they will call him Immanuel" [Isaiah 7:14]—which means, 'God with us.'"
>
> Matthew 1:20-23

Thus, within Adam, the first son of God whom the Creator forms from the dust of the ground, is the first y-chromosome, which his male descendants all share. This precludes third-party *fallen angels* introducing y-chromosomes unrelated to Adam. One exception to this is the Son of God, born of a *virgin*. His y-chromosome, as with Adam's originally, is of divine origin, since he is the *Second Adam* (1 Corinthians 15:45-47):

> The [I-SHALL-BE] had said to Abram, "Leave your country, your people and your father's household and go to the land I will show you.
>
> "I will make you into a great nation and I will bless you; I will make your name great, and you will be a blessing.
>
> "I will bless those who bless you, and whoever curses you I will curse; and all peoples on earth will be blessed through you." Genesis 12:1-3
>
> The fear of the [I-SHALL-BE] is the beginning of wisdom, and *knowledge of the Holy One* is understanding. Proverbs 9:10
>
> I am the most ignorant of men;
>   I do not have a man's understanding.
> I have not learned wisdom,
>   nor have I *knowledge of the Holy One.*
> Who has gone up to heaven and come down?
>   Who has gathered up the wind in the hollow of his hands?
> Who has wrapped up the waters in his cloak?
>   Who has established all the ends of the earth?
> What is *his name*, and the *name of his son?*
>   Tell me if you know!   Proverbs 30:2-4

a *7ʰ-Gift* page

The one who comes from above is above all; the one who is from the earth belongs to the earth, and speaks as one from the earth. The one who comes from heaven is above all. He testifies to what he has seen and heard, but no one accepts his testimony. The man who has accepted it has certified that God is truthful. For the one whom God has sent speaks the words of God, for God gives the Spirit without limit. The Father loves the Son and has placed everything in his hands. John 3:31-35

What Did the *Woman* Actually Say?

In Genesis 3:17, Adam *listens* as the [I-SHALL-BE] rebukes him with the following:

To Adam he said, "Because you *listened* to your wife *and* ate from the tree about which *I commanded* you, 'You must not eat of it,'

"Cursed is the ground because of you; through painful toil you will eat of it all the days of your life."

The word "listen" occurs in multiple proverbs, and always involves hearing *spoken* words:

> *Listen*, my son, to a father's instruction;
>> pay attention and gain understanding.
>>> Proverbs 4:1

> My son, pay attention to what I say;
>> *listen* closely to my words.
> Do not let them out of your sight,
>> keep them within your heart;

for they are life to those who find them
　　and health to a man's whole body.
　　　　　　　　　　　　　　　　Proverbs 4:20-22

"Now then, my sons, *listen* to me;
　　blessed are those who keep my ways.
Listen to my instruction and be wise;
　　do not ignore it.
Blessed is the man who *listens* to me,
　　watching daily at my doors,
　　waiting at my doorway.　　　　Proverbs 8:33,34

So, what did the *woman*, Adam's wife, actually *say* prior to Adam receiving the fruit from her, and then eating it? At the time she gives the fruit to Adam [Genesis 3:6], the Scripture does not disclose the *woman* saying *anything* to Adam. The *only* words spoken by her are in an earlier response to the serpent's query. And *that* response—her first response and the only spoken words that Scripture discloses she says prior to Adam eating the fruit—is in *harmony with the Word of God*, as noted in Genesis 3:2,3:

> The woman said to the serpent, "We may eat fruit
> from the trees in the garden, but *God did say*, 'You
> must not eat fruit from the tree that is in the middle
> of the garden, and you must not touch it, or you
> will die.'"

Genesis 3:6 makes clear Adam is with the *woman* when she *speaks* this, out loud. *If* this is what Adam *listens* to, referenced in Genesis 3:17, then the [I-SHALL-BE] acknowledges the *woman's initial*, spoken response as consistent with the *Word of God*. Furthermore, in doing so, the [I-SHALL-BE] establishes the precedent of two witnesses, prevalent throughout Scripture, regarding the authority to judge:

"But if I do judge, my decisions are right, because I
am not *alone.* I stand with the Father, who sent me.
In your own Law it is written that the testimony of
two men is valid. I am one who testifies for myself;
my other witness is the Father who sent me."

John 8:16-18

The *Word of God* that Genesis 3:2,3 discloses informs the
woman's spoken response, contrasts with Scripture's silence
about anything else she says prior to judgment. This finding,
regarding her initial, spoken response, affirms the primacy of
the *Word of God,* independent of who gives notice, whether male
or female, husband or wife, and that the *woman* has standing
as a witness to the *Word of God.* It is her *spoken* response that
the [I-SHALL-BE] acknowledges, consistent with the proverb:

The [I-SHALL-BE] detests the way of the wicked,
but he loves those who pursue righteousness.

Proverbs 14:9

He who leads the upright along an evil path will fall
into his own trap, but the blameless will receive a
good inheritance. Proverbs 28:10

The [I-SHALL-BE] attaches no merit to any action or
statement by the serpent, after the *woman* gives righteous
notice to the serpent about the *Word of God.* The serpent then
violently attacks her. Adam later discloses to the [I-SHALL-BE]
that *he* eats the fruit, with no further disclosure. He is silent
about the serpent's involvement in any way with the *woman*—
the *woman* whom only the [I-SHALL-BE] creates:

The man said, "The *woman* you put here with me—she
gave me some fruit from the tree and *I* ate it."

Genesis 3:12

Thus, without the *woman's* disclosure about the serpent deceiving her, it appears Adam simply wants to eat some fruit, which his wife obediently provides him, to do as he please. Condemnation rests only with Adam—until the *woman* declares, "The *serpent* deceived me, *and I ate*." [Gen 3:13]

> Then the [I-SHALL-BE] said to the *woman,* "What is this you have done?"
>
> The *woman* said, "The *serpent* deceived me, and I ate." [Gen 3:13]

Through this, the [I-SHALL-BE] accepts that the serpent spoke to the *woman,* who first *resists* the serpent with the *Word of God*. When the *woman* resists with the *Word of God,* the serpent violently and angrily escalates his attack on the *woman* by proclaiming to her what "*God knows*":

> "You will not surely die," the serpent said to the *woman.* "For *God knows* that when you eat of it your eyes will be opened, and you will be like God, knowing good and evil." [Gen 3:4,5]

Declaring what "*God knows,*" in this way, is violent iniquity. Through this deception, the serpent claims he is only being truthful and sharing what "*God knows,*" to help the *woman* fulfill her desire, including her desire as a wife. The deception is that knowing what "*God knows,*" about good and evil, will give the *woman* an advantage in fulfilling her own desire. This is at the heart of murder itself, *from the beginning,* as it is written:

> [H]e was a murderer *from the beginning,* not holding to the truth, for there is no truth in him. When he lies, he speaks his native language, for he is a liar and the father of lies. John 8:44

The *woman* is always faithful to the [I-SHALL-BE] and to Adam, prior to this. Both love her, knowing that she is a creation of the [I-SHALL-BE], himself, from Adam's side. There is no reason for Adam not to trust her, or impose upon her freedom to act as she will—before eating the fruit. Adam has the command to keep, not to eat from the Tree of the Knowledge of Good and Evil, as well as his vow, that the *woman* is "bone of my bones and flesh of my flesh." If the *woman* eats from the tree, and he does not, their flesh and bones will become different. It will be a witness against Adam's vow, through words he speaks to the [I-SHALL-BE] and the *woman*. The proverb states:

It is a trap for a man to dedicate something rashly, and only later consider his vow. Proverbs 20:25

The man said,

"This is now bone of my bones and flesh of my flesh; she shall be called 'woman,' for she was taken out of man."

Genesis 2:23

Adam eats the fruit, visiting whatever effects on his own flesh and bones that the fruit will have on his wife's flesh and bones. This gives him standing to speak for both of them, rather than leaving her to suffer the fruit's effects and answer, alone, for her actions. He keeps his vow through the words he declares to the [I-SHALL-BE] and to the *woman*, though it will surely cost him his life. Adam keeps the *woman* as his wife, and answers inquiries on her behalf. Later, they fulfill the promise of being fruitful, together, across the generations, with a seed of the *woman's* to become the salvation of the whole world, including for Adam and his wife.

The Garden of Eden, Love, and the 7th-Gift

The Garden of Eden is where the [I-SHALL-BE] first reveals the glory of his love between heaven and earth:

o   Love's bond of trust between the [I-SHALL-BE] and MAN.
o   Love's unity between husband and wife.
o   Love's gift to bless 7th-days with rest.

All of the above the [I-SHALL-BE] anchors, first, in the Garden of Eden, with MAN as male and female. The serpent's deception, in one stroke, attempts to defile the [I-SHALL-BE's] glory of love, and therefore defile the [I-SHALL-BE] or, if possible, to destroy the [I-SHALL-BE] himself—the serpent's fondest desire.

Genesis 3:6-8 reveals the devastation the serpent causes in the Garden of Eden:

> [Gen 3:1] "[D]id *God really say* you must not eat from any of the trees in the garden."
>
> [Gen 3:2] The woman said to the serpent, "We may eat fruit from the trees in the garden, [Gen 3:3] but *God did say*, 'You must not eat fruit from the tree in the *middle* of the garden, and you must not touch it, or you will die.'"
>
> [Gen 3:4] "You will not surely die," the serpent said to the *woman.* [Gen 3:5] "*For God knows* that when you eat of it your eyes will be opened, and you will be like God, knowing good and evil."
>
> [Gen 3:6] When the woman saw that the fruit of the tree was pleasing to the eye and good for food, and also desirable for gaining wisdom, she took

a *7th-Gift* page

some and ate it. She also gave some to her husband, who was with her, and he ate it. [Gen 3:7] Then the eyes of both of them were opened, and they realized they were naked; so they sewed fig leaves together and made coverings for themselves. [Gen 3:8] Then the man and his wife heard the sound of the [I-SHALL-BE] God as he was walking in the garden in the cool of the day, and they hid from the [I-SHALL-BE] God among the trees of the garden.

[Gen 3:9] But the [I-SHALL-BE] God called to the man, "Where are you?"

[Gen 3:10] He answered, "I heard you in the garden and I was afraid because I was naked; so I hid."

Relative to running *away* from the [I-SHALL-BE] or running *to* him, it remains written:

The name of the [I-SHALL-BE] is a strong tower; the righteous run *to* it and are safe. Proverbs 18:10

Adam and his wife are running *away* from the [I-SHALL-BE], defiling the splendor, majesty and trust of his love to come and be with them, compelling him to ask, "Where are you?" The question means there is agreement to be together *on that day*, the [I-SHALL-BE] fulfilling the time to be with them. The [I-SHALL-BE] is there, at the appointed time, but not Adam and his wife. Adam and his wife run away from the *middle* of the garden, where the Tree of Life and the Tree of the Knowledge of Good and Evil are. They run away from those trees to hide among the "trees of the garden," away from meeting with the [I-SHALL-BE], as it is written:

> [Gen 2:9] And the [I-SHALL-BE] God made all
> kinds of trees grow out of the ground—trees that
> were pleasing to the eye and good for food. In the
> *middle* of the garden were the tree of life and the
> tree of the knowledge of good and evil.

They run away from the *middle* of the garden *after* they sew *fig leaves* together to make coverings for themselves. They flee from meeting with the [I-SHALL-BE] in the *middle* of the garden. Thus, the tree from which they gather leaves, to make coverings for themselves, is in the *middle* of the garden. The Tree of the Knowledge of Good and Evil is one of the two trees in the *middle* of the garden. And it is from that tree Adam's wife takes fruit, before she and Adam gather *fig leaves* from *a* tree in the *middle* of the garden to make coverings for themselves. The fig tree is repeatedly evocative of indifference to divine will:

> Then [I-SHALL-BE-SALVATION] said to the tree,
> "May no one ever eat fruit from you again[.]"
>   Mark 11:14

> In the morning, as they went along, they saw the
> *fig tree* withered from the roots. Peter remembered
> and said to [I-SHALL-BE-SALVATION], "Rabbi,
> look! The *fig tree* you cursed has withered!"

> "Have faith in *God*," [I-SHALL-BE-SALVATION]
> answered. Mark 11:20, 21

> I watched as the Lamb opened the sixth seal. There
> was a great earthquake. The sun turned black
> like sackcloth made of goat hair, the whole moon
> turned blood red, and the stars in the sky fell to the
> earth, as late *figs* drop from a *fig tree* when shaken by
> a strong wind. Revelation 6:12,13 [In other words,

some fruit cling to the *fig tree*, unless shaken off by a strong wind, just as some cling to "the tree of knowledge," to the detriment of faith, until events prove such clinging a poor choice, when clinging precludes acting by faith.]

The next day, [ישוע] decided to leave for Galilee. Finding Philip, he said to him, "Follow me."

Philip, like Andrew and Peter, was from the town of Bethsaida. Philip found Nathanael and told him, "We have found the one Moses wrote about in the Law, and about whom the prophets also wrote—[ישוע] of Nazareth, the son of Joseph."

"Nazareth!" Can anything good come from there?" Nathanael asked.

"Come and see," said Philip.

When [ישוע] saw Nathanael approaching, he said of him, "Here is a true Israelite, in whom there is nothing false."

"How do you know me?" Nathanael asked.

[ישוע] answered, "I saw you while you were *still* under the *fig tree* before Philip called you."

Then Nathanael declared, "Rabbi, you are the Son of God; you are the King of Israel."

[ישוע] said, "You believe because I told you I saw you under the *fig tree*. You shall see greater things than that." John 1:43-50

Answering the call by faith, Philip follows the Son of God. However, Nathanael stays under a fig tree, convinced by his own knowledge that he can ignore the one to whom others are responding. The Son of God does not go to the fig tree, himself, but reaches out to Nathanael through Philip, who exhorts Nathanael to come, based on Moses and the prophets. But Nathanael's knowledge *still* keeps him under the fig tree, because of his objection that Nazareth is contrary to his understanding that the "real" Son of God would avoid such a place.

It is only through Philip's personal appeal that Nathanael comes from under the fig tree, though he is *still* very skeptical, until he reaches the Son of God, who tells him, "I saw you while you were *still* under the *fig tree* before Philip called you." To Nathanael, only the Son of God, the King of Israel could know about the fig tree, Nathanael's heart, and what kept him, until Philip helps him to act by faith instead of presumed knowledge.

Similarly, Adam and his wife remain in the proximity of the tree from which they gather the leaves they are wearing, until they hear the sound of the [I-SHALL-BE] God walking in the garden in the cool of the day—at the time he and they are to be together, in the cool of that day. Acting from the knowledge of good and evil, rather than by faith, they avoid him and flee from the *middle* of the garden, wearing the *leaves* they gather there. And so, they hide among the "trees of the garden," that is, among the trees *not* in the *middle* of the garden. The fig leaves they gather, therefore, aren't from the "trees of the garden" they hide among.

Thus *knowledge* compels them to flee from the love of the [I-SHALL-BE], wearing coverings of fig leaves as evidence of their transgression, consistent with the serpent's desire to sever the loving bond of trust between the [I-SHALL-BE] and MAN, for the first time, and for all time, if possible. Additionally, the serpent desires to defile love's gift of rest on 7ᵗʰ-days—the day the [I-SHALL-BE] blesses for the [I-SHALL-BE] and MAN to rest and be together, free from all their labors. Knowing when and where they meet together, and the time and day, allows the serpent to attack these expressions of love, and the blessing thereof, between heaven and earth.

What remains is love's unity between husband and wife, which the serpent also desires to destroy, but fails, because Adam keeps his "bone of my bones and flesh of my flesh" vow. Adam and his wife are wearing the same coverings, and are both enduring, together, the effects that the fruit from the Tree of the Knowledge of Good and Evil visits upon their bodies. And yet, they are still together. Adam and his wife, hiding among the "trees of the garden," appear before the [I-SHALL-BE], whose garden's *middle* has the Tree of Life and the Tree of the Knowledge of Good and Evil. And the [I-SHALL-BE] sees they are both wearing the same coverings, from fig leaves they gather while in the *middle* of the garden, before meeting with him when he calls out, "Where are you?"

To protect his purity and righteousness, the [I-SHALL-BE] could simply depart and return to heaven, and end love's bond of trust with MAN, and withdraw love's gift, on earth, and take back the blessing of rest on 7th-days and its origin from heaven. Doing so would prevent the serpent from ever defiling the [I-SHALL-BE's] glory of love again, removing any similar vulnerability the serpent may exploit to reach into heaven and molest the [I-SHALL-BE's] righteousness and purity. The [I-SHALL-BE] can take his glory, his love, and his majesty and return to heaven and leave MAN and earth behind, for the serpent to do with whatever he chooses. Or the [I-SHALL-BE] can simply destroy all of them, along with the earth.

The [I-SHALL-BE] does neither. He does not retreat, but rather rebukes the trespasses against him and his righteousness. Mercifully, he sets a future judgment for the serpent, allowing him to live, and places a curse upon him to crawl on his belly and eat the dust of the earth all the days of his life, for deceiving the *woman*. The serpent's blasphemous "*God knows*" declaration about fruit from the Tree of the Knowledge of

Good and Evil undermines the *woman's* understanding about "God's will," since the serpent is risking his future existence if this isn't "true".

This creates a perception that the serpent is being "truthful," and that for the *woman* not to act on what "*God knows*" could place her at odds with "God's will." Being perceived as "truthful" is the only way the serpent can attack the *woman's* faithfulness, when she rebuffs and angers the serpent with the *Word of God*: "But *God did say* you must not eat . . ." When she resists him, the serpent's violent, angry and desperate act is to immediately declare "*God knows*" otherwise, his creatibility being that God would destroy him if this is false. Such *new* knowledge of what "*God knows,*" contrary to the *Word of God,* without patient humility to ask God himself, is at the very heart of partaking fruit from the Tree of the Knowledge of Good and Evil.

In encouraging her to do so, based on what "*God knows,*" the serpent desires to destroy the *woman's* fertility, or defile or kill any *future-seed-descendant.* Either way, the serpent could prove the [I-SHALL-BE] can't keep the promise to be "fruitful and multiply." The serpent conceals this by telling the woman she will have everything she desires by eating fruit from the Tree of the Knowledge of Good and Evil, and that "*God knows*" this.

The [I-SHALL-BE] does fulfill his promise with Adam and his wife, that they will be fruitful and multiply. He does so by separating their offspring from those of the serpent's, after the serpent's "*God knows*" deception:

> So the [I-SHALL-BE] God said to the serpent, "Because you have done this,
>
> "Cursed are you above all the livestock and all the wild animals. You will crawl on your belly and you will east dust all the days of your life. And I will put enmity between you and the woman, and between

a *7ʰ-Gift* page

your offspring and hers; he will crush your head
and you will strike his heel."

Gen 3:14, 15

Thus, even as the [I-SHALL-BE] condemns the serpent, he
is reassuring the *woman* that she will have offspring, meaning
she will live to have them, and that Adam will live, since
her desire can not be for a dead husband. Such words are
comforting for Adam and his wife to hear, as they await their
judgment, listening as the [I-SHALL-BE] mercifully delays
final judgment on the serpent. And when the [I-SHALL-BE]
speaks to Adam and his wife, directly, he affirms that they
are husband and wife, and they will be fruitful and multiply.
Through his love he clothes them, and watches over them
and their descendants, across the generations, as he fulfills
his promise of love through salvation. Knowledge, alone,
without the [I-SHALL-BE], can never fulfill such promise. As
they repent and acknowledge their transgression, they receive
through faith what only love can give, freely, and generously,
generation after generation.

Through love, the [I-SHALL-BE] sustains the bond of trust
between himself and MAN. The love in the unity between
husband and wife glorifies the promise of a *future seed*, as Adam
and Eve remain together their entire lives, as a wonderful
precedent for generations to come, to which the Son of God
gives testimony:

> Some Pharisees came to [I-SHALLL-BE-
> SALVATION] to test him. They asked, "Is it lawful
> for a man to divorce his wife for any and every
> reason?"

"Haven't you read," he replied, "that at the *beginning* the Creator 'made them male and female' [Gen 1:27] and said, 'For this reason a man will leave his father and mother and be united to his wife, and the two will become one flesh'? [Gen 2:24] So they are no longer two, but one. Therefore what God has joined together, let man not separate."

"Why then," they asked, "did Moses command that a man give his wife a certificate of divorce and send her away?"

[I-SHALL-BE-SALVATION] replied, "Moses permitted you to divorce your wives because your hearts were hard. But it was not this way *from the beginning*. I tell you that anyone who divorces his wife, except for marital unfaithfulness, and marries another woman commits adultery."

                                        Matthew 19:3-9

Adam maintains his vow, as a husband, his entire life, and is the one to whom the *Second Adam* gives witness, regarding the *from the beginning* declaration. And love's gift to bless 7th-days with rest, magnifies the wisdom in the [I-SHALL-BE's] love, as he keeps the promise of rest for those who love him, across generations, to this very day.

The serpent repeatedly attacks the [I-SHALL-BE], through the glory of his love, to force him to withdraw his love, and retreat from keeping MAN, as male and female, in his love through trust, unity, and rest. The attacks across the generations are often violent, and painful, and yet the [I-SHALL-BE], with MAN, patiently endures the suffering, pain, tears, mourning, death, and every affront by the serpent to exploit MAN, to kill, steal, and destroy what the [I-SHALL-BE] gives through the love that never fails, and patiently endures.

The Ten Commandments, Exodus 20:1-17, given much, much later through Moses, reveal how the serpent exploits MAN to kill, steal, destroy *and change* that which the [I-SHALL-BE] gives through love. The Ten Commandments remind everyone of what has been lost, as they also guard and hold open the way back to that love, as powerfully as parting the sea, to walk between two walls of water, and fulfill the promise of love that only is from the [I-SHALL-BE].

Out of love, the [I-SHALL-BE], himself, sets forth the blessing of rest on the 7ᵗʰ-day. Why follow anyone else?

> By the 7ᵗʰ-day God had finished the work he had been doing; so on the 7ᵗʰ-day he rested from all his work. And God blessed the 7ᵗʰ-day and made it holy, because on it he rested from all the work of creating that he had done.
>
> Genesis 2:2,3

This blessing, with the *7ᵗʰ-Gift*, is a gift of love. The serpent frightens Adam and his wife to run away from the [I-SHALL-BE] and the *7ᵗʰ-Gift* he gives them—a gift of inestimable value, a restful gift reaching from heaven's throne, through the Garden of Eden, to today. And through grace, the [I-SHALL-BE] allows them to keep this gift, even after departure from the Garden of Eden—a gift for all subsequent generations from the [I-SHALL-BE]. The entire book of Revelation is given through the *7ᵗʰ-Gift*, on the [I-shall-be's] day:

> Do not move an ancient boundary stone set up by your forefathers.
>
> Proverbs 22:28

> On the [I-shall-be's] day I was in the Spirit . . . . And when I turned I saw 7 golden lampstands, and among the lampstands was someone "like a son of

man," dressed in a robe reaching down to his feet
and with a golden sash around his chest.

Rev 1:10-13

From the throne came flashes of lightning,
rumblings and peals of thunder. Before the throne,
7 lamps were blazing. These are the 7-fold Spirit
of God.

Rev 4:5

Then I saw a Lamb, looking as if it had been slain,
standing in the center of the throne . . . He had 7
horns and 7 eyes, which are the 7-fold Spirit of God
sent out into all the earth.

Rev 5:6

"For the Son of Man is [I-shall-be] of the Sabbath."

Matthew 12:8

He who has an ear, let him hear what the Spirit says
to the churches. To him who overcomes, I will give
the right to eat from the *tree of life*, which is in the
paradise [or garden] of God.

Rev 2:7

To him who overcomes, I will give the right to sit
with me on my throne, just as I overcame and sat
down with my Father on his throne. He who has
an ear, let him hear what the Spirit says to the
churches.

Rev 3:21, 22

Appendix ז

The Ten Commandments and
the "Ten" Kinds of Fruit of the Spirit

What follows, the writer considers only for his benefit, from committing the entire book of Proverbs to memory, by chapter and verse recall, for all 31 chapters. This has been one of the most fulfilling endeavors ever pursued by this writer. It is prerequisite to becoming an ISB∴Institute™ *Emeritus* Fellow, chapter 31 completed during two workouts, each constituting 24-mile walks. The writer has always found Proverbs embracing, admiring its discrete structure and similarity to instruction-sets of computers. If one understands the instruction-set of a computer, one appreciates the architecture of the whole system. Proverbs helps develop a similar appreciation and respect for the unity of all of Scripture, through discrete precepts, statutes, commands, promises, and the *name*, the *word, words,* the *law* and *laws*.

Though the initial undertaking may seem daunting, it becomes less and less so with progress, verse by verse—not proceeding to the next one until the prior one is confidently recalled from memory; and chapter by chapter—not proceeding to the next chapter until all verses of the prior one are confidently recalled from memory. One of the most uplifting verses, regarding this type of endeavor, comes from Proverbs itself:

> Pay attention and listen to the sayings of the wise;
> apply your heart to what I teach, for it is pleasing
> when you *keep them in your heart* and *have all of them*
> *ready on your lips.*
>
> Proverbs 22:17,18

After committing all 31 chapters to memory, the writer revisited a numeration question regarding Scripture: Why are there *nine* kinds of fruit of the Spirit? Why *nine*? How does *nine* harmonize with other Scripture?

> But the fruit of the Spirit[1] is *love*[2], *joy*[3], *peace*[4],
> *patience*[5], *kindness*[6], *goodness*[7], *faithfulness*[8], *gentleness*[9]
> and *self-control*[10]. Against such things there is no *law.*
>
> Galatians 5:22

At Caltech, one learns to apply *substitutions* or *offsets* or *re-numerations* to resolve or simplify complex, mathematical phenomena into more understandable forms.

Exodus 20 contains the *Ten* Commandments, and Galatians 5:22 has the *nine* kinds of fruit of the Spirit—that is, there is one less kind of fruit of the Spirit than commandments. The Son of Man gives witness to the Ten Commandments, and law thereof, with just *two* commandments, a regression that helps grasp the unity of all the Ten Commandments:

> Hearing [יִשְׁע] had silenced the Sadducees, the
> Pharisees got together. One of them, an expert in

a *7th-Gift* page

the law, tested him with this question: "Teacher, which is the greatest commandment in the Law?"

[ישוע] replied: "'Love the [I-shall-be] your God with all your heart and with all your soul and with all your mind.' [Deuteronomy 6:5] This is the first and greatest commandment. And the second is like it: 'Love your neighbor as yourself.' [Leviticus 19:18] All the law and the Prophets hang on these two commandments."

Matthew 22:34-40

Similarly, how might the *nine* kinds of fruit of the Spirit be witness to the *Ten* Commandments? The first of the Ten Commandments establishes the primacy of the One who writes them with his own finger. If primacy is given to the Spirit himself, as the source of his fruit, then there is a one to one correspondence between the Ten commandments and "Ten" Kinds of Fruit of the Spirit, the first being the source of the fruit and the primacy of the Spirit himself.

To the writer, Proverbs and other books thereof, that the writer commits to memory, are witnesses to these *tens*—both the commandments and the kinds of fruit—by the grace of the Name of the writer's Father, the [I-SHALL-BE]; and the writer's Father's Son, the [I-SHALL-BE-SALVATION]; and the Spirit of truth himself. Herein, the kinds of fruit of the Spirit are numbered [F1] . . . [F10], and the commandments [C1] . . . [C10], consistent with the ordination of א . . . י in Psalms 119:1-80:

א Fruit

א Commandment

[F1] *The Spirit*[1]: "But when he, *the Spirit of truth* comes, he will guide you into all truth. He will not speak on his own; he will speak only what he hears, and he will tell you what is yet to come. He will bring glory to *me* by taking from what is mine and making it known to you. All that belongs to my Father is mine. That is why I said *the Spirit* will take from what is mine and make it known to you.

[John 16:13-14]

[C1] "You shall have no other gods before *me.*"

[Exodus 20:3]

ב Fruit   ## ב Commandment

[F2] *Love*[2]: This is how we know that we *love* the children of God: by *loving* God and carrying out his *commands*. This is *love* for God: to *obey* his *commands*. And his *commands* are not burdensome, for everyone born of God overcomes the world. This is the victory that has overcome the world, even our faith. [1John 5:2-4] If you *obey* my *commands*, you will remain in my *love*, just as I have *obeyed* my Father's *commands* and remain in his *love*. [John 15:10]

[C2] "You shall not make for yourself an idol in the form of anything in heaven above or on the earth beneath or in the waters below. You shall not bow down to them and worship them. For I, the [I-SHALL-BE] your God, am a jealous God, punishing the children for the sin of the fathers to the third and fourth generation of those who hate me. But showing *love* to a thousand generations of those who *love* me and *keep* my *commandments*. [Exodus 20:4-6]

ꙩ Fruit

ꙩ Commandment

[F3] *Joy*³: Until now you have not asked for anything in my *name*. Ask and you will receive and your *joy* will be complete. [John 16:24] And I will do whatever you ask in my *name*, so that the *Son* may bring glory to the *Father*. [John 14:13] I have told you this so that my *joy* may be in you and your *joy* may be complete. [John 15:11]

[C3] You shall not misuse the *name* of the [I-SHALL-BE] your *God*, for the [I-SHALL-BE] will not hold any one guiltless who misuses his *name*. [Exodus 20:7]

7 Fruit

7 Commandment

[F4] *Peace*[4]: "'Observe my Sabbaths[.] . . . If you follow my decrees and are careful to obey my commands . . . I will grant *peace* in the land, and . . . no one will make you afraid.'" [Lev 26:3-6] "I have told you these things, so that in me you may have *peace* . . ." [John 16:33] "For the Son of Man is [I-shall-be] of the *Sabbath*." [Matt 12:8] When he opened the *7th* seal, there was *silence* in heaven for about half an hour. [Rev 8:1] The *7th* angel poured out his bowl into the air, and out of the temple came a loud voice from the throne, saying, "It is *done!*" [Rev 16:17]

[C4] Remember the *Sabbath* day by keeping it holy. Six days you shall labor and do all your work, but the *7th* day is a *Sabbath* to the [I-SHALL-BE] your God. On it you shall not do any work, neither you, nor your son or daughter, nor your manservant or maidservant, nor your animals, nor the alien within your gates. For in six days the [I-SHALL-BE] made the heavens and the earth, the seas, and all that are in them, but he *rested* on the *7th* day. Therefore the [I-SHALL-BE] blessed the *Sabbath* day and made it holy. [Exodus 20:8-11]

And God blessed the 7th day and made it holy, because on it he *rested* from all the work of creating that he had *done*. [Gen 2:2,3] Early on the 1st day of the week, while it was still dark, Mary Magdalene went to the tomb and saw that the stone had been removed from the entrance. [John 20:1] God called the light "day," and the darkness he called "night." And there was evening, and there was morning—the 1st day. [Gen 1:5]

a *7ʰ-Gift* page

ה Fruit   ## ה Commandment

[F5] *Patience*[5]: He who has   [C5] *Honor* your *father* and
a *wise* son delights in him;   your *mother,* that you may
the *father* of a *righteous* man   live long in the land the
has great joy. May your *father*   [I-SHALL-BE] your God is
and *mother* be glad; may she   giving you. [Exodus 20:12]
who gave you birth rejoice.
[Prv 23:24,25]. A man's
wisdom gives him *patience,* it
is to his glory to overlook an
offense. [Prv 19:11] My son,
keep your *father's* commands
and do not forsake your
mother's teaching. [Prv 6:20]
Better a *patient* man than a
warrior; a man who controls
his temper than one who
takes a city. [Prv 16:32] *Honor*
the [I-SHALL-BE] with your
wealth, with the first fruit of all
your crops. Then your barns
will be filled to overflowing
and your vats will brim over
with new wine. [Prv 3:9,10]
My son, do not despise the
[I-SHALL-BE's] discipline
or resent his rebuke; for the
[I-SHALL-BE] disciplines
those he loves, as a *father* the
son he delights in.[Prv3:9-12]

ו Fruit   ### ו Commandment

[F6] *Kindness*[6]: He who   [C6] You shall not *murder*.
oppresses the poor shows   [Exodus 20:13]
contempt for their maker,
but whoever is *kind* to the
needy honors God. [Prv
14:31] He who despises his
neighbor *sins*, but blessed is
he who is *kind* to the needy.
[Prv 14:21] But if you show
favoritism, you *sin* . . . [f]or
he who said, "Do not commit
adultery," also said, "Do not
murder." [James 2:9-11]

† Fruit

† Commandment

[F7] *Goodness*[7]: A *good* man obtains favor from the [I-SHALL-BE], but the [I-SHALL-BE] condemns a crafty man. [Prv 12:2] He who finds a wife finds what is *good* and receives favor from the [I-SHALL-BE]. [Prv 18:22] Houses and wealth are inherited from parents, but a *prudent* wife is from the [I-SHALL-BE]. [Prv 19:14] But a man who commits *adultery* lacks judgment, and whoever does so destroys himself. [Prv 6:32] The mouth of an *adulteress* is a deep pit; he who is under the [I-SHALL-BE's] wrath will fall into it. [Prv 22:14] Her house is a highway to the grave, *leading down* to the chambers of death. [Prv 7:27]

[C7] You shall not commit *adultery*. [Exodus 20:14]

ח Fruit   ח Commandment

[F8] *Faithfulness*[8]: The *faithful* man will be richly blessed, but one eager to get rich will not go unpunished. [Prv 28:20] . . . [G]ive me neither poverty nor riches . . . Or I may become poor and *steal, and* so dishonor the name of my God. [Prv 30:8-9] Do not those who plot evil go astray? But those who plan what is good find love and *faithfulness*. [Prv 14:22] The *faithless* will fully *repay* for their ways, and the good man *rewarded* for his. [Prv 14:14] "Will a man *rob* God? Yet you *rob* me . . ." [Mal 3:8] He who *robs* his father or mother and says it is not wrong—he is partner to him who destroys. [Prv 28:24] He who conceals his sins *does not prosper,* but whoever confesses and renounces them finds *mercy.* [Prv 28:13]

[C8] You shall not *steal.* [Exodus 20:15]

ט Fruit

ט Commandment

[F9] *Gentleness*[9]: A *gentle* answer turns away wrath, but a harsh word stirs up anger. [Prv 15:1] Through patience a ruler can be persuaded, and a *gentle* tongue can break a bone. [Prv 25:15] Like a club or a sword or a sharp arrow is the man who gives *false testimony* against his neighbor. [Prv 25:18] *Do not accuse* a man for no reason—when he has done you no harm. [Prv 3:30] *Do not say,* "I will do to him as he has done to me; I'll pay that man back for what he did." [Prv 24:29] *Reckless* words pierce like a sword, but the tongue of the wise brings *healing.* [Prv 12:18]

[C9] You shall not give *false testimony* against your neighbor. [Exodus 20:16]

' Fruit

' Commandment

[F10] *Self-control*[10]: Let your eyes look *straight* ahead; fix your gaze *directly* before you. [Prv 4:25] *Seldom* set foot in your neighbor's house; too much of you and he will hate you. [Prv 25:16] Do not wear yourself out to get rich; have the wisdom to show *restraint.* [Prv 23:4] A fool gives full vent to his anger, but a wise man keeps himself *under control.* [Prv 29:11] It is not good to eat *too much* honey; nor is it honorable to seek one's own honor. [Prv 25:27] Do not join those who drink *too much* wine or *gorge* themselves on meat . . . you will be like one *sleeping* on the high seas, *lying* on top of the rigging. [Prv 23:20-34] Like a city whose walls are broken down is a man who lacks *self-control.* [Prv 25:28]

[C10] You shall *not covet* your neighbor's house. You shall *not covet* your neighbor's wife, or his manservant or maidservant, his ox or donkey, or *anything* that belongs to your neighbor. [Exodus 20:17]

a *7th-Gift* page

Do not envy a wicked man or *choose* any of his ways. [Prv 3:31] The fear of the [I-SHALL-BE] teaches a man wisdom, and *humility* comes before honor. [Prv 15:33] Death and destruction are *never satisfied*, and neither are the eyes of man. [Prv 27:20] Do not lie like an *outlaw against* a righteous man's *house*, do not raid his *dwelling place*, for though a righteous man falls seven times, he rises again, but the wicked are brought down by calamity. [Prv 24:15,16] Do not *envy* wicked men, do not desire their company; for their hearts plot violence, and their lips talk about making trouble. [Prv 24:1,2]

I

Shall

Be

Index

"a 7th-*Gift* page"

7, 14, 21, 28, 35, 42, 49, 56, 63, 70, 77, 84, 91, 98,
105, 112, 119, 126, 133, 140, 147, 154, 161, 168,
175, 182, 189, 196, 203, 210, 217, 224, 231, 238,
245, 252, 259, 266, 273, 280, 287, 294, 301, 308,
315, 322, 329, 336, 343, 350, 357, 364, 371, 378,
385, 392, 399, 406, 413, 420, 427, 434, 441, 448,
455, 462, 469, 476, 483, 490, 497, 504, 511, 518,
525, 532, 539 (77th entry)